EAI/Springer Innovations in Communication and Computing

Series editor
Imrich Chlamtac, European Alliance for Innovation, Ghent, Belgium

Editor's Note

The impact of information technologies is creating a new world yet not fully understood. The extent and speed of economic, life style and social changes already perceived in everyday life is hard to estimate without understanding the technological driving forces behind it. This series presents contributed volumes featuring the latest research and development in the various information engineering technologies that play a key role in this process. The range of topics, focusing primarily on communications and computing engineering include, but are not limited to, wireless networks; mobile communication; design and learning; gaming; interaction; e-health and pervasive healthcare; energy management; smart grids; internet of things; cognitive radio networks; computation; cloud computing; ubiquitous connectivity, and in mode general smart living, smart cities, Internet of Things and more. The series publishes a combination of expanded papers selected from hosted and sponsored European Alliance for Innovation (EAI) conferences that present cutting edge, global research as well as provide new perspectives on traditional related engineering fields. This content, complemented with open calls for contribution of book titles and individual chapters, together maintain Springer's and EAI's high standards of academic excellence. The audience for the books consists of researchers, industry professionals, advanced level students as well as practitioners in related fields of activity include information and communication specialists, security experts, economists, urban planners, doctors, and in general representatives in all those walks of life affected ad contributing to the information revolution.

Indexing: This series is indexed in Scopus, Ei Compendex, and zbMATH.

About EAI

EAI is a grassroots member organization initiated through cooperation between businesses, public, private and government organizations to address the global challenges of Europe's future competitiveness and link the European Research community with its counterparts around the globe. EAI reaches out to hundreds of thousands of individual subscribers on all continents and collaborates with an institutional member base including Fortune 500 companies, government organizations, and educational institutions, provide a free research and innovation platform. Through its open free membership model EAI promotes a new research and innovation culture based on collaboration, connectivity and recognition of excellence by community.

More information about this series at https://link.springer.com/bookseries/15427

Sanjay Misra • Amit Kumar Tyagi
Editors

Blockchain Applications in the Smart Era

Springer

EAI
RESEARCH MEETS INNOVATION

Editors
Sanjay Misra
Department of Computer Science and Communication
Østfold University College (HIOF)
Halden, Norway

Amit Kumar Tyagi
Vellore Institute of Technology
Tamil Nandu, India

ISSN 2522-8595 ISSN 2522-8609 (electronic)
EAI/Springer Innovations in Communication and Computing
ISBN 978-3-030-89545-7 ISBN 978-3-030-89546-4 (eBook)
https://doi.org/10.1007/978-3-030-89546-4

This Springer imprint is published by the registered company Springer Nature Switzerland AG
The registered company address is: Gewerbestrasse 11, 6330 Cham, Switzerland

Preface

Blockchain is the transformational technology that will define a paradigm shift from the traditional ways we used to carry out our activities. We are approaching an era where virtually everything human activity will now be undertaken by machines automatically. This era is expected to extend beyond the activities carried out by robots in factory product lines. This change is envisaged to be a combination of artificial intelligence, the Internet of Things, virtual reality, and data analytics. Blockchain is expected to serve as a bridge between the digital and the real world. Blockchain applications are already gaining prominence in various aspects of global economy. For instance, blockchain currently serves as a distributed ledger technology on which Bitcoin and other cryptocurrencies are relying upon as it has the potential to sustain what is referred to as the programmable economy.

It is envisaged that the blockchain technology, which will be a combination of AI, robotics, and smart contracts, as mentioned before, will in the nearest future lead to a transaction between a buyer and seller that requires minimal human intervention. Also, it will be possible with Blockchain technology to directly relate an order to an automated factory for manufacturing, with collection by a self-driving truck, loading by a port's automated crane, transfer by an unmanned ship, and final delivery by an aerial drone. There is also the consideration of 3D printing that will make factories obsolete for certain products. Blockchain technology will, in no doubt, extend to boundaries we can only imagine, right. The coverage of blockchain has extended to smart homes, where it is now used to support automation of lighting, climate, entertainment, and even the ordering of groceries using the smart refrigerator. Blockchain technology is estimated to cover a wide range of sectors in the global economy.

Given all of the above, it is logical to begin to focus on developing practical applications that will power the new era of technological transformation, thereby serving a good foundation for the new digital economy with its attendant benefits.

Therefore, it is our pleasure to present to you this book: ***Blockchain Applications in the Smart Era.*** The book provides adequate insights on the concepts and fundamentals of blockchain technology. The book consists of 14 different contributions from authors that cut across different continents working in the domain of blockchain technology. We present a brief summary of the contributions of these authors to stimulate readers' interest in studying the entire textbook.

Yamada et al., in their chapter titled **"Perspectives of Blockchain in Digital Health in Brazil,"** suggested the need to automate patients' data management processes in government hospitals. They proposed an electronic patient record (EPR) that is secured based on blockchain capability. Their study provided the first theoretical contact in an academic context for implementing a blockchain system for digital transformation involving mainly EHR.

In the chapter titled **"Edge Information Systems Based Blockchain Frameworks for Industrial IoT: A Novel Approach,"** Parimala et al. proposed novel approaches in building efficient frameworks for Industrial IoT through blockchain-based edge information systems, thus providing findings that pave the way for authorizing and easing the researches in this field, and supported decision-makers in blockchain espousal and investment in Industry 4.0 and IIoT space.

Olaniyi et al., in their chapter titled **"A Secure Electronic Voting System Using Multifactor Authentication and Blockchain Technologies,"** presented a distributed e-voting system that solves the problems of vote-rigging, voter impersonation, and vote falsification, all of which are prevalent in traditional paper ballot systems. The study employed a combination of multifactor authentication (MFA) and blockchain techniques to secure electronic voting. MFA hampers the compromising of voters' identities and allows for easy verification, while blockchain technology protects the integrity of the votes and ensures the verifiability of the cast votes. They combined a facial recognition algorithm and RFID for authentication and authorization of voters to participate in the election process while a smart contract implemented on an Ethereum network provided the required measures of integrity and verifiability for secure e-voting.

In the chapter titled **"Enhanced Security and Privacy Issue in Multi-tenant environment of Green Computing using Blockchain Technology,"** Ogundokun et al. presented various forms of green computing, including the need for green computing, benefits, and security concerns of multi-tenancy in cloud environments. The study also proposed blockchain technology as a method to deal with the security and privacy challenges in a multi-tenant environment. The study implemented blockchain in a cloud environment with the use of Ganache and meta-mask to create a dummy secure account for each cloud tenant.

Ajao et al., in their chapter titled **"Application of Crypto-Blockchain Technology for Securing Electronic Voting Systems,"** proposed integration of security countermeasure to enhance the adoption of smart electronic voting systems for election processes using blockchain technology with bi-factors authentication methods. They adopted the private blockchain technique to provide immutability, avoid records tampering, and ensure records transactions' integrity and reliability. Paillier homomorphic encryption algorithm was applied to the decentralized system-based blockchain and electronic voting system through bi-factors certification and secrecy (Iris and Fingerprint). Their crypto-blockchain technique using Paillier homomorphic encryption was found to perform well with moderate latency during certification execution and data retrieval timing.

In the chapter titled "**Enhanced Hash Values and Public Keys Infrastructure Generations for Blockchains using Sooner Lightweight Cryptography,**" Alfa et al. proposed a lightweight cryptosystem known as Sooner-C to reduce the complexity while improving the encryption and hashing used in blockchain technology. The proposed technique was found to be more effective when compared to the traditional blockchain Technology's cryptosystem using encryption time (1672.2 secs to 18385.3 secs), decryption time (618.8 secs to 806 secs), ciphertext size (32-bits to 79-bits), and a number of rounds (15–10), respectively.

Arowolo et al., in their chapter titled "**A Prediction Model for Bitcoin Cryptocurrency Prices,**" proposed a predicting model for blockchain bitcoin cryptocurrency prices and its profitability trading strategies using machine learning algorithms (ICA-Firefly and SVMs). For the prediction analysis of Bitcoin cryptocurrency data, this study combined ICA-Firefly with SVM algorithms. The model was tested on a large dataset of 2,194 samples, and its performance was analyzed in terms of evaluation metrics. The ICA-Firefly with the SVM model was found to be most viable for financial system sustainability management strategy.

Awotunde et al., in their paper titled "**Blockchain-Based Framework for Secure Medical Information in the Internet of Things System,**" explained the important role played by blockchain in securing medical data. They also proposed an architecture based on blockchain to secure medical data in an Internet of Things–based healthcare system. In their opinion, the concern of security and privacy in the healthcare system can be reduced if blockchain technology is used in the healthcare system.

In the chapter titled "**Blockchain Technology and Organizational Practices: The Case of Nigerian Academic Libraries,**", Ojobor et al. discussed the extent of applicability of blockchain technology towards enhancing library practice. They also pointed out the challenges of applying blockchain technology to library practice. The study engaged 133 professional librarians while adopting a descriptive survey design. Questionnaire and focus groups discussions were used for data collection.

Saheed et al., in their chapter titled "**A Comparative Study of Regression Analysis for Modelling and Prediction of Bitcoin Price,**" proposed six regression models for Bitcoin price prediction based on historical data from 2014 to 2020. The study employed six different regression models, namely Cat Boost regressor, Gradient Boosting regressor, Extra Tree regressor, Ada Boost regressor, K-Neighbor regressor, and the Theil-Sen regressor. The models' performance was evaluated using coefficient of determination (R2), mean absolute error (MAE), mean square error (MSE), root mean square error (RMSE), root mean square logarithmic error (RMSLE), and mean absolute percentage error (MAPE).

In the chapter titled "**Adoption of Blockchain Technology in the Indian Business Market: Obstacles and Opportunities,**" Litoriya et al. presented a summary of their findings and potential application and use cases of blockchain technology to help companies eliminate risks.

Ogundokun et al., **in their chapter titled "Machine Learning, IoT and Blockchain Integration for Improving Process Management Application Security,"** employed the combination of blockchain technology and machine learning to protect network communications and manage datasets to eliminate counterfeit effect datasets. To bring about and evaluate the gathered dataset, big data procedures were employed. Likewise, the fault diagnosis forecast aspect was evaluated on the predictive ML approach proposed, which is the improved ensemble learning (IEL) classification ML technique. The system was implemented using the traditional ensemble learning (TEL), and the improved ensemble learning (IEL) and performance matrices like accuracy, precision, sensitivity, and false-positive rate (FPR) were used to evaluate the system performance.

In the chapter titled **"A Blockchain-Powered Energy Monitoring System,"** Swain et al. proposed a blockchain-based application that integrated advanced metering infrastructure and smart contracts leading to a peer-to-peer monitoring system that ensures transparency in energy usage and automation of the billing process and detect energy theft in the power lines. Major issues associated with the existing infrastructure of energy monitoring and billing systems like the manual intervention of the billing process, overdue bill payment, illegal tapping of power lines, and unethical selling of energy units by the electricity board were considered. Their proposed smart system is capable of executing the smart contract functionality in automated billing and storing transactional data in the blockchain.

In the final chapter titled **"Multifactor IoT Authentication System for Smart Homes Using Visual Cryptography, Digital Memory and Blockchain Technologies,"** Thompson et al. proposed a multifactor authentication mechanism that is based on digital memory, visual cryptography, and blockchain technologies with a view to securing the digital memories by incorporating an efficient (2, 2) visual cryptography scheme (VCS) in the digital memory authentication service (DMAS) in smart homes. Blockchain technology ensures decentralized identity storage and sensitive data management. While this authentication system prevents unauthorized access to smart doors, its mobile application permits authorized users remote access to the smart door(s) in smart homes. The proposed strategy offers about 50% improvement in speed over conventional public key crypto-based authentication systems.

In conclusion, blockchain technology is expected to combine AI, robotics, and smart contracts to define the future programmable economy, which includes a transaction between a buyer and seller that requires minimal human intervention. It is, therefore, logical to begin to focus on developing practical applications that will power the new era of technological transformation, thereby serving as a good foundation for the new digital economy, which has great potential to disrupt and improve the new global economy.

Østfold University College (HIOF), Norway Sanjay Misra

Contents

About the Editors

Sanjay Misra a senior member of IEEE and ACM Distinguished Lecturer, is a professor at Østfold University College(HIOF), Halden, Norway. Before coming to HIOF, he was a professor at Covenant University (400–500 ranked by THE(2019)) Nigeria for 9 years. He holds a PhD in information and knowledge engineering (software engineering) from the University of Alcala, Spain, and MTech (software engineering) from MLN National Institute of Technology, India. As per SciVal (SCOPUS- Elsevier) analysis (on 01.12.2021), he has been the most productive researcher (no. 1) in Nigeria since 2017 (in all disciplines), no. 1 in computer science in the country and no. 2 in the whole of Africa. He has published a total of around 600 articles (SCOPUS/WoS) with 500 coauthors worldwide (110 JCR/SCIE) in the core and application areas of software engineering, Web engineering, health informatics, cybersecurity, intelligent systems, and AI. He is among the top 2% of scientists in the world (published by Stanford University) for the last 2 consecutive years and also got several awards for outstanding publications (2014 IET Software Premium Award(UK)) and from TUBITAK-Turkish Higher Education and Atilim University). He has delivered more than 100 keynote/invited talks/public lectures in reputed conferences and institutes (traveled to more than 60 countries). Sanjay is one of the editors in 58 LNCSs, 4 LNEEs, 1 LNNSs, 3 CCISs, and 10 IEEE proceedings, editor of 10 books, editor-in-chief of IT Personnel and Project Management series and the *Int J of Human Capital & Inf Technology Professionals* (IGI Global), and editor of various SCIE journals.

Amit Kumar Tyagi is an assistant professor (senior grade) and senior researcher at Vellore Institute of Technology (VIT), Chennai Campus, India. His current research focuses on machine learning with big data, blockchain technology, data science, cyber physical systems, smart and secure computing, and privacy. He has contributed to several projects such as "AARIN" and "P3-Block" to address some of the open issues related to the privacy breaches in vehicular applications (such as parking) and medical cyber physical systems. He received his PhD degree from Pondicherry Central University, India. He is a member of the IEEE.

Perspectives of Blockchain in Digital Health in Brazil

Talita de Oliveira Vargas Yamada and Fernanda Nascimento Almeida

1 Introduction

According to a survey carried out by the United Nations (UN), the world population reached 7.6 billion inhabitants in 2017. This population representation has tremendous impact and importance for developing new technologies in several sectors, among them health. Currently, in response to the unfolding of the modern technological era in this segment, essential aspects have emerged involving Health 4.0, mainly issues related to Digital Health. Health 4.0 comes from the need to integrate different computational platforms, which is impacted mainly with the provision and management of health services, and the production of medical devices for the medical field. In this way, the term Health 4.0 is designed to ensure the incorporation of information technology (IT) with Digital Health, encompassing online service and information logistics (medical images from clinical examinations, genetic and clinical data, etc.) [8, 20]. According to the World Health Organization (WHO), the interoperability guaranteed by Digital Health in the health sector happens with information and communication technologies (ICT) and is the main recommendation made for its Member States. So Health 4.0 becomes an innovative and challenging era to propose the development and implementation of

Supported by organization UFABC

T. de Oliveira Vargas Yamada · F. N. Almeida (✉)
Bioinformatics and Health Informatics Group (BHIG), Center for Engineering, Modeling and Applied Social Sciences (CECS), Federal University of ABC (UFABC), São Bernardo do Campo, São Paulo, Brazil
e-mail: talita.yamada@aluno.ufabc.edu.br; fernanda.almeida@ufabc.edu.br

S. Misra, A. Kumar Tyagi (eds.), *Blockchain Applications in the Smart Era*,
EAI/Springer Innovations in Communication and Computing,
https://doi.org/10.1007/978-3-030-89546-4_1

processes and systems management to ensure security and interoperability in ICT use [8].

In Brazil, the Public Health System was established in 1988 by the Federal Constitution, and it is currently known as the Sistema Único de Saúde (Unified Health System), better known by the acronym SUS. The SUS has the function of regulating, inspecting, controlling, and executing health services and actions. In 2015, a survey by the Ministry of Health carried out in partnership with the Brazilian Institute of Geography and Statistics (IBGE) found that 71.1% of the Brazilian population sought public health facilities to be served. Searching for a health facility means participating in scheduling appointments, care, or health procedures. These processes generate records with clinical and administrative information that make up the patient's record. With this scenario, we can determine that Digital Health has a significant role in providing proportional access to quality health and information integrity [7, 8, 27]. According to the SUS computer department, in 2016, about 76% of basic health units used the patient's history on paper. Thus, one of the innovative strategies applied in the system using information and communication technologies was the implementation of Electronic Health Records (EHR), as a need for solutions and improvements for the management and storage of data in the public health system, such an implementation has become an ideal technological tool of great importance for the current scenario [4, 8]. Indeed, the EHR is a tool that has continuous improvements for the computerization of the Unified Health System, which favors a more excellent transaction of information on the history of clinical conditions related to the patient. In addition, the EHR is a digital document that facilitates the communication process with a multidisciplinary health team, involving from the health institution to a health agent (nurses, instrumentalists, doctors). However, one of the most significant risks of this technology is that it is linked to data security's ethical and legal aspects. In order to avoid risks to the EHR database, it must be determined in legal terms that all the contents of the medical record belong to the patient and are confidential. In addition, institutions and health professionals cannot use this information for corporate purposes or their interests, promoting that every patient has the right to access any information contained in their medical records. Thus, it is the responsibility of health institutions to ensure a secure system for protecting the content of the EHR [20, 21].

In recent years, information technology has emerged to transform and exploit advances in data and information sharing. With technological advancement, a revolutionary and impactful digital transformation took place in several areas, such as financial markets, smart cities, education, and health. Among the various sectors, significant investments are being directed to health, specifically for digital health, involving mainly electronic medical records (EMR) and electronic health records with information on prevention, diagnosis, treatment, and rehabilitation of the patient [16]. Digital health or e-Health is the interface between information technology, the Internet, communication, and medicine. According to [13, 31], digital health can be defined as an emerging field at the intersection of medical informatics, public health, and business, related to health services and information transmitted over the Internet and related technologies. In a broader sense, the term characterizes not only a technical development but also a state of mind, a way of thinking, an attitude and a commitment to the network, the global thinking, to improve health care at the local, regional level and worldwide using information and communication technology. Blockchain aims at the existence of possible cyber threats in digital health, and this method emerges as an emerging technology to contemplate data security applications and mainly the integrity and confidentiality of the system. The use of blockchain favors an environment of data sharing in a secure way aiming at the decentralization of transactions. This technology was well known in the era of digital currency, bitcoin [12, 16]. The blockchain stores its transactions in an orderly manner, linked to the previous block, divided into header and transactions and stored detailed information according to existing transactions. This technology uses Peer-to-Peer (P2P) networks, being characterized mainly in a decentralized consensus network. Thus, in a P2P network, users of the system are equal and responsible for the functioning of the network [14].

In Brazil, the entire data digitalization and information integration of SUS users are managed through the e-SUS AB System. All the system modeling and software engineering is currently under development by the Department of Primary Care, the Department of Health Care, and the Ministry of Health, organizations that have committed themselves to restructure the Health Primary Care Information. Inside the System e-SUS AB, two software are responsible for data collection: (i) Simplified Data Collection (SDC): through this system, it is possible to transfer the information in the printed form to the system, SDC is indicated for Basic Health Units (UBS) that do not have computerization; (ii) system with Electronic Health Record (EHR): software used to computerize the Basic Health Unit (currently known as the UBS - Unidade Básica de Saúde), individually registering all the actions related to the patient's health, that is, it is the electronic record containing the entire consultation history of an individual [10, 24]. It is of interest to understand deeper into the Software Engineering landscape behind EHR specifically for this work. However, the EHR for being precisely the electronic document to support computational platforms within the health scenario in Brazil will be presented throughout this work. Considering the barriers brought by the operability of digital health and the safety of data in the EHR, this work aims to analyze the applicability of blockchain as a solution to privacy and anonymization of health records [2, 29].

This work is structured in seven sections. In Sects. 2, 3, and 4, a description is made of those covered in this research. The Electronic Patient Record used in SUS is written throughout Sect. 2 and is for understanding the Digital Healthcare scenario in Brazil. Section 3 presented the blockchain concept to ensure interoperability and guarantee security during the transactions of information between institutions, health professionals, and users. The topic of Ethereum and Hyperledger Fabric was covered in Sect. 4, and essential concepts were related to two technology platforms developed around blockchain concerns. Section 5 talks about the applicability of blockchain in the Healthcare Sector and digital health. In Sect. 6, there was a discussion on using blockchain as an essential tool to solve systemic problems of data privacy in the digital health sector. Moreover, it was compared Ethereum and

Hyperledger Fabric tool and introduced the best mechanism to apply in the Brazilian Healthcare scenario; finally, Sect. 7 concludes the article.

2 Electronic Health Records

It is known that the health care of a patient can occur in different clinics and hospitals, such as infirmary, ambulatory, surgical center, ICU, nursing home, among others. Technological and scientific development in the information sector also provided that a multidisciplinary team with doctors, nurses, nutritionists, psychologists, physiotherapists, and health managers become agents crucial in this sector [23, 30].

The Brazilian Federal Council of Medicine described the minimum components needed for management records in the EHR, which are: patient identification and anamnesis, visual record of vital signs, progress chart/average prescription, progress chart for laboratory test results (and other methods additional diagnostics), discharge summary and other documents for the recording of reports on surgical, anesthetic, obstetrics, and monitoring of the management of time spent in the operating room.

The EHR is a fundamental element for informatization in the health care sector. EHR was an element developed by doctors and nurses so that all information related to the patient could be recorded and, also, to promote the integrity of clinical facts. In this way, the EHR allows communication between all professionals involved in the health care process of the patient. One of the most well-known models of the EHR was proposed in 1997 by the United States Institute of Medicine (IOM). This model represented a record electronic with user support information, bringing personal data (name, age, gender, address, etc.), reminders, doctor's clinical opinions, among other information about the patient. However, before computerization in the health care sector, the medical record of the patient was registered and filed in physical documents with fields filled in manually. Over the years, the EHR has emerged as an extremely important tool to assist in handling and management of each patient's data within sectors in hospitals, allowing greater support for clinical research, epidemiological studies, evaluation of the quality, and clinical trials. The EHR has become an electronic document and decision support for health agents and their users, making available complete and updated data regarding the patient, the clinical status, trajectory of diagnosis involving examinations, prescribed treatment, and all kinds of sensitive information [6, 30].

2.1 Electronic Citizen Record Template in e-SUS AB

The PEC available on e-SUS AB is a fully electronic document for the registration of clinical practice in Primary Care (AB). Introducing the same concepts previously illustrated on the EHR, the PEC is also an electronic medical record capable of compiling information and ensuring a history of care for any patient [15]. The development of this software in the Brazilian health scenario was based on the requirements of the Problem-Oriented Clinical Record (POCR) model and with the Classification of Primary Care (ICPC). A crucial point of this project in SUS is the integration of a system meeting the requirements of the POCR model with the implementation of the SOAP method (subjective, objective, evaluation, and plan) for the clinical records of patients [15].

2.2 Model POCR: Problem-Oriented Clinical Record

The POCR model is a method based on the problem-oriented medical record (POMR) proposed by the American physician Lawrence Weed in the 1960s in his study Medical Records That Guide and Teach [28]. The functionality of this model is closely related to the objectives of making access to patient data an agile process with information control for decision-making, in addition to organizing and obtaining data on an ongoing basis. The structure of the POCR is composed of [28]: a database that is a component for the identification of the patient during the first care in a primary care unit. This base is subdivided into two categories, namely, identification block and background block. The list of Problems (Cover Sheet) is a component that precedes the patient's clinical record, being a cover sheet. In it there are three categorizations of problems, being they are active, inactive, and resolved. In addition, within the inactive problems, it is determined that there was an exposure of latent problems, which are those that even being inactive, require vigilance because they can compromise the health of a patient, for example, ex-smoker or cardiovascular disease. The cover page is extremely important for the EHR because it also contains additional information in blocks about past contacts, a list of allergies and adverse reactions, and a list of medications assets in use. For the item Evolution, according to the SOAP method, every clinical record of the patient is indicated with the SOAP model; it contains information categorized as subjective and under evaluation. This component contains all the information on Clinical Examinations. And finally, the item Follow-up Sheet is the summary of all the clinical follow-ups of the patient, determining the clinical evolution of a particular pathology or symptom. The accompanying sheet is categorized by flowcharts or spreadsheets, which are also called program sheets.

2.3 Model ICPC: Classification of Primary Care

Bringing a different concept from the POCR model, but with a similar purpose in proposing a clinical history of the patient, the ICPC model captures questions of health related to the individual and not only in diagnoses and pathologies. In this model, classification by health professionals is done by capturing a context of causes and effects of why the patient is seeking clinical help for symptom resolution, assessment, and diagnosis.

3 Blockchain: Interoperability and Data Privacy

One of the biggest concerns currently related to EHR is security and the interoperability of information between institutions, health professionals, and health and users. In this way, blockchain is a technology with the viability of addressing these concerns, and then one must understand the functionality and applicability of this emerging technique to Digital Health, so that we can analyze the impact of blockchain on data security [2, 28, 29]. Some concepts related to blockchain are of great importance to give continuity of the study, being they:

- Asset: It is any data that needs security and interoperability.
- Peer-to-Peer (P2P): This operation allows the decentralization of the information, causing the data to be distributed.
- Transactions: It is the modification of the state of an asset and the registration of that action in the blockchain. A transaction is performed by a peer.
- Ledger: This is the immutable record of the blockchain, known as the ledger. Therefore all transactions carried out by the participants are recorded in it.

As this record cannot be deleted by any participant all codes in this technology are considered pure codes (smart contract) and that all peers only have access to information authorized by the system.

3.1 Implementation and Functionality System

The blockchain acts as a decentralized database containing immutable and incorruptible transactions and it is possible to store a transaction log base through encryption in your system. In this section, we will cover the implementation system. For this, it is important to understand the theoretical concept of blockchain and how the use of this data privacy system works together with encryption [3, 18]. The block is made up of a header and transactions, with the header consisting of essential metadata for block identification and the transactions are represented by a listing of transactional shares within the block. So that we can proceed with the theoretical knowledge of the blockchain and its functionality, the structure of the block header must be identified and also how transactions influence this technology [3, 18]. One of the parts that make up the block is the header; in this part, there is the following metadata:

- Version: The version number of the software used.

- Previous block hash: Identifier code and that is linked to the block previous.
- Merkle root: Delimits in numerical quantity the existing transactions in the block.
- Timestamp: Date and time of block creation.
- Difficulty target: Delimiter of difficulty and complexity in the creation of the block.
- Nonce: Algorithm counter.

Transactions, as previously mentioned, are composed of a list of actions and are divided into inputs and outputs. The concept of entry and outgoing transactions in the block is essential, as the input represents transactions from a subject that is providing information, and the output represents a subject's transactions that are receiving this information [3, 18].

A crucial point of identification in this section is that in the header, the "previous block hash" ensures a data privacy system that works in conjunction with the encryption SHA-256 hash, where there is the storage of a hash reference with 64 characters in a block. The information transition occurs in blocks that are linked by a hash from the previous block. Therefore, we realize that the last hash block reference always contains information of the prior block hash of the previous block, and this allows the blocks to be all interconnected and remain intact due to cryptographically linked data [18].

Currently, the large number of blocks available for viewing on the internet are blocks that contain transactions in the world of cryptocurrencies. The check of these blocks can be found, for example, on sites like blockchain.info or etherscan.io [3].

Understanding the composition of the elements of a block is fundamental for we can apply them to EHR information transaction concepts and, mainly, in data security. In this case, information transactions patients, such as medical histories or exams, can occur securely as long as all transactions are validated and authorized during transactions categorized as incoming and outgoing. For these transactions to be validated, cryptography will be covered in the next section.

3.2 Data Security and Privacy Encryption

After understanding the functionality of the blockchain in the previous section, it is an in-depth investigation of how all this technology can offer data security. It was identified that the authorization of shares of information occurs within transactions, where these transaction listings depend on a user who initiates the transaction (outgoing) and a user who receives all this information (input). For this information transaction to occur safely, cryptography is used to secure the blockchain, particularly asymmetric encryption [29]. Within cryptography, two categories are defined: symmetric and asymmetric. For symmetric encryption, encrypting and decrypting a message is the same secret key used in communications between the user who initiates the transaction (outbound) and the user who receives the transaction (inbound). In the blockchain, asymmetric encryption is used, and there are processes to encrypt and decrypt messages using public and private keys. For the concept asymmetric cryptography is easy to assimilate to the reader, an example contextualizing how this category of cryptography would work within the EHR. One of the essential points that must be considered in cryptography asymmetric is that the private key does not allow the user who receives the transaction to encrypt the final message and enable another user who has critical public access to the document. And it is precisely this block in the transaction that guarantees the security of information exchange within asymmetric cryptography.

Currently, in the category of asymmetric cryptography, there is Rivest Shamir Adleman (RSA), and Elliptic Curve Cryptography (ECC) [11, 25]. The ECC encryption method is being widely studied and used in blockchain, and this is because the main advantage of ECC is that the keys to public and private data generated in this category are shorter in length and offer the same security provided, for example, by RSA. The ECC does develop in an algorithmic method based on discrete logarithm problems on elliptical curves. For the ECC to generate a key pair is required 256 bits, while in the RSA to generate the same key pair is the required size of 3072 bits [19]. However, both the RSA method and the ECC are asymmetric encryption processes that guarantee confidentiality, integrity, and availability [19].

4 Ethereum and Hyperledger Fabric

As previously seen, blockchain emerged as a tool to assist in financial transactions. As of 2013, new features were developed with the premise of expanding the service to various types of transactions, among which stand out: Ethereum and Hyperledger.

4.1 Ethereum

The development of Ethereum occurred due to the need to implement more robust techniques beyond which the blockchain allows, mainly so that the transactions were recorded in a decentralized manner and that could cover other transactions besides cryptocurrencies. Vitalik Buterin scientifically presented the Ethereum in 2013 through a technical article highlighting a new proposal for blockchain-based open-source that allows for more evolved transactions involving smart contracts (smarts contracts) [26, 34].

The main feature of the Ethereum platform is that it allows you to store the source code used for each transaction carried out within this system. The execution of smart contracts is performed by the Ethereum Virtual Machines (EVM), where a set of protocols can be established, allowing for a scope of transactions that are not limited only to the world of cryptocurrencies [34]. In addition, the EVM protocols can be implemented in programming languages like JavaScript and Python. However, for programmers and developers, there is the possibility to program in a more advanced language known as Solidy, which is also similar to JavaScript. One of the fundamental concepts within the Ethereum platform is two accounts: externally owned accounts and contract accounts. These accounts are managed by agents responsible for certain transactions, in the case of externally owned stores are accounts that can be controlled by any user and using the private key cryptography, on the other hand, contract accounts are controlled totally by the protocols generated within the smart contracts [26].

The difference between these two accounts available on Ethereum is that in EOAs, there is no code implemented, and there is no data storage. With this, it is only possible to control the transactions within the EOAs, which is why it is an account dependent on users. On the other hand, contract accounts not being managed by users but by codes present in smart contracts can have data storage and do not have private keys, so you have the communications made through messages or transactions [26].

For transactions to occur within the EVM, every protocol added in another block must be proven with proof of work, known as Proof-of-Work (PoW). This computational effort required during mining has a cost that is defined as Gas.

In Ethereum, there is also the use of blockchain; accounts have information similar to what was presented earlier in this bibliographic review. Ethereum also uses a blockchain. However, it has the following components defined uniquely in its account: nonce; ether required for the transaction to occur; smart contract code; and storage. All transactions carried out within an EVM are carried out by the user defined as a customer; however, for each block transaction and to maintain network security, it is necessary to know the responsibilities of three available nodes. For each node, there is a pre-defined functionality that performs different transactions. Being:

- Storage nodes (full node): Stores all the blockchain information of the smart contract. Checks and validates all transactions in a block and states this transaction. It is one of the most important nodes, as it manages to prove blockchain integrity. In addition, it provides the user with all the information and bills.
- Mining nodes (archive node): Keeps stored all the information that the storage nodes can evidence, and with this information records a history of the transaction status of each block. In addition, it is able to create new blocks and use PoW.
- Collector nodes (light node): These are nodes that verify transactions, and that in most cases, devices with little capacity to exert effort computational and storage are much lower than the system.

In this way, all transactions that occur through externally owned accounts or contract accounts need to be done via nodes. In Ethereum, a customer can be a user or a smart contract, and depending on your transaction, you need to have a public key to be identified in the system and a private key to make the transaction.

4.2 Hyperledger Fabric

The Hyperledger Fabric was founded in 2015 and emerged as a need to create smart contracts that meet the needs of the corporate world, and that did not need a currency of exchange for each transaction [32]. It is a platform with a flexible architecture, and one of its main characteristics is that networks are not anonymous.

The development of Hyperledger Fabric was based on the greatest needs of corporate use, where users cannot be anonymous; all networks need to be granted permission, guarantee high-performance transactions, and guarantee privacy and data security during transactions. It is a tool that had its development based on distributed logging technology (DLT), which means that there is a ledger that cannot be immutable but that all transactions are stored on multiple computers, or nodes [32]. The development of Hyperledger Fabric was based on the greatest needs of corporate use, where users cannot be anonymous, all networks need to be granted permission, guarantee high-performance transactions, and guarantee privacy and data security during transactions. It is also guaranteed to use consensus protocols like Crash Fault Tolerance or Byzantine Fault Tolerance, different protocols from PoW [17, 32].

The exposure of Hyperledger Fabric to an environment of authorized networks guarantees to the system a governance model with more evidence on the security of transactions that occur within that system. Smart contracts on the Hyperledger Fabric platform are called chain code. In this case, it provides the same functionalities as the one formalized and developed on Ethereum. A chain code does not necessarily need to contain a code to prove a transaction made from the ledger. It may contain a code that determines guidelines, rules, and definitions so that a user has permission to carry out the transaction and that it can be validated effectively [32]. It is crucial to point out that all communications from transactions made within the Hyperledger Fabric take place within the channels. Thus only users who are part of a channel can be aware of all the transactions from the ledger. It ensures the safety and privacy of the information handled within the platform since if the user does not be part of the channel, he will also have no access to the information entered, transferred, and generated within that protocol.

Understanding the Hyperledger Fabric architecture is necessary to identify components of that software and have specific responsibilities for the management and transactions of chain codes. Whether they are peers, orderers, certification authorities, and customers, in this way, the architecture of the Hyperledger Fabric does promote through the construction of a network and that there is an integrated system of channels where authorized users authorize transactions to occur [32]. One of the most significant issues involved with the customer is whether it could somehow generate cyber-attacks or carry out transactions maliciously. The fact is that the architecture of Hyperledger Fabric already predicts this type of failure, and this ensures security in the system with the implementation of protocols and rules in the network settings [17].

5 Applicability of Blockchain in Health Sector

In recent years, Blockchain is not just revolutionizing the financial sector with digital currencies. Its purpose of provide decentralized technology that can guarantee interoperability and data security. This technology also provides studies and implementations in managing clinical records, tracking medicines in the supply chain, and can to guarantee the validity of information transactions in the health area. The development of new technologies and tools in Blockchain in health is being extensively studied, investigated, and tested by its main features. These characteristics are in line with the needs of health companies [22].

In Brazil, in 2020, the national platform National Data Network on Health (RNDS) was created to guarantee the interoperability of data in the health system. This project has the participation of states and municipalities. It is still an embryonic project, and the Blockchain is present as a security service [1].

In 2016, Asaph Azaria proposed a platform called MedRec that guaranteed the interoperability of information from the Electronic Medical Record of patients. This system aims to solve the fragmentation of records and the lack of continuous control over the patients' clinical history information. The architecture of this system was published by [5] and has three contracts smartphones: registration contract, summary contract, and the patient–provider relationship. The big issue with the MedRec proposal is that as it is introduced on the Ethereum platform, there are no guarantee that users will be anonymous, which makes the system more vulnerable and data identification possible in sensitive patients [5].

On the other hand, there is an excellent movement of technology companies toward developing systems and applications that use the Blockchain. An important example is the IBM corporation, part of the consortium of blockchain technologies. At the end of 2020, IBM announced that with SalesForce, the application would be used to check the vaccination of users and make available that information to establishments ensuring data security and preventing the improper sharing of users' data. This way, it will be possible for institutions to verify test results, records of vaccination, and temperature checks of their employees, customers, and visitors. All transactions that occur within this application have been developed by IBM Blockchain Technology.

The biggest issue concerning EHR interoperability is the importance of protecting the patient's data. Given the ease of communication between health agents and users by the EHR, the critical point about data confidentiality was studied. In this regard, it is essential to note that only authorized entities are allowed access to provide data quality and accuracy of their representation for any information about the EHR.

Data confidentiality is the biggest concern for this electronic system in digital health because bringing the patient's personal information to multiple distributed users can generate cyber-attacks and compromise this exchange of sensitive data. Figure 1 was constructed to analyze the possible interoperability of the EHR between healthcare services.

It was observed that the interoperability of the EHR within the health sector is one of the most sensitive issues to be questioned, evidencing the question of the extent to which a health agent can access the patient's sensitive information. As previously said, health agents are doctors, nurses, physiotherapists, managers of health institutions (public and private), nutritionists, or psychologists.

Therefore, the impact of using Blockchain on digital health and data security was studied and discussed. In the yellow box of Fig. 1, we indicate the use of this technology in EHR to provide privacy and interoperability in the system.

Then a solution was proposed to ensure that the patient feels secure in reporting his or her sensitive data to an institution and that only the authorized health worker can access this information about the patient.

Blockchain is a shared transactional network that emerged to solve problems related to system insecurity and ensuring interoperability. Given what has already been discussed, the EHR currently contains several difficulties and instabilities: data security and interoperability of information between institutions, health professionals, and users.

Figure 2 was created to illustrate the main idea of a distributed communication network from the EHR with decentralized information. In the case of the schematization presented in Fig. 2, it is essential to emphasize that every patient has the

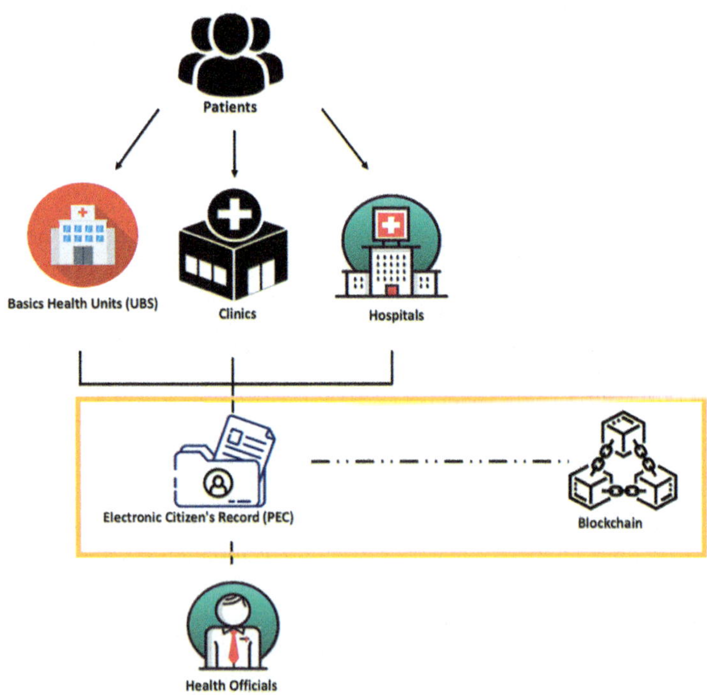

Fig. 1 Indication of the use of blockchain in the interoperability at EHR

right to access any information contained in his/her medical record, so the patient is the main focus of this flowchart; in addition, a safe system to protect the content of the EHR is under the responsibility of health institutions. In this way, Blockchain emerges as an effective tool for the secure communication of patient information between health agents and institutions as a solution. The concern was to understand the functionality and applicability of this emerging technique for digital health so that it was possible to analyze the impact of Blockchain on data security [28].

In order to better understand the use of Blockchain in EHR, the following scenario was proposed: We have a patient who is currently seeking primary health care for consultation with health professionals to retract a complaint of sudden illness. However, after this consultation, this patient was referred for exams, returned to the same primary health care, and was seen by another health professional. With his clinical attendance history, the patient ended up going directly to a hospital or specialized clinic. In this scenario, the EHR is unique and should contain information on all consultations, clinical results, and patient diagnosis.

The patient only has a single EHR (asset). However, access to data must occur in a decentralized manner and with interoperability. Thus, as indicated in Fig. 1, the Blockchain will assist in the secure information transmission process, ensuring system interoperability.

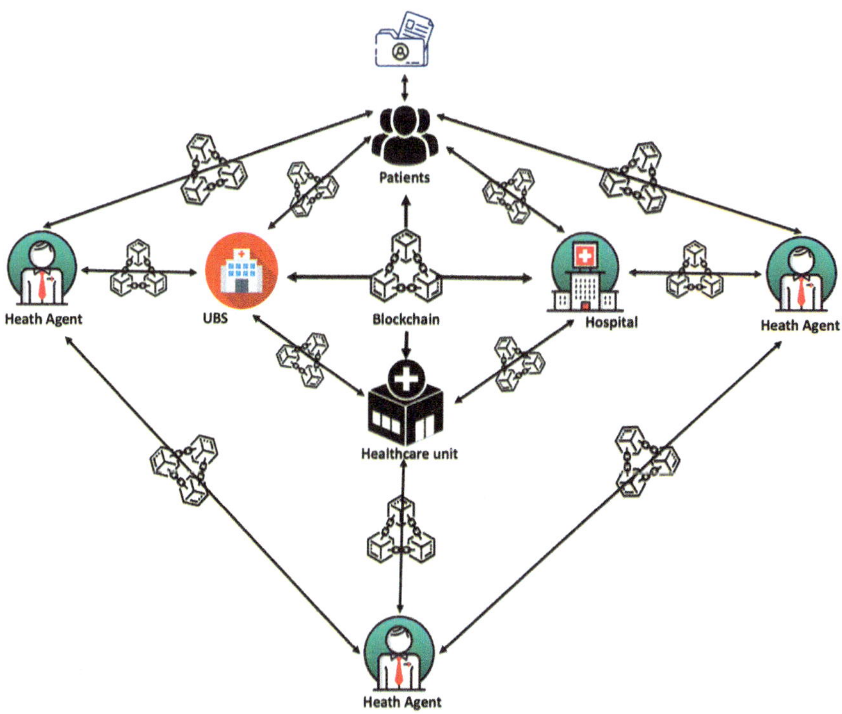

Fig. 2 Network distribution of blockchain in relation to the patient data

6 Discussions

After understanding the Blockchain functionality, an in-depth investigation into data security using symmetric and asymmetric cryptography during information transactions was necessary. Then, it was identified that the authorization of information sharing occurs within the transactions, where these transaction listings depend on a user who initiates the transaction (outgoing) and a user who receives all this information (incoming). For this information transaction to occur securely, cryptography is used within the Blockchain, specifically asymmetric cryptography.

6.1 Symmetric and Asymmetric Encryption

Analyzing the context of the EHR, suppose that a patient initially goes to primary health care to report on some chest pain. After being examined and medicated, this patient is seen in primary health care and is then referred to the institution to a specialized clinic. In this way, primary health care (outbound) will share the EHR with the clinic (inbound), and for this transaction of information from the EHR to take place securely, asymmetric cryptography is then applied.

For asymmetric cryptography to occur, primary health care is enough to encrypt the patient's EHR information with the specialized clinic's public key. Once primary health care allows this transaction, only the specialized clinic with its private key will decipher the entire content of the EHR.

6.2 EHR Domain Model with Blockchain

The proposal for the EHR domain model with Blockchain will be presented according to the schematics in Fig. 1. We have a patient who can go to a hospital, clinic, or primary health care. Moreover, during the service, the institution has access to the patient's EHR, and this document can be managed by health agents (administrators, doctors, and nurses), and ideally, it should contain the entire clinical history.

That is why it is essential to identify who will be the network's founder to allow users to communicate and information transactions. As stated in the previous section, it is ideal to have an information security management system to manage and be responsible for the system. In the rules presented in the previous section, the person responsible for managing and retaining knowledge of the system's functioning is part of the Senior Management.

After the definition of the founding originator, some criteria must be followed for the construction of the EHR domain model with Blockchain, which are:

- Definition of members for the network, meeting the criteria and policies stipulated and determined in the configuration of the channel.
- Create an order service in the system that defines transactions.
- Develop criteria and policies within the channel configuration.
- Define how many channels will be available in the system and which users are allowed to access the information.
- Develop chain codes and link them correctly to the channels defined by the founding originator.

It is recommended that the EHR domain model be built within the Hyperledger Fabric tool since it is a free domain and guarantees the auditing of the information inserted in that system.

Currently, Hyperledger Fabric is increasingly gaining evidence in its functionality and being a tool that adapts according to the user's needs. Furthermore, it is for this reason that for the construction of chain code on this platform, it is essential to re-establish an ISMS in hospitals, health clinics, and primary health care [33].

The criterion established here is just a proposal for constructing the EHR domain model in the Hyperledger Fabric tool, and its implementation is of future interest to ensure the functionality and applicability of the ideas presented in this study.

During the development of this study, it was found that Blockchain is a relatively new technology for digital health and is being widely studied by several researchers around the world. Currently, essential publications in the area can be found in the scientific journal Blockchain in Healthcare Today. This is the first worldwide peer-review journal that brings together several 2018 articles and contributions to the development of Blockchain in healthcare.

At Blockchain in Healthcare Today, it is possible to find research related to case studies of the use of Blockchain for health technologies, present new business models and proposed tools as a solution, and avoid cyber-attacks on systems, data governance, and information in digital health mainly in EHR, and studies for continuing education about Blockchain.

As presented in the bibliographic review, in Brazil, all the technologies behind the RNDS project for the security of patient clinical information transactions in the e-SUS environment are being filed and developed in a blockchain pilot project. According to the IBGE, this fact is significant for ensuring that all federations in Brazil can count on a secure system that can be audited since 71.1% of Brazilians were served in the public sector. However, as it is a relatively new subject and with few experts in the country, there are no materials that expose the blockchain technology used in the RNDS.

On the other hand, Blockchain is not just a technology that is being disseminated in digital health directly related to EHR, and it is not just about this applicability in question. Furthermore, with this study, it was possible to become aware of several projects around the world that directly impact digital health, such as traceability of medical equipment in medical units and even the control of the drug distribution network as a result of the supply area in the pharmaceutical sector.

One of the most practical examples found in journals within Blockchain in Healthcare Today was that the Drug Supply Chain Security ACt (DSCA) is an agency responsible for managing the distribution of medicines in the USA [9]. This agency aims to ensure data traceability to guarantee confidentiality, immutability, integrity, authorization, and verification throughout the process.

7 Conclusion

More and more, we are generating health data through various technological resources, such as wearables, the Internet of Things (IoT), Artificial Intelligence, and electronic medical records. Patients can manage this data, so we control our financial resources via the web. It is essential to learn how to manage our health data, primarily if this benefits us directly. The Blockchain enables the data owners to use them for various personal and collective purposes and can share this data for studies of diseases, population health, new medicines, etc.

Throughout the development of this work, it was observed that the impact of the use of Blockchain on digital health and data security is a very interdisciplinary, challenging, and innovative subject. This study provided the first theoretical contact for implementing a blockchain system for digital transformation involving mainly EHR and possibly one of the first steps toward implementing and developing a system that meets the data security requirements within digital health in an academic context.

To implement and execute the model proposed here is crucial to involve an interdisciplinary team that meets technology and information requirements, software engineering, law, and administration of public healthcare systems. Biomedical Engineering professionals can mainly assist in database management, software development and support decision-making according to the regulations in force in the country.

In addition, the topics presented and discussed here are aligned with the need for legal compliance with local laws and regulations that have been implemented in recent years, including the implementation of Brazil's General Data Protection Law (Lei Geral de Proteção de Dados Pessoais—LGDP). It is essential to highlight that the development of new technologies to promote digital transformation in SUS or the private sector in the country must, mandatorily, comply with all jurisprudence.

It is also essential to emphasize and show that data protection is critical within digital health. This data protection is because, of course, within this area, it is necessary to handle and store personal data and sensitive data, especially in the EHR, which has such a complexity of information, and the traceability of a patient's clinical information is possible.

The prospects regarding the Blockchain for digital health in Brazil are pretty positive, one of the indications that this is an excellent way to guarantee data security and interoperability in the health data of citizens using SUS based on Blockchain.

References

1. Rede nacional de dados em saúde. (2021). https://rnds.saude.gov.br/
2. Abayomi-Zannu, T. P., Odun-Ayo, I., Tatama, B. F., & Misra, S. (2020). Implementing a mobile voting system utilizing blockchain technology and two-factor authentication in Nigeria. In P. K. Singh, W. Pawłowski, S. Tanwar, N. Kumar, J. J. P. C. Rodrigues, & M. S. Obaidat (Eds.), *Proceedings of First International Conference on Computing, Communications, and Cyber-Security (IC4S 2019)* (pp. 857–872). Singapore: Springer.
3. Agbo, C. C., Mahmoud, Q. H., & Eklund, J. M. (2019). Blockchain technology in healthcare: A systematic review. *Healthcare, 7*(2), 3–9.
4. Ajayi, P., Omoregbe, N., Misra, S., & Adeloye, D. (2018). Evaluation of a cloud based health information system. In M. F. Kebe, C. Gueye, & A. Ndiaye (Eds.), *Innovation and interdisciplinary solutions for underserved areas* (pp. 165–176). Cham: Springer International Publishing.
5. Asaph, A., Ariel, E., Thiago, V., & Andrew, L. (2016). Medrec: Using blockchain for medical data access and permission management. In *2nd International Conference on Open and Big Data (OBD)* (pp. 25–30). IEEE.
6. Atherton, J. (2011). Development of the electronic health record. *American Medical Association Journal of Ethics: Virtual Mentor, 13*(3), 186–189.
7. Ayeni, F., & Misra, S. (2014). Overcoming barriers of effective health care delivery and electronic health records in Nigeria using socialized medicine. In *2014 11th International Conference on Electronics, Computer and Computation (ICECCO)* (pp. 1–4).
8. Ayeni, F., Omogbadegun, Z., Omoregbe, N., Misra, S., & Garg, L. (2018). Overcoming barriers to healthcare access and delivery. *EAI Endorsed Transactions on Pervasive Health and Technology, 4*(15), 156515.
9. Chien, W., de Jesus, J., Taylor, B., Dods, V., Alekseyev, L., Shoda, D., & Shieh, P. B. (2020). The last mile: DSCSA solution through blockchain technology: Drug tracking, tracing, and verification at the last mile of the pharmaceutical supply chain with BRUINchain. *Blockchain in Healthcare Today, 3.*
10. Committee, B. I. S. (2019). *Survey on the use of information and communication technologies in Brazilian healthcare facilities: ICT in Health 2018.* Brazil, by Brazilian Network Information Center - NIC.br and Brazilian Internet Steering Committee - CGI.br
11. Diffie, W., & Hellman, M. E. (1976). New directions in cryptography. *IEEE Transactions on Information Theory, 22*, 644–654.
12. Ejaz, W., & Anpalagan, A. (2019). *Blockchain technology for security and privacy in internet of things* (pp. 47–55). Cham: Springer International Publishing.
13. Eysenbach, G. (2001). What is e-health? *Journal of Medical Internet Research, 3*(2), e20.
14. Gayvoronskaya, T., & Meinel, C. (2021). *Where does the hype end, and where does the innovation of blockchain technology begin?* (pp. 35–68). Cham: Springer International Publishing.
15. Gontijo, T. L., Lima, P. K. M., Guimarães, E. A. D. A., Oliveira, V. C. D., Quites, H. F. D. O., Belo, V. S., & Cavancante, R. B. (2021). Computerization of primary health care: The manager as a change agent. *Revista Brasileira de Enfermagem, 74*, e20180855.
16. Hira, A. Y. (2012). Digital health: New paradigm of convergence of information technologies focused the attention of health. Ph.D. thesis, Escola Politécnica, University of São Paulo.
17. Kumar, M., & Chand, S. (2021). MedHypChain: A patient-centered interoperability hyperledger-based medical healthcare system: Regulation in covid-19 pandemic. *Journal of Network and Computer Applications, 179*, 102975.
18. Li, P., Nelson, S. D., Malin, B. A., & Chen, Y. (2019). DMMS: A decentralized blockchain ledger for the management of medication histories. *Blockchain in Healthcare Today, 2.*
19. Mahto, D., & Yadav, D. K. (2017). RSA and ECC: A comparative analysis. *International Journal of Applied Engineering Research, 12*(19), 9053–9061.
20. Marrone, P. V. (2015). Saude 4.0 – Propostas para Impulsionar o Ciclo das Inovações em Dispositivos Médicos (DMAs) no Brasil. ABIIS

21. Martins, C., & Lima, S. M. (2014). Advantages and disadvantages of electronic health record for health institutions. *RAS, 16*.
22. McGhin, T., Choo, K. K. R., Liu, C. Z., & He, D. (2019). Blockchain in healthcare applications: Research challenges and opportunities. *Journal of Network and Computer Applications, 135*, 62–75.
23. Organization, G. W. H. (2009). *WHO guidelines on hand hygiene in health care: First global patient safety challenge clean care is safer care*. Geneva: World Health Organization (WHO).
24. Pellison, F. C., Rijo, R. P. C. L., Lima, V. C., Crepaldi, N. Y., Bernardi, F. A., Galliez, R. M., Kritski, A., Abhishek, K., & Alves, D. (2020). Data integration in the Brazilian public health system for tuberculosis: Use of the semantic web to establish interoperability. *JMIR Medical Informatics, 8*(7), e17176.
25. Rivest, R. L., Shamir, A., & Adleman, L. (1978). A method for obtaining digital signatures and public-key cryptosystems. *Communications of the ACM, 21*, 120–126.
26. Sagar, V., & Kaushik, P. (2021). Ethereum 2.0 blockchain in healthcare and healthcare based internet-of-things devices. In M. Dave, R. Garg, M. Dua, & J. Hussien (Eds.), *Proceedings of the International Conference on Paradigms of Computing, Communication and Data Sciences* (pp. 225–233). Singapore: Springer.
27. Santos, L. (2017). Healthcare regions and their care networks: An organizational-systemic model for SUS. *Ciência & Saúde Coletiva, 22*, 1281–1289.
28. da Saúde, M., & de Atenção à Saúde, S. (2018). e-SUS Atenção Básica: Manual do Sistema com Prontuário Eletrônico do Cidadão PEC, vol. Versão 3.1. Ministério da Saúde and Secretaria de Atenção à Saúde and Secretaria-Executiva. http://dab.saude.gov.br/portaldab/esus.php
29. Schaar, P. (2010). Privacy by design. *Identity in the Information Society, 3*, 267, 274.
30. Slee, V. N. (2000). The endangered medical record: Ensuring its integrity in the age of informatics. [National Health Informatics Collection], Triaga Press, St. Paul, 1st edn.
31. Soyemi, J., Misra, S., & Nicholas, O. (2015). Towards e-healthcare deployment in Nigeria: The open issues. In R. Intan, C. H. Chi, H. N. Palit, & L. W. Santoso (Eds.), *Intelligence in the era of big data* (pp. 588–599). Berlin/Heidelberg: Springer.
32. Stamatellis, C., Papadopoulos, P., Pitropakis, N., Katsikas, S., & Buchanan, W. J. (2020). A privacy-preserving healthcare framework using hyperledger fabric. *Sensors, 20*(22), 1–14.
33. TIFAC-CORE, S. P. (2018). On blockchain applications: Hyperledger fabric and ethereum. *International Journal of Pure and Applied Mathematics, 118*, 2965–2970.
34. Yu, H., Sun, H., Wu, D., & Kuo, T. T. (2019). Comparison of smart contract blockchains for healthcare applications. *AMIA Annual Symposium Proceedings, 2019*, 1266–1275.

Blockchain Based Edge Information Systems Frameworks for Industrial IoT: A Novel Approach

M. Parimala Devi, Mani Deepak Choudhry, R. Nithiavathy, G. Boopathi Raja, and T. Sathya

1 Introduction

To bridge the gap between the physical and digital environment, an intellectual milieu is rendered for the different sector to reduce the cost of production by using IoT, which are capable of sensing and communicating over the Internet [1]. IoT can be habiliment, automated device growing in numbers; the immense volume of data is congregated, managed effectively. IoT is a networked device that communicates to progress the upcoming applications transversely in all domains. In IoT, security becomes a significant apprehension as it deals with huge data from various domains and services [2, 3]. Early the IoT was in data processing and storage in the cloud environment, it deployed in the healthcare sector, industries who were involved in manufacturing monitored [4]. There are security breaches and time sensitivity in the IoT data process as the number of systems interconnected. Lately, blockchain technology is advanced, which is capable of handling applications that may assimilate peer-to-peer distributed storage and encryption along with various technologies.

The main fundamental technology of block chaining is digital currency, such as Ethereum and Bitcoin [5]. The decentralized system is a prime solution of the trust among the nodes by consent and verification of nodes [6]. Blockchain includes

M. Parimala Devi (✉) · G. Boopathi Raja · T. Sathya
Department of ECE, Velalar College of Engineering and Technology, Erode, Tamil Nadu, India

M. D. Choudhry
Assistant Professor, Department of Information Technology, KGiSL Institute of Technology, Coimbatore, Tamil Nadu, India

R. Nithiavathy
Department of Computer Science & Engineering, Sri Krishna College of Technology, Coimbatore, Tamil Nadu, India

S. Misra, A. Kumar Tyagi (eds.), *Blockchain Applications in the Smart Era*, EAI/Springer Innovations in Communication and Computing,
https://doi.org/10.1007/978-3-030-89546-4_2

confidentiality in the transaction with security, suitability, invariance of data, and fault tolerance widely used in finance, transportation, and encryption technology, etc. [7–14]. The democratization of the network of things and the IoT system is emphasized in blockchain by some companies such as IBM [15–17]. There is no need for third party in IoT, where the blockchain by itself proposes storage distribution and generation of data that are recorded in an incontrovertible and certifiable manner [18].

All the action traced in the IoT network, which is triggered according to the timestamp decision to fulfill the governing acquiescence and operations in the system. The IoT transactions are secure in sharing the data, and the blockchain yields steady and trustworthy environment. Though the blockchain and IoT are emerging technologies, there are deficiencies in approaches which consider management of data, and its provision in the architectural design results in developing IoT blockchain system. The chapter flow in the next section describes the work carried out to understand different blockchain systems for industrial IoT. The following section of related work deals with an approach or framework that solves the current issues and challenges posed. The final section of the chapter deals with the experimental results and concluding remarks.

2 Related Works

With limited use of memory, battery resources, and computing proficiencies, sensors and actuators are the Internet of Things including mobiles, home appliances, and vehicles that are connected in the network of devices where the data are collected and exchanged for an application such as in the medical field, industries, transport, etc. [19–21].

For usage of a huge volume of data, they use cloud servers as a centralized system, which leads to high latency and bandwidth ingesting in the network. The data manipulation, e.g., altered and interfered by outsiders, in the cloud leads to a security breach of sensitive and vital information stored in the cloud storage [22].

The issue in the trust information system proposed by Satoshi Nakamoto [5] provides two perceptions: Bitcoin and blockchain. A bitcoin maintains value without the entity of financial centralized authority. The coin is held by a decentralized peer-to-peer network with actors where they validate. Blockchain is a mechanism that involves transaction monitored by actors. It holds an auditable ledger that is translucent, absolute, and protected. The blockchain protocol assemblies the data in a chain of blocks, where each one of the blocks holds a set of Bitcoin dealings completed at a specified time. Blocks are related and organized by a reference to the preceding block, creating a chain. To fund and activate the blockchain, system peers have to deliver the below functionality wallet services and mining routing technique and storage space [6].

Using the blockchain technology and the manufacturing controller system leads to lowering the cost of the process, resources supervision, and vigorous use, thereby avoiding threats and attacks. A peer-to-peer network is established by blockchain, which allows to share the maintainable cost by allocating the storage and computing needs in a centralized cloud. There is a problem in communication at a single point of a fiasco. This can be addressed by IoT working along with block chaining to maintain privacy using encryption algorithms. A tamperproof ledger [23] is used to resolve the dependability issue in the Internet of Things. Arisas et al. have highlight confidentiality, safekeeping, and performance issues in employing the blockchain technology in IoT [24].

For the cyber-physical social services, a cloud-based framework focuses on the trust mechanism and optimized performance by employing blockchain along with IoT applications [25]. A structured network that is heterogeneous allows data protection from information processing and communication protocol through sensors [25, 26]. There are some delay and issues in access control in a distributed manner to handle the classified information. To afford substantiation and secrecy, IPsec along with TLS is used, where it does not satisfy all requirements such as computing devices with limited resources and high cost. To the above problem, the solution is done by block validation along with the consent methods, but for complex blockchain it is difficult. They did, however, build on the constraints of recollection and computational features, power, and industrial Internet submission obligations to cloud computing in data layer management.

For example, we can take food safety guarantee by tracing the various food products which involve many members like a producer, nourishing, handling, circulation, etc., when there is an intruder or a break the chain in part of the blockchain leads to data leak results in slow down the process of finding out infected part may affect the lives of people, economical fall in markets in the circumstance of foodborne outburst [27]. A healthier regulator in these zones would surge food safety, cultivating the data distribution among contestants and dropping the exploration time in the case of a food-borne outbreak, which will save a lot of lives. The usage of blockchain technology and IoT deals with secure and reliable data. Together, the IoT and blockchain are implemented in smart cities, cars, etc. by adding new members in the environment and providing better-quality services and their adoption [28]. Scalability, confidentiality, and consistency problems related to the IoT paradigm can be tackled by blockchain technology. Combination of IoT and blockchain yields the following scalable decentralization, moving the central construction to P2P by eliminating the vital reason for failure and bottleneck [29]:

- **Uniqueness:** Using mutual blockchain system, members can recognize all solitary devices. Statistics provide and feed the scheme, which is absolute and uniquely classifies definite data that was provided by a device. Furthermore, blockchain can offer reliable circulated confirmation and agreement of devices for IoT applications [30].
- **Self-sufficiency:** Blockchain technology provides future application features, manufacture conceivable the growth of smart independent assets and hardware as a facility [29, 30].

- **Services:** The construction of an IoT network of facilities and data marketplaces can be hastened by blockchain, where communications among peers are plausible without authorities. Microservices can be simply positioned, and micro-payments can be made with full protection in a faithless environment [31–33]. Protected code positioning: enchanting the benefit of blockchain secure-immutable storage, encryption can provide security devices [27, 34].

The benefits of blockchain and the benefits of current IoT communications, such as fog computing, can be leveraged to balance the confines of blockchain and the IoT. For example, fog computing comprises rarer computational partial strategies such as gateways and where mining is done similarly as enterprises that employ IoT [35, 36].

Today's manipulators can necessitate many benefits using blockchain in several applications, i.e., using shared infrastructure while maintaining a level of security and privacy in a system. It makes blockchain unique, but creating blocks using the blockchain concept comes not easy; it requires a lot of mathematical and cryptographic operation [37–39]. It also requires a lot of time to compute such an operation. This is a disadvantage of this novel concept. Cybersecurity is also playing an important role in industrial IoT [40, 41], and the need for recent techniques for ICT was explained well [42].

3 Novel Tier-Based EIS Framework of Blockchain for Industrial IoT

Internet of Things (IoT) is a pioneering, novel, and groundbreaking computing model, which empowers every device (IoT) with communication, computation, and storage competence to link traditional Internet. IoT applications are found in various fields since it is a massive field. IoT devices in the healthcare sector reside on the patient interact straight to the healthcare organization through a communication network. These IoT devices unceasingly diffuse substantial information to the healthcare organization, such as blood pressure and heart rate. In the transportation domain, IoT is evolving as IoV (Internet of Vehicles), which is mainly responsible for safe traffic, less fuel consumption, and optimization of travel times. In a nutshell, IoT is an upsurging innovation that has an enormous impact on various stakeholders.

Blockchain (BC) is a technique initially planned for cryptocurrency, and the financial industry can be exploited within IoT-enabled networks to attain anticipated security and confidentiality. The key knowledge behind the BC is to extant a communal and public "open ledger," where every contributing node can get anticipated information without any necessity on a third party. So, a fully distributed method is espoused in BC, which, as a result, upsurges the effectiveness of the network in terms of security and transparency as identical update and precise and reliable evidence are comprised of every node.

In a BC network, solving complex consensus algorithm was carried out by miners, the nodes participating in the network which can add or modify data in BC. Many existing algorithms, such as Proof of Work (PoW), Proof of Stake (PoS), etc., use consensus algorithms where nodes are allowed to access and modify information in BC only if it can solve the PoW algorithm. This is a huge challenge in the modern day of IoT-based framework design in industries.

Blockchain methodology assimilates consensus, distributed storage, encryption, peer-to-peer transmission, and other technologies. The key values of BC are attracting much research and development in a wide range of industrial Internet of Things (IIoT) and become an important topic of discussion among researchers. With limited storage, computing, and bandwidth, IoT smears to a large number of edge sensor devices. In the field of industrial Internet device, layer requires hundreds of data sources. The disseminated and enormous data traffic retorting to quality of service (QoS) requirements becomes tailbacks. Blockchain approach as the essential methodology of digital currency, such as Ethereum [35] and Bitcoin, resolves the delinquent trust-building among nodes of a devolved system through the substantiation and distributed node consensus method, thus effecting value transfer while disseminating information and comprehending the noteworthy conversion of current network architecture from "information Internet" to "value Internet."

Generally, IIoT comprises edge computing devices that are resource inhibited. The minimum levels of computing power, battery capacity, and memory are the physiognomies of an edge computing IIoT. Thus, to balance computing and resource consumption, the system requires a trivial algorithm. It is tough to comprehend the valuable data interconnection because IIoT lacks operative data sharing method.

The overall organization of this chapter is structured and demonstrated as per guidelines in [42]. The contribution of this chapter is to propose a novel, efficient modified blockchain framework for industrial IoT and its edge applications. The framework can be explained by defining the three-tier architecture along with an improved consensus algorithm, and its performance results are compared with the existing approach.

3.1 Blockchain Architecture for Industrial IoT

In this section, we provide the layered architecture [43], which is found effective and efficient in many industrial applications. The architecture proposed in Fig. 1 consists of three distinct or unique domains:

1. ***IoT device domain:*** IoT device data transmission through a wireless channel is carried out through this domain. It will be helpful in the generation of useful information.
2. ***Communication domain:*** It acts as a message gateway between tiers 1 and 3. The relay nodes are important as it relays information to BS, which improves latency and delays.

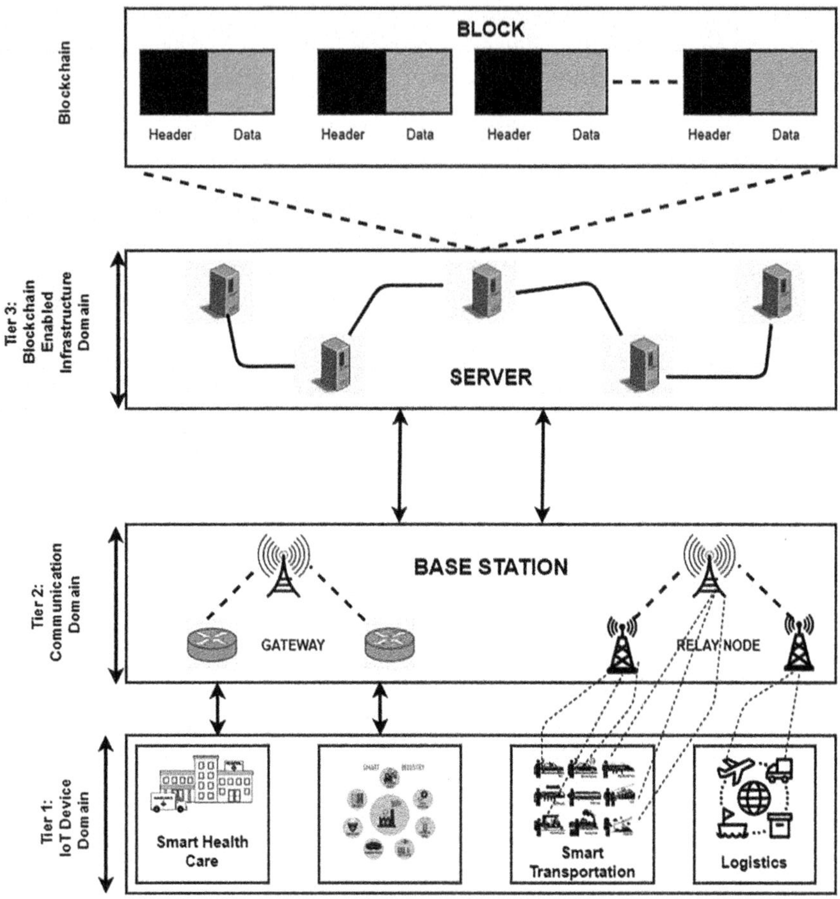

Fig. 1 Layered architecture of blockchain for industrial IoT

3. ***Blockchain-empowered infrastructure domain:*** Blockchains are in the layered architecture, which is represented by the back-end domain. The block has two major components: block header and block data, to validate the information and solve the PoW algorithm.

3.2 Operations in the Layered Blockchain IIoT Architecture

The main tedious process is linking IoT devices and blockchains. To mainly address this issue, the layered architectures of blockchain are proposed to provide connectivity. It operates in three parallel phases:

Phase 1: *Generation of information of IoT devices.*
Phase 2: *Dissemination of information.*
Phase 3: *Add or modify the record.*

The operation of layered architecture is thus straightforward, and various IoT applications are making use of it.

3.3 System-Level Architecture Model

The system-level model in Fig. 2 gives a detailed view of the task or process involved in each domain of the layered architecture. It is a sectional architecture, where designers shall substitute or enhance any new component as each layer is dissociated from other layers without distressing the other parts of the system.

The physical layer of IoT system-level architecture model with the capabilities of communication, computing, and data storage embraces abundant linked devices. Self-organization is essential because physical devices don't have any common Internet protocols (IP), such that routing management is the key mechanism provided by the connectivity layer. The other functionalities, such as services, managing networks, security maintenance, and breaking of messages, are provided by this layer.

The final layer is all about servicing blockchain, to provide several mechanisms, such as blockchain mechanisms, managing identity, consensus, and peer-to-peer (P2P) communication, by establishing common services with all modules.

The participants of the BC network have a self-copy of the ledger, which is carried out by the distributed ledger, that is, the consensus of shared, harmonized, and replicated digital data that are widespread along with this network. To handle the configuration of devices and data detection of physical sensors, safe storage spaces are offered by this layer. The replication of any modification is done in all copies within seconds. The ledger can be either permission or permission less, concerning if a peer can be run to validate transactions by anybody or only authenticated members.

The system-level model contains big data analytics function which is a powerful method that empowers the BC for online storage of data. The transactional data is an impeccable source for further investigation, which is stored as an organized form of ledgers. The components involved are authenticated to access all details in a single network.

The smart contract is used to realize access and modifications in the ledger, which is instantiated by the client as a code and is installed in each peer network.

The module which manages the event directs it every time either to fulfill the precondition of a smart contract or when a ledger gets a new block. The accessibility and management of the network can be done through the BC network service provider as services are exposed by the API interface. The data from physical

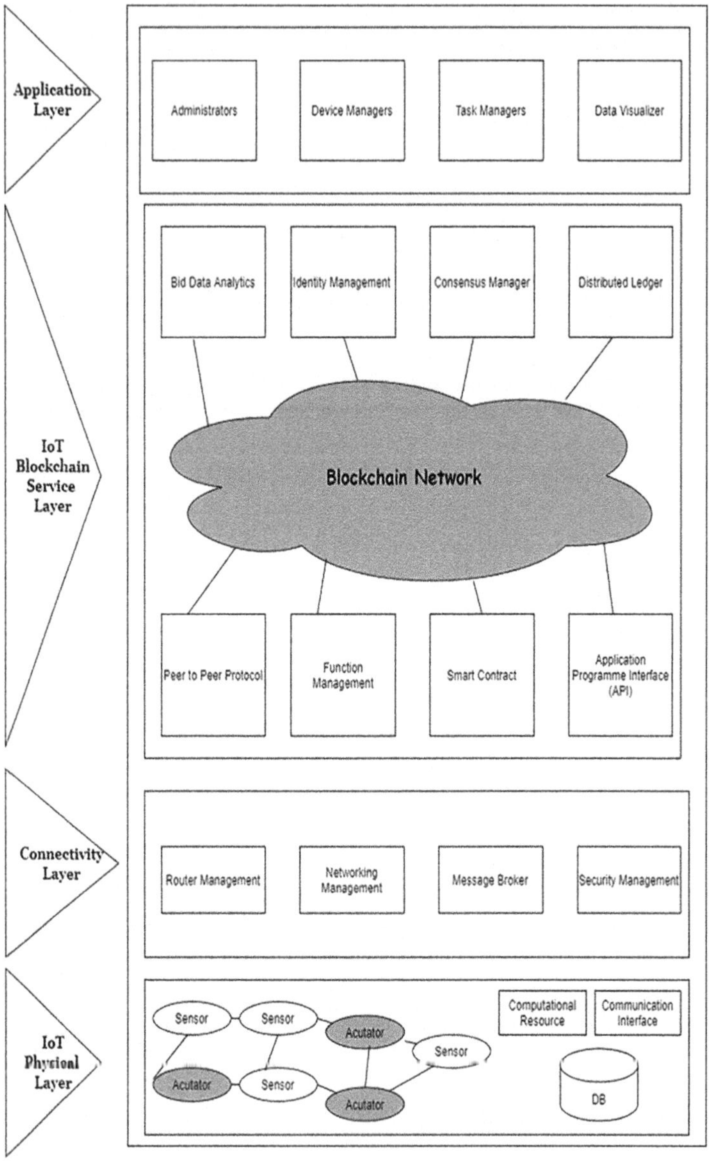

Fig. 2 System-level layered architecture for blockchain-based IoT framework

devices, controlling and deployment of the devices are foreseen by the application layer, which is the topmost layer of the model through various interfaces [43].

3.4 *Interaction Model for the Proposed Layered Architecture for Blockchain-Based IoT Framework*

The app client is used to deliver intense services, such as enrollment of the user, registration of the device, and generation of task services, and submit the proposals of transactions to the BC network through an instinctive. The registration is done before submitting a transaction where an explicit participant is provided with a certificate, which encompasses private keys to sign it is an essential part.

Figure 3 defines the workflow-based interaction diagram and gives a clear understanding of its components. It incorporates not just a user service outline but also a methodological setup, in which the smart contract and dispersed ledger are disclosed to the application.

The IoT server is responsible for generating a new job or device, which is carried out through the device owner. In turn, to achieve some secured operations, the BC network accepts the request from the server and processes it. In real time, the device that has acquired data that is sensed or the modification in the status can be sent back and also instruct to handle the job request from the client.

The individual owner who is related to the physical device is allowed to submit transactions right away to the BC network as the identity of the device owner is authentic.

The threshold defined by the smart contract compares the data which is detected or the status affixed in the ledger. The device owner is notified by generating a warning if the compared value exceeds the threshold defined.

3.5 *Transaction Flow of BC Framework*

The comprehensive transaction performance process of the BC network is illustrated in Fig. 4. To submit proposals of the transaction, the client application has to get the authorized permission for it. The permission can be obtained through credentials issued by the service which manages the identity.

The clients who wish to join the network are validated by the identity manager that clasps user IDs. In the BC network, the client application sends transaction tenders to peers. The BC network and the client application between communications supervene the SDK of the application. The peers can be of two types: either endorses or committers. Endorsers can feign and sign transaction offers, retort to conceding, or repudiate endorsements; committers authenticate results of transaction and write the block of the transaction once to the ledger. Each peer of the endorser invokes the smart contract to receive and implement the offer of transaction in its simulated environment. The results of execution won't be reproduced in the ledger. The endorser peers record the read data from the present state at the time of simulation of transaction and write data after the execution of the transaction by simply apprehending the RW sets. For authorization review, ciphering of the RW

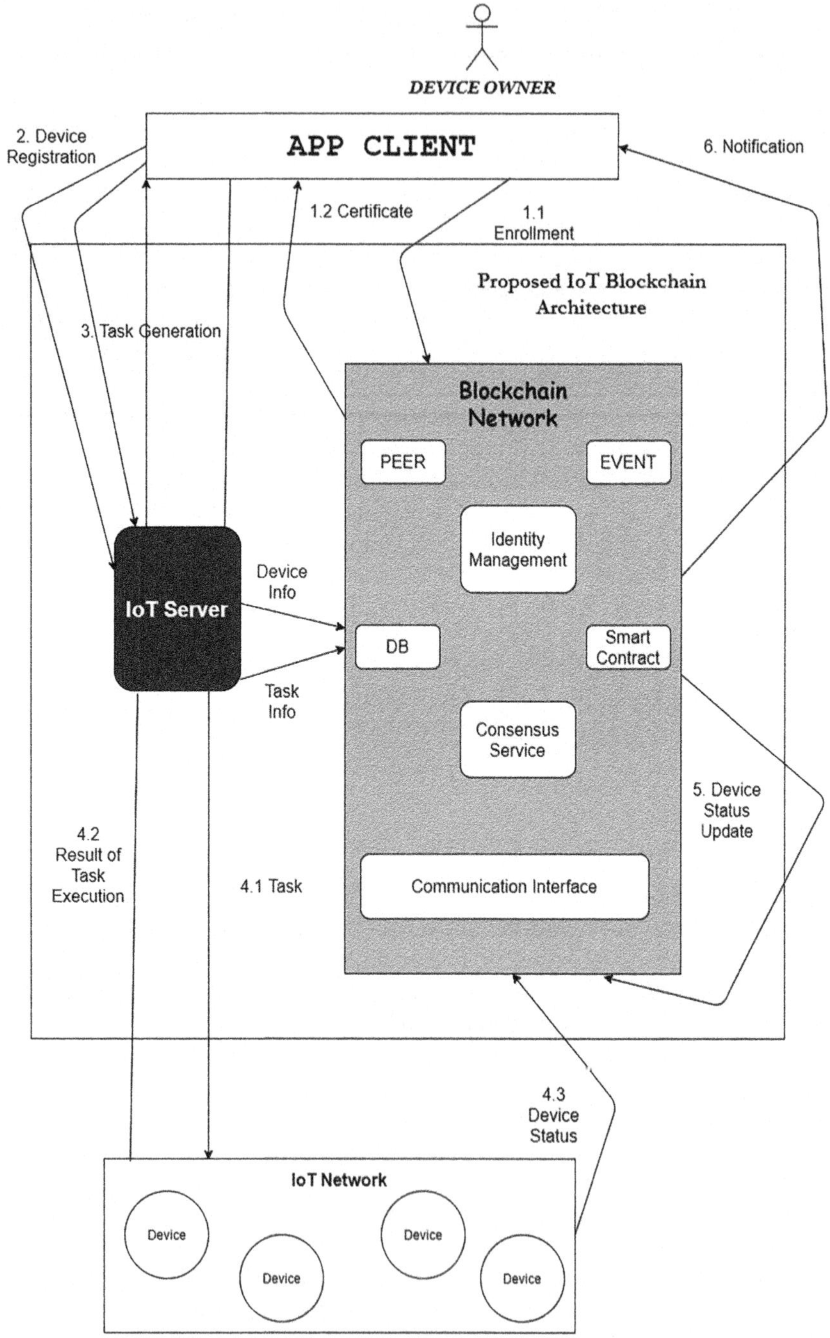

Fig. 3 Proposed IoT BC framework workflow diagram

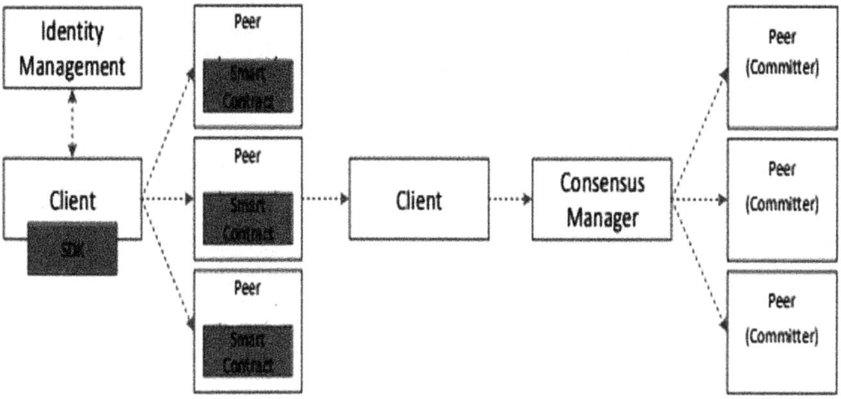

Fig. 4 Workflow implementation: a detailed diagram

sets and sending back the proposal rely on the client application was carried out by the peers of the endorser.

The signatures are to be approved by the client to validate and oversee if the stated endorsement policy has been fulfilled. The client packages the transaction along with the RW sets, and it is combined and submitted to the consensus manager. In parallel, with the help of signed transactions across the network, consensus occurs and RW sets are submitted, and peers of committer are distributed with the block of this order. The legalization of the transaction is carried out by each peer of the committer by comparing present state and RW sets to find if they are matching. If the simulation of the endorser results in the same as the present state, the data of reading is still available.

After the transaction is authorized by a peer of the committer, the state can be updated with the write data from RW set accordingly based on the transaction written to the ledger. Finally, the client application is notified at different times whether the transaction submitted is a success or failure by the peers of the committer asynchronously. Each peer of the committer notifies event occurrence to the client application which occurs when enrolled for events.

3.6 Improved Consensus Mechanism for Blockchain

Figure 5 demonstrates the PoW technique of the BC network, in which a PoW puzzle was created first by each miner. Secondly, broadcasting of the created puzzle was carried out by a node in BC, which is observable and reachable to each contributing node. But embracing, accessing, and modifying data in BC can be carried out only by the nodes which resolve the PoW mechanism.

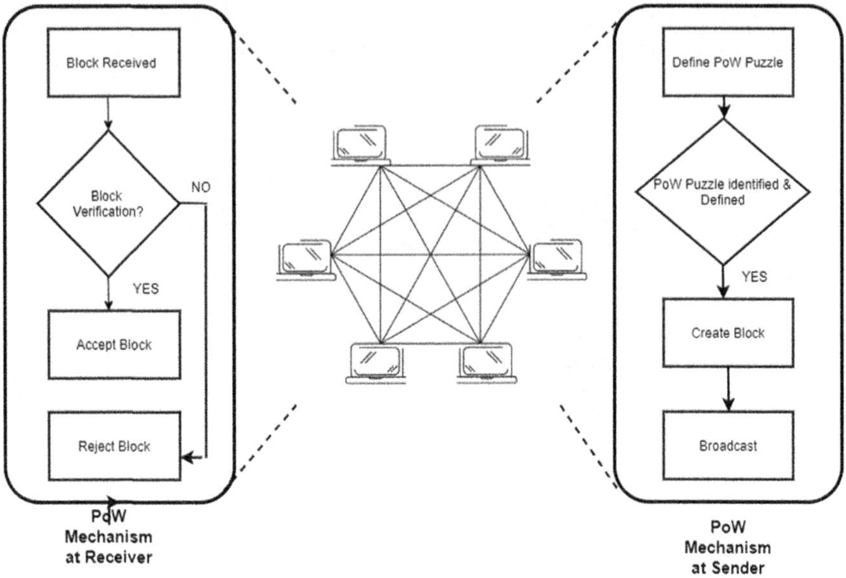

Fig. 5 Consensus mechanism for blockchain

The central administration doesn't have BC. Network miners independently generate the blocks. The same inference is reached and fabricated the identical common record of each node, thus attaining global consensus by a single node that uses information that is transmitted through the apprehensive connection. The whole chain of blocks was managed by the nodes that are complete, which validate it. In the main chain, if numerous nodes have identical blocks, then it can be concluded that consensus had reached.

The steps involved in the consensus technique are block authentication and the most widespread chain assortment. The node achieves these steps autonomously. In the network, firstly, the blocks are broadcasted, and when a new block is acquired by each node, it retransmits to the neighboring nodes. To ensure that only legal blocks are broadcasted, the blocks perform block authorization before this retransmission. The following is the explicit checklist to be followed:

(i). Block design (structure).
(ii). Verifying hash of header with difficulty established (met or not).
(iii). Limit of block size.
(iv). All transaction verification.
(v). Timestamp validation.

BC defines one parent for one block, but, in some circumstances, at a single point in time, miners produce new blocks, leading to have one parent with many children. This divergence resulted in the chain. Selecting blocks to be part of the main chain while rejecting others is the final step.The communication domain

```
Consensus Algorithm:
Input Transaction(X)
Output: True or False
Verification of Requester:
if (hash((X.requester)=X(output[2]) then
      return F
else
      if (X.requester-PK redeem X.requester-Signature) then
              return F
      end if
end if
Output Verification:
if(X.output[0] - (X- 1).output[0]) + (X.output[1])-(X- 1).output[1]>1) then
      return F
end if
Verification of Requestee:
if (X.requestee-PK redeem X.requestee-Sign) then
      return T
end if
```

Fig. 6 Algorithm for consensus

produces transient communication in IoT devices, which is the main feature of the IoT networks. In constrained time the delicate information which might proliferate must be shared among nodes. Thus, data can be made accessible for other nodes at a particular point in time by providing an optimized consensus algorithm in miners. So, improved consensus mechanism was deployed in the architecture, which helps applications with resource-constrained environments.

A time reliability algorithm is proposed to eliminate the problems faced by the old mean exhaustive procedure. The block creator is arbitrarily nominated for this procedure. In each block the CH waits for T time randomly before producing a new block. The block generations carry over a while, and when the number of blocks exceeds the threshold (based on the environment of the network and performance necessities), the CH will discard the blocks. So, to verify the block before attaching it to the chain, Fig. 6 shows the improved consensus algorithm, which will do the process:

The requester hash is compared with output transaction by CH, and if the requester agrees, output [0] will increase to 1 or else output [1] will be improved to 1.

In the transaction verification process, CH checks for output [0] fruitful transaction or failure transaction output [1]. Then, the requested signature is verified.

4 Experimental Simulation

In this section, the performance of the Lightweight Acquired Blockchain Framework (LABF) blocks is assessed to examine the viability of IIoT BC platforms. 20 nodes CH is the default arrangement for simulation. So, to assess the feasibility, different nodes were installed on different platforms. The framework proposed the cluster node as responsible for block generation, verification, and consensus because CH nodes have the authority to generate blocks. According to the data rate of sensor and delay of data gathering and block production, fine-tuning can be done in the gateway block production time. In Fig. 7 the comparison of results with the typical starting point based on the single simulation, which took 120 s and competed ten times, is illustrated. The left axis signifies the CH block verification time, and the right axis denotes the percentage of the transactions authenticated (PTA).

Due to the ungainly common trust by CH, the indulgence time of the two approaches was the same when LABF block substantiation starts. Unreliable trust was established between CH as many blocks were generated and validated by Resource Constrained Layer Block Chain (RCLBC). The processing time was reduced by LABF in comparison to baseline as LABF requires only a small communication part in the block, which is a newly generated block. Additionally, there are increases in a subsidiary conviction of CH and in block verification and a progressive reduction in transaction volume.

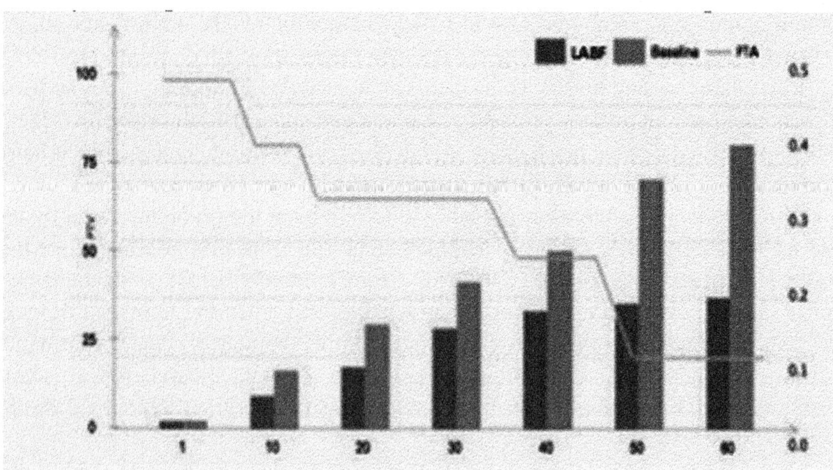

Fig. 7 Block verification performance evaluation

The integration of trust mechanism with block substantiation scheme accomplished by CH reputation module enhanced the authentication process.

Through the above experiment, the intricacy and network overload of this algorithm are analyzed. The analyzed results of the experiment show that there is an increase in the effectiveness of BC with network scale. Data flow and transaction flow are separated by the BC system. The transmission dormancy and lower packet outlay were taken care of by LABF. When there is an increase in CH, package outlay increases. Mutual trust is established between the blocks when CH produces more blocks as time goes by. Inversely, when the validation of blocks increases, the transaction validation of numbers decreases. Thus, there will be fixed processing time and verification of several transactions to be carried out.

To provide a complete solution, numerous investigational tests were carried out with the help of different performances.

$$\begin{aligned}\text{Service Execution Time} = {} & \text{Transaction Appeal Time for transmitting} \\ & + \text{length of ack acquired by Web Client}\end{aligned} \tag{1}$$

As per Eq. 1, it is understood that the execution time of service is the total appeal time of the transaction for transmission plus Web client acknowledgment length. Figure 8 illustrates the analyzed cost of service execution time on registration of device, which is the first study. To undergo this study, devices are segregated into four groups as 50, 150, 200, and 250, and their information is given to the proposed framework. Through Hyperledger Caliper [38], implementation was carried out. With the help of indicators, set users are permitted to configure particular execution of BC use case script. The implementation time was recorded as Min, Avg, and Max to perform this transaction in the proposed BC platform. The four groups of devices recorded different min, max, and average times. The group which has 50 devices shows 2262 ms as the minimum time, 2286 ms as the average time, and 2375 ms as the maximum time. The second group which has 150 devices recorded the min time as 2257 ms, the avg. time as 2335 ms, and the max time as 2801 ms. The third group of 250 devices delivered execution time as 2254 ms for the min time, 2585 ms for the avg. time, and 3004 ms for the max time. Finally, the fourth group, which has 500 devices, recorded the transaction execution time as 2267 ms for the min time, 2923 ms for the avg. time, and 4013 ms for the max time.

The next study is carried out to evaluate the execution time of service in storing data that are sensed in the BC network. The devices can appeal to Representational State Transfer (REST) server for API Sensor Reading as they have HTTP Client. REST server drew the implementation results from the BC network and sends them back to the device when sensed data is added to the BC. Figure 9 shows the estimation results of the execution time on accomplishing sensor reading transactions.

In the third study, evaluation of the BC network system performance is carried out when records that are sensed are stored in the distributed ledger. In Fig. 10 querying of data records in BC and their implementation time is measured. The min,

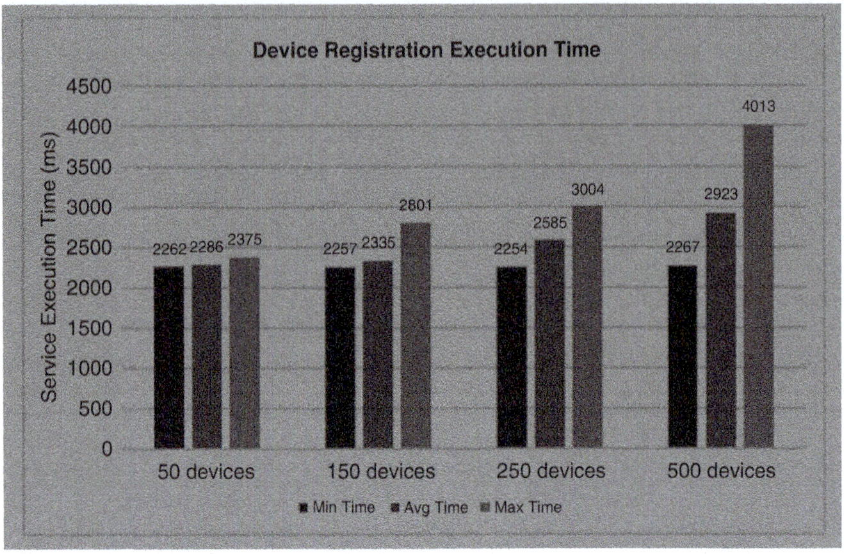

Fig. 8 Device generation performance study graph

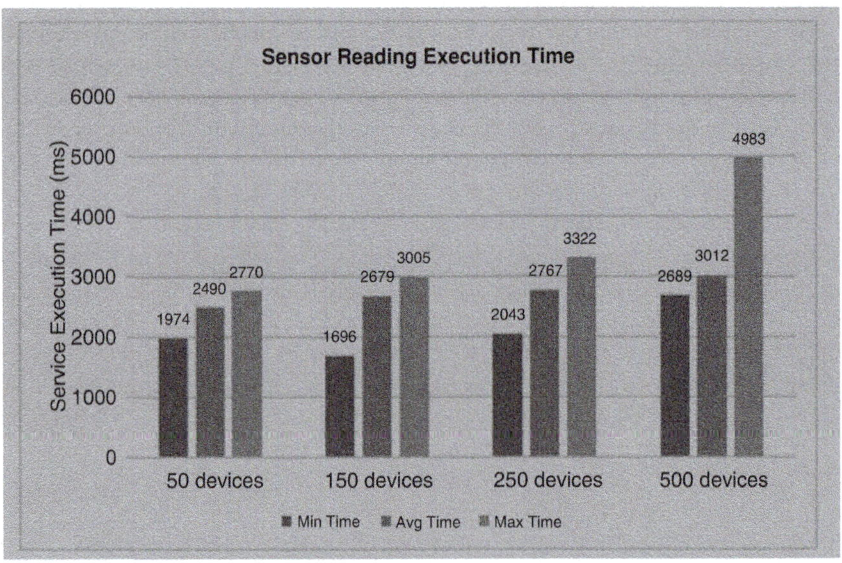

Fig. 9 Graph for sensor reading performance study

avg., and max delay times in ms taken by the proposed framework to repossess the sensing records were noted down ten times at arbitrarily selected system resource consumption levels.

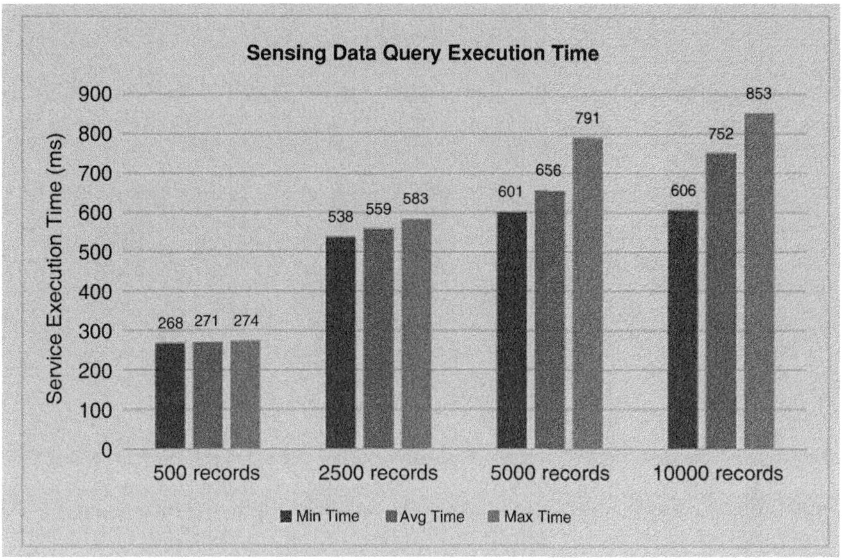

Fig. 10 Sensing data query performance graph

The proposed approach shows BC platform performance is significantly highlighted by the properties that play an essential role are considered for this study to compare the overfed platforms.

The selected system [39] setup was utilized for the analysis of this study. Through 50 peers, simulation was run for 60s during which 950 transactions were executed. Figure 11 illustrates processing overhead evaluation. As depicted in the graph, when compared to selected system processing overhead, processing time is reduced by 18% by changing the number of blocks from 10 to 60.

Permission-less BC network allows anyone to participate who is unidentified, through which most systems are developed. This depicts that there is neither privacy in contract nor privacy in a transaction that is produced. These systems issue their tokens to incent exclusive mining or to trigger the execution of smart contracts to alleviate nonexistence privacy. The transaction rate and rapidity can be significantly affected by undesirable links with cryptocurrencies.

In addition, as the token used in both systems must be unified, the BC network hinders the interaction with other distributed systems. In contrast, the proposed system lessens the peril of malicious code presentation through a smart contract intentionally by a participant where it is built on a permission network. All the activities of the participants are recorded on the BC in terms of affirmation policy given for the network and type of transaction as participants are known to each other. Additionally, in IoT devices, many systems simply deploy the full nodes to attain time-consuming mining as they are lagging in resource-constrained IoT devices. The consensus algorithm is limited to work with constraints, so the IoT resource-

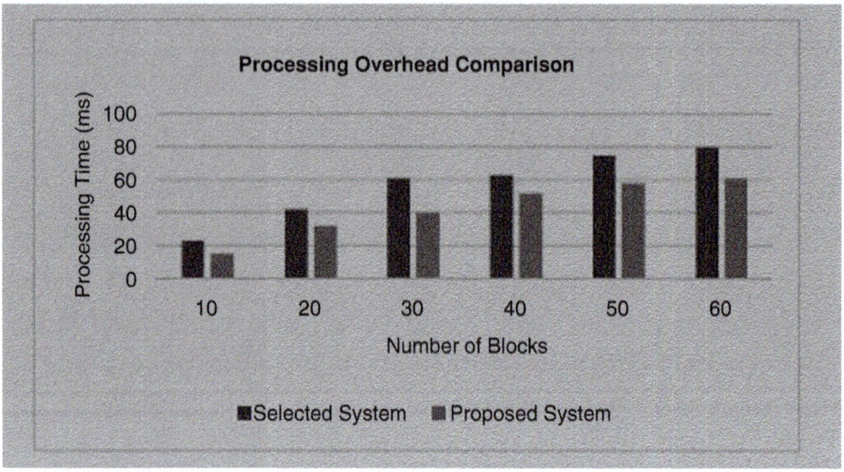

Fig. 11 Comparison of processing overhead and its performance analysis graph

constrained architecture has always been the chief interference in assimilating IoT with the blockchain. But present works deploy such huge algorithms on other devices, which are part of IoT-like gateways. However, these have limited storage space.

To authenticate transactions and block full nodes with the whole BC should be deployed on the gateway but many BC frameworks don't provide these requirements. If this is provided, gateways act as self-reliant targets and stand as the first line of defense as they link between the Internet and device. An Light Weight (LW) solution neglects the integration of the BC platform into IoT devices, and no modification is required and is proposed.

In the proposed model, the BC acts as a peripheral service to make the network provide secure storage, which is consistent. In addition, without extracting the whole BC, legalization of transactions produced by IoT devices is carried out. With limited capabilities, the proposed solution can be utilized in a wide range of IoT scenarios. Additionally, through Web service API BC network communicates with IoT devices, thus permitting cross-platform communication, which helps in the integration of the proposed and existing systems.

To prove the feasibility of the system proposed, a real-life smart space case study was implemented in the experiment. The proposed platform can be easily prolonged to numerous domains as it is built on a modular architecture.

The IoT sensors can be consigned to someone as it is linked to any product with remotely sensed data. Each party of a supply chain is allowed to access the ledger as it is made for sharing, such that recording of all the processing steps and storing it on the BC, including audit certificates, test evaluation results, and digital compliance documentation, can be carried out. This chapter aims to solve all the mentioned problems, and the demand for such IoT BC applications increased due to its various offers, such as permissioned network, user-friendly API, flexibility in architecture, and the latency of transaction is low, while the throughput of the transaction is high.

5 Conclusion

In this chapter LABF for edge computing-based IIoT applications was proposed. With a variety of resource capabilities, the BC operation is designed for edge devices. A time reliability algorithm is designed for limiting the generation of different blocks in the consensus cycle and also to diminish the asynchronous block operation in network delay. Accumulation of other node evidence is done by each node based on the generation of original blocks. To evaluate the effectiveness of BC, a high-throughput administration method is proposed. The highlight is the integration of BC with IoT, which is not an easy task that requires more attention from all directions.

Universal challenges are faced by IoT as millions and millions of devices are available online. Factors and manufacturers are always different from connected devices as they are diverse. Thus, in a secure way, uniqueness and interoperability are to be ensured. To enable IoT devices to have trusted interoperability for data and e-commerce, the BC platform gives innovative infrastructure and protocol for security. This paper provides an intuitive approach to address uniqueness and challenges in the data security of BC networks by delivering a decentralized IoT platform.

In the future adaptive block verification scheme can be implemented to verify and diminish the calculation cost of the block authentication process, thus improving the scalability and trust architecture delay. Furthermore, the blockchain architecture can be implemented through Resource Constrained Layer (RCL), Resource Extended Layer (REL), and cloud layers for expanding the areas, and quantitative researches can be formalized for investigation of functioning competence.

References

1. Díaz, M., Martín, C., & Rubio, B. (2016). State-of-the-art, challenges, and open issues in the integration of internet of things and cloud computing. *Journal of Network and Computer Applications, 67*, 99–117.
2. Fan, L., et al., (2018). *Investigating blockchain as a data management tool for IoT devices in smart city initiatives*. In Proceedings of the 19th annual international conference on digital government research: Governance in the data age, 2018, p. 100.
3. Razzaque, M. A., Milojevic-Jevric, M., Palade, A., & Clarke, S. (Feb. 2016). Middleware for internet of things: A survey. *IEEE Internet of Things Journal, 3*(1), 70–95.
4. Botta, A., De Donato, W., Persico, V., & Pescapé, A. (Mar. 2016). Integration of cloud computing and internet of things: A survey. *Future Generation Computer Systems, 56*, 684–700.

5. Nakamoto, S. *Bitcoin: A peer-to-peer electronic cash system*. Available online: https://bitcoin.org/bitcoin.pdf.
6. Antonopoulos, A. M. (2014). *Mastering bitcoin: Unlocking digital crypto-currencies*. O'Reilly Media, Inc..
7. Huang, X., Xu, C., & Wang, P. (2018). LNSC: A security model for electric vehicle and charging pile management based on blockchain ecosystem. *IEEE Access, 6*, 13565–13574.
8. Dorri, A., Steger, M., & Kanhere, S. S. (2017). BlockChain: A distributed solution to automotive security and privacy. *IEEE Communications Magazine, 12*, 119–125.
9. Lei, A., Cruickshank, H., & Cao, Y. (1832–1843). Blockchain-based dynamic key management for heterogeneous intelligent transportation systems. *IEEE Internet Things, 2017*, 6.
10. Kang, J., Yu, R., & Huang, X. (2017). Enabling localized peer-to-peer electricity trading among plug-in hybrid electric vehicles using consortium Blockchains. *IEEE Transactions on Industrial Informatics, 6*, 3154–3164.
11. Li, L., et al. (2018). CreditCoin: A privacy-preserving Blockchain-based incentive announcement network for Communications of Smart Vehicles. *IEEE Transactions on Intelligent Transportation Systems, 19*, 2204–2220.
12. Yang, Z.; Zheng, K.; Yang, K.; Leung, V. (2017). *A blockchain-based reputation system for data credibility assessment in vehicular networks*. In Proceedings of the 2017 IEEE 28th annual international symposium on Personal, Indoor, and Mobile Radio Communications (PIMRC), Montreal, QC, 8–13 October 2017.
13. Wang, J., et al. (2018). A Blockchain based privacy-preserving incentive mechanism in Crowdsensing applications. *IEEE Access, 6*, 17545–17556.
14. Tian, F. (2016). *An agri-food supply chain traceability system for China based on RFID & blockchain technology*. In Proceedings of the IEEE 13th International Conference on Service Systems and Service Management (ICSSSM), Kunming, China, 24–26 June 2016.
15. Abayomi-Zannu, T. P., Odun-Ayo, I., Tatama, B. F., & Misra, S. (2020). Implementing a mobile voting system utilizing Blockchain technology and two-factor authentication in Nigeria. In *Proceedings of first international conference on computing, communications, and cybersecurity (IC4S 2019)* (pp. 857–872). Springer.
16. Lu, Z., et al. (2018). A privacy-preserving trust model based on Blockchain for VANETs. *IEEE Access*.
17. Brody, P., & Pureswaran, V. (2014). *Device democracy: Saving the future of the internet of things*. IBM.
18. Zheng, Z., Xie, S., Dai, H., Chen, X., & Wang, H. (2017). An overview of blockchain technology: Architecture, consensus, and future trends. In *IEEE international congress on Big Data (BigData Congress)* (pp. 557–564).
19. Atzori, L., Iera, A., & Morabito, G. (2010). The internet of things: A survey. *Computer Networks, 54*(15), 2787–2805.
20. Gubbi, J., Buyya, R., Marusic, S., & Palaniswami, M. (2013). Internet of things (IoT): A vision, architectural elements, and future directions. *Future Generation Computer Systems, 29*(7), 1645–1660.
21. Hammoudi, S., Aliouat, Z., & Harous, S. (2018). Challenges and research directions for internet of things. *Telecommunication Systems, 67*(2), 367–385.
22. Atzori, M. (2017). *Blockchain-based architectures for the internet of things: A survey*. University College of London.
23. Gao, J., Asamoah, K. O., Sifah, E. B., Smahi, A., & Xia, Q. (2018). Grid monitoring: Secured sovereign blockchain based monitoring on smart grid. *IEEE Access, 6*, 9917–9925.
24. Arias, O., Wurm, J., Hoang, K., & Jin, Y. (2015). Privacy and security in internet of things and wearable devices. *IEEE Transactions on Multi-Scale Computing Systems, 1*, 99–109.
25. Wang, X., Yang, L. T., Xie, X., & Jin, J. R. (2017). A cloud-edge computing framework for cyber-physical-social services. *IEEE Communications Magazine, 55*, 80–85.
26. Zhang, G., Gao, Y., Luo, H., Sha, N., Wang, S., & Xu, K. (2019). Security performance analysis for relay selection in cooperative communication system under Nakagami-m fading channel. *IEICE Transactions on Communications, E102-B*, 603–612.

27. Buzby, J. C., & Roberts, T. (2009). The economics of enteric infections: Human foodborne disease costs. *Gastroenterology, 136*(6), 1851–1862.
28. Malviya, H. (2016). *How Blockchain will defend IOT*. Available online: https://ssrn.com/abstract=2883711.
29. Veena, P., Panikkar, S., Nair, S., & Brody, P. (2015). Empowering the edge-practical insights on a decentralized internet of things. In *Empowering the edge practical insights on a decentralized Internet of Things* (Vol. 17). IBM Institute for Business Value.
30. Gan, S. (2017). *An IoT simulator in NS3 and a key-based authentication architecture for IoT devices using Blockchain*, Indian Institute of Technology Kanpur.
31. Chain of things. (2017). Available online: https://www.blockchainofthings.com/. Accessed 1 Feb 2018.
32. Filament. (2017). Available online: https://filament.com/. Accessed 1 Feb 2018.
33. LO3ENERGY. (2017). Available online: https://lo3energy.com/. Accessed 1 February 2018.
34. Aigang. (2017). Available online: https://aigang.network/. Accessed 1 Feb 2018.
35. My bit. (2017). Available online: https://mybit.io/. Accessed 1 Feb 2018.
36. Samaniego, M., & Deters, R. (2016). Hosting virtual IoT resources on edge-hosts with blockchain. In *2016 IEEE international conference on Computer and Information Technology (CIT)* (pp. 116–119). IEEE.
37. Ethembedded. (2017). Available online: http://ethembedded.com/. Accessed 1 Feb 2018.
38. Raspnode. (2017). Available online: http://raspnode.com/. Accessed 1 Feb 2018.
39. Sawal, N., Yadav, A. Tyagi, A. K., Sreenath, N., & Rekha, G. (2019). *Necessity of Blockchain for building trust in today's applications: An useful explanation from user's perspective* (May 15, 2019). Available at SSRN: https://ssrn.com/abstract=3388558 or https://doi.org/10.2139/ssrn.3388558.
40. Hyperledger Caliper (2019) Available online https://www.hyperledger.org/projects/caliper. Accessed on 15 Jan 2019
41. Awotunde, J. B., Ogundokun, R. O., Jimoh, R. G., Misra, S., & Aro, T. O. (2021). Machine learning algorithm for cryptocurrencies Price prediction. In S. Misra & A. Kumar Tyagi (Eds.), *Artificial intelligence for cyber security: Methods, issues, and possible horizons or opportunities* (Studies in computational intelligence) (Vol. 972). Springer. https://doi.org/10.1007/978-3-030-72236-4_17
42. Misra, S. (2021). A step by step guide for choosing project topics and writing research papers in ICT related disciplines. In S. Misra & B. Muhammad-Bello (Eds.), *ICTA 2020, CCIS 1350* (pp. 727–744). Springer Nature. https://doi.org/10.1007/978-3-030-69143-1_55
43. Parimala Devi, M., Choudhry, M. D., Boopathi Raja, G., & Sathya, T. (2022). A roadmap towards robust IoT-enabled cyber-physical systems in cyber industrial 4.0. In *Handbook of research of internet of things and cyber-physical systems: An integrative approach to an interconnected future*. Apple Academic Press. [In Press].

A Secure Electronic Voting System Using Multifactor Authentication and Blockchain Technologies

O. M. Olaniyi, E. M. Dogo, B. K. Nuhu, H. Treiblmaier, Y. S. Abdulsalam, and Z. Folawiyo

1 Introduction

Voting is a fundamental component of a consensus-based society practicing a democratic system of governance. Citizens' voting rights must be confidential and strictly based on the "one person, one vote" principle exercised through either traditional or electronic voting systems [1]. Historically, most elections in developing countries are manipulated, and the announced results are frequently based on a nontransparent underlying electoral system [4, 5]. The electoral process is frequently characterized by problems ranging from ballot stuffing to bribery, manual counting errors, problems in the delivery of election materials from central locations to polling centers, external interference by agents handling election materials or voting database management, inconclusive ballots, high election-related costs, as well as time-consuming and nontransparent processes in general [5–8]. Therefore, voters are concerned whether their preferred choice in the electoral process will count and whether the votes recorded and collated truly represent the general interest of the populace [2].

In Nigeria, for instance, recent elections have adopted a semiautomated paper ballot system to address the challenges associated with the previous paper ballot system. However, despite these efforts by the electoral body in Nigeria, many

O. M. Olaniyi (✉) · E. M. Dogo · B. K. Nuhu · Z. Folawiyo
Department of Computer Engineering, Federal University of Technology, Minna, Nigeria
e-mail: mikail.olaniyi@futminna.edu.ng

H. Treiblmaier
Department of International Management, Modul University, Vienna, Austria

Y. S. Abdulsalam
DNA Lab, Department of Computer Science and Engineering, University Mohammed VI Polytechnic, Ben Guerir, Morocco

S. Misra, A. Kumar Tyagi (eds.), *Blockchain Applications in the Smart Era*,
EAI/Springer Innovations in Communication and Computing,
https://doi.org/10.1007/978-3-030-89546-4_3

of the challenges associated with conducting free and credible elections persist [3]. To provide a competitive advantage over the traditional paper ballot voting system, an electronic voting system requires security measures both during the authentication and vote casting processes [9]. In this regard, electronic voting machines have been shown to have technical and socio-technical vulnerabilities [9]. To achieve a competitive advantage, electronic voting systems must meet technical security requirements, such as eligibility, coercion freeness, availability, anonymity, integrity, correctness/accuracy, robustness, fairness, receipt-freeness, voter verifiability, and universal verifiability [10]. Huge varieties of security measures are suggested in the scholarly literature to meet these requirements, including biometrics, security firewalls, cryptography, smart cards, steganography, and cryptography (i.e., the combination of cryptography and steganography) [10, 11].

Existing centralized trust-based systems such as secure electronic voting in [13, 19, 33] are vulnerable to distributed denial of service (DDoS) and Sybil attacks from malicious users and provide no mechanism to track possible compromises of the electoral process by either internal or external actors [4]. Furthermore, they lack real-world deployment. To avert these possible electoral frauds, we propose a multifactor authentication (MFA) mechanism in combination with a public blockchain network that ensures the required integrity of a vote in a decentralized database environment on a cloud/edge computing [14] architectural arrangement. Facial recognition and radio frequency identification (RFID) techniques confirm voter identification, and verification averts possible insecurities through the authentication of invalid voters. Blockchain technology can help avert possible integrity, verification, and auditing issues, both during and after electoral processes. The proposed public blockchain contains transactions in the form of blocks, whereby each block is linked with the previous block using a cryptographic hash algorithm. The hash contained in the blocks makes use of the SHA-256 algorithm, and all blocks are distributed to each node on the network to avoid a central point of attack, which is a common weakness of existing electronic voting mechanisms.

This chapter presents the development of a secure decentralized electronic voting system using MFA and blockchain techniques. MFA is a security approach that uses more than one means of authentication from independently available credentials to accredit a person's eligibility to vote [10]. It is widely recognized as the most secure method for authenticating access to data or a specific application [10, 12, 38]. The more authentication factors exist to determine a subject's identity, the greater the authenticity trust. This chapter specifically addresses security flaws of semiautomated electronic voting systems that frequently occur in developing countries [3]. Existing problems that motivated this research to secure electronic voting systems are as follows: (1) centralized data at a single location, (2) vulnerabilities to cyber-security attacks, (3) the problem of validating voters' identity, and (4) lack of transparency, trust, and forgery during the electioneering process. Applying the proposed security mechanism will help increase the robustness in the authentication phase of future electronic voting systems and guarantee an uninfluenced, fair, and transparent election during and after the voting process.

The remainder of this chapter is organized according to the thought of author in [40] as follows: Section 2 gives a brief overview of similar works in the problem domain. Sections 3 to 5 present the materials, methods, and findings from the study. Section 6 contains the performance evaluation. Section 7 presents the security analysis, and Sect. 8 concludes the chapter and suggests future research endeavors.

2 Review of Related Works

Several electronic voting systems that include various security mechanisms have been proposed in the academic literature, some of which are based on blockchain. Table 1 shows a synthesis of previous approaches. Over the years, blockchain-based electronic voting systems have emerged widely and replaced paper ballot systems for securing and providing trust to ensure transparent e-voting. Several papers have demonstrated the use of blockchain by using different consensus protocols such as proof of stake (PoS) and proof of work (PoW). Hardwick et al. [26] proposed a blockchain-based decentralized system that offers voters a dynamic way of updating and changing votes during e-voting. Their approach supports complex voting situations but does not provide auditability, consistency, and user privacy. Kshetri and Voas [27] proposed an e-voting system that allows voters to pay a certain amount to cast votes without the problem of double-spending. This scheme, however, lacks scalability due to the excessive workload on nodes during simultaneous executions.

Bartolucci et al. [28] proposed an Ethereum-based blockchain system that implements the circle shuffle technique for registering. Their proposed system provides a trusted environment for transparent voting processes but necessitates the use of a trusted authority. The limitation of their proposed system is that if at any point the trusted authority goes malicious, then the entire system becomes compromised. Giving the sensitivity of information during a voting process, issues of susceptible rogue parties are to be avoided at all costs. Thuy et al. [29] proposed the Votereum blockchain-based voting system on Ethereum, ensuring security and privacy. Their proposed solution supports requirements such as verifiability and robustness but lacks resisting coercion and receipt-freeness. Yavuz et al. [30] proposed a voting application that uses smart contracts on the Ethereum blockchain and is based on an android platform. However, their proposed scheme lacks robustness and receipt-freeness.

Other blockchain platforms such as Hyperledger Fabric have also been used to ensure transparency during e-voting. Hyperledger Fabric is a private permissioned network that does not rely on the use of smart contracts or cryptocurrency. Previous research illustrated the use of Hyperledger Fabric for ensuring end-to-end privacy during e-voting, providing correctability and detectability, but also exhibits a lack of coercion resistance [31, 32]. Oke et al. [10] developed an MFA technique (i.e., a biometric fingerprint combined with a cryptographically secured smart card) to

Table 1 Synthesis of recent related works

Reference	Work description	Limitations
[12]	Mechanism for securing an e-voting system using MFA and cryptographic hash functions	The authentication mechanism proposed is a single factor that can easily be compromised
[2]	This voting system applies RFID and fingerprint technologies for voters' authentication	No extra layer of protection is added to the RFID technique, thus posing an open door for masquerading voters
[18]	Enhanced stegano-cryptographic model for a secure electronic voting system in the voting station	Neglects key requirements of an electronic voting system, such as checking the identity of the voters
[19]	Unimodal fingerprint biometrics and advanced encryption standard-based wavelet-based crypto-watermarking approach	The system stores the vote cast in a centralized server that a malicious third party can compromise
[7]	Applies an RFID reader module which senses the RFID tags with unique identity that is serially controlled by an embedded system	Similar to the limitation of Ref. [2]
[10]	MFA technique via biometric fingerprint and cryptographically secured smart card to secure an e-voting authentication process	Fails to secure the integrity of the cast votes stored in the database
[16]	A secure private blockchain-based electronic voting system for a university election	The system fails to address the issue of authentication to verify the voter's identity
[3]	Proposes blockchain technology to replace an existing manual or semi-digitized e-voting system	Neglects several key requirements of an electronic voting system, such as repudiation, confidentiality, and privacy
[8]	Multilayer security scheme based on a hybrid RSA algorithm and AES algorithm with a least significant bit steganographic algorithm	The scheme lacks design consideration for averting possible impersonation of ineligible erring voters through proper identification and verification measures
[9]	Addresses the voter eligibility problem through the development of a fingerprint biometric authentication system for secure electronic voting machines	The scheme design consideration fails to observe the integrity and verifiability of the vote
[17]	Explores the use of biometric smart cards for voter verification and identification. Adopting this method will enhance the electoral process by ensuring that only registered voters can cast votes	This approach does not address the issue of confidentiality of the vote cast by the user
[33]	Presentation of a secure and verifiable polling system (SeVEP) scheme that implements MFA and well-known cryptographic techniques to achieve privacy, verifiability, and authorized multiple voting and prevents double voting	The proposed system lacks scalability and usability in a real-world deployment

secure the e-voting system's authentication. An enhanced Feistel block cipher is used to secure confidential data on voters' smart cards, and a first-moment feature extraction technique secures the voter's fingerprint template. This system deals with issues encountered during authentication but fails to secure the integrity of the cast votes once stored in the database.

Ashok et al. [2] applied RFID and fingerprint technologies for authentication in an electronic voting system. Each voter has an ID in the form of an RFID tag and has his/her fingerprints scanned for comparison with the ones stored in the user's profiles. While overcoming voter authentication issues, this system also fails to protect the integrity of the vote once cast. In Fusco et al. [6], the authors propose methods to improve the traceability and auditing of voting operations using blockchain technology. Their system, however, does not present any means for authenticating the user for the election.

The security mechanisms presented in academic literature such as [4, 6, 7, 9, 15–18] solve either authentication or confidentiality issues surrounding e-voting, and some even manage to solve both problems, but none meets the multiple security requirements of authentication, confidentiality, integrity, and verifiability, all of which are crucial to delivering credible electronic democracy through e-voting. This research solves these critical security requirements by proposing MFA using facial recognition and RFID cards combined with a public blockchain. Table 1 shows the synthesis of related works in this domain.

3 Preliminaries

3.1 Blockchain in E-Voting

Blockchain has emerged as a trustless system used in several domains to ensure data integrity. It has been implemented in e-voting systems and has become an important option in overcoming various security challenges [34]. Blockchain-based e-voting systems have been predicted to be the next generation of modern e-voting due to their decentralized and distributed nature. A blockchain network is suitable for e-voting because transactions are time-stamped when recorded and cannot be modified after being validated. Also, certain blockchains offer programmability via smart contracts and are secure through encryption. Most importantly, blockchain is a distributed ledger technology, where all participating full nodes in the network maintain a copy of the ledger to ensure transparency.

A blockchain is a linear combination of blocks representing different data elements. These blocks are linked using a cryptographic collision-resistant hash function to form a chain of connected blocks (see Fig. 1). To concatenate each block or transaction data in a blockchain, a hash pointer links a block to a previous block. This pointer also creates an integrity check, allowing only verified blocks to be included in the blockchain [37].

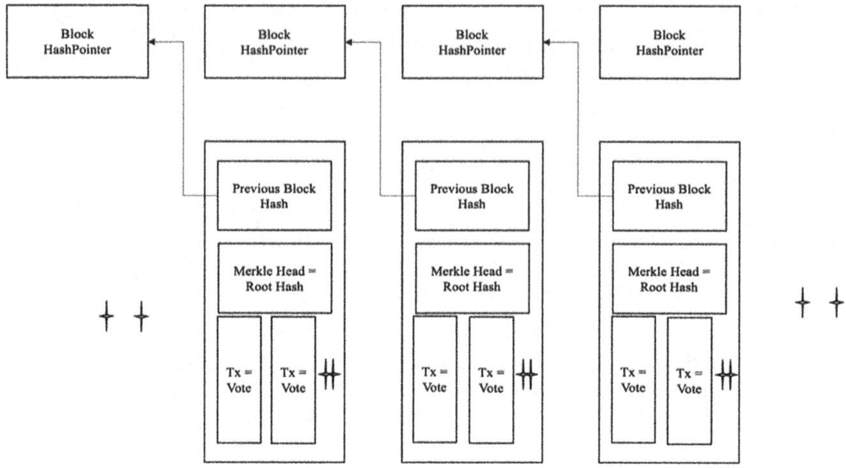

Fig. 1 Representation of a blockchain

Blockchain networks can be classified into private and public networks and hybrid solutions that combine both types. A permissioned setting allows only designated peers to participate in the consensus agreement protocol [34], and only authorized users can contribute and modify block information. A public and permissionless blockchain does not limit the number of peers who can participate in the consensus protocol. All participants can record block information. The most common public blockchain types available include the bitcoin network [35] and Ethereum [36]. In this type of blockchain, its decentralized public nature serves as a distributed ledger to immutably record transactions between participants.

A distributed ledger is inherently resistant to modification and verifiable by authorized users. In our proposed scheme, we deploy the Ethereum blockchain to build a secure e-voting system. The Ethereum blockchain is easily accessible and makes use of a state transition system. The different states make it possible for new blocks to be easily verified when they are added to the blockchain. When a vote has been cast and verified in our proposed scheme, a transaction is hashed and added to the blockchain. We made use of the SHA-256 hash function, which is a collision resistance one-way function.

3.2 *Multifactor Authentication*

MFA is a way of authenticating end users (voters) in two or more different ways that establishes access control and identity. MFA includes three different ways of authentication: something you have (e.g., a smart card), something you know (e.g., passwords in the forms of tokens), and something you are (e.g., biometric or face recognition). In our proposed scheme, a two-factor authentication is used to verify the entire e-voting process. The first level of authentication is microcontroller data verification. The microcontroller compares the data newly supplied by the RFID module with that stored in the database during authentication. Suppose the microcontroller confirms that the data matches its counterpart in the database. In that case, it sends a string of data to the software application to grant the user access to navigate to the second phase of the authentication. Facial recognition is implemented during the second phase of the authentication. The software application contains a facial recognition Web interface that takes a picture of the user's face and compares it with one already stored in the database. In case of a match, the user is granted access to the voting page.

4 System and Threat Model

The system model of our proposed design consists of three main participants described as follows:

1. Voters: These are all eligible voters denoted as $V = \{v_1, v_2, v_3, \ldots, v_n\}$, where n is the total number of eligible voters.
2. Voting Authority: This contains a set of all election administrators (EA) = 1, responsible for the management and verification of voters' identity during the election.
3. Auditors: Agents responsible for inspecting EA compliance to election norms and monitoring the power of the EA.

The framework of the proposed blockchain voting system contains the participants = {voters}, EAs = {poll sites under the districts}, auditors = {EA representative}, hash algorithm = {SHA-256}, and voting server.

4.1 Threat Model

In an e voting system, a malicious user can exploit different attack scenarios, as summarized in Table 2. When using blockchain for e-voting, issues such as double voting can arise in which an authenticated malicious voter can attempt to cast multiple votes without being detected. Voter coercion can occur by persuading a voter to vote for a particular option. This can be accomplished only when a voter provides the coercer with his/her voting credentials, such as the private key. Voting modification or interruption by a malicious voter or device can also occur as a result of an infected malware or by being controlled by an attacker.

Table 2 Threat model scenario in an e-voting system

Threat scenario	Forged private key	Malicious auditor with access to storage	Unauthorized network provider	Rogue voting device	Malware infected operating system
Double voting	•	•	•	•	
Unauthorized administrative access	•	•	•		
System modification	•	•	•	•	•
Vote coercion	•	•		•	•
Audit log tampering	•	•	•		
Transparency	•	•	•	•	•
Biometric attacks		•	•		

In all these cases, the option selected by the voter can be inadvertently modified before submission, which can result in falsely counting polling votes. In terms of coalition attacks, voters can collude with the voting authority to affect the transparency of the voting experience, and they can also form a coalition to affect the polling option or even modify ballot options. When using biometrics for authentication, the security of the biometric templates can be undermined through attacks using keystrokes and voice patterns stored in the database. Storing biometric templates in a plain format without encryption can result in gaining access by an unauthorized attacker. Also, records stored in the database can be modified or stolen by any malicious individual, granting them access to enrolling a voter.

5 Proposed E-Voting System

In this section, we present the mechanisms and procedures, as well as the selected hardware subsystems and the software design considerations used in the realization of the proposed secure electronic voting system. The block diagram of the system is shown in Fig. 2 and the proposed system architecture in Fig. 3. They outline the structure of a decentralized database to store the encrypted votes, in essence making it more difficult to modify or alter a vote once cast. The architecture is robust with a two-way authentication, which helps prevent unauthorized users from accessing the system or casting a vote.

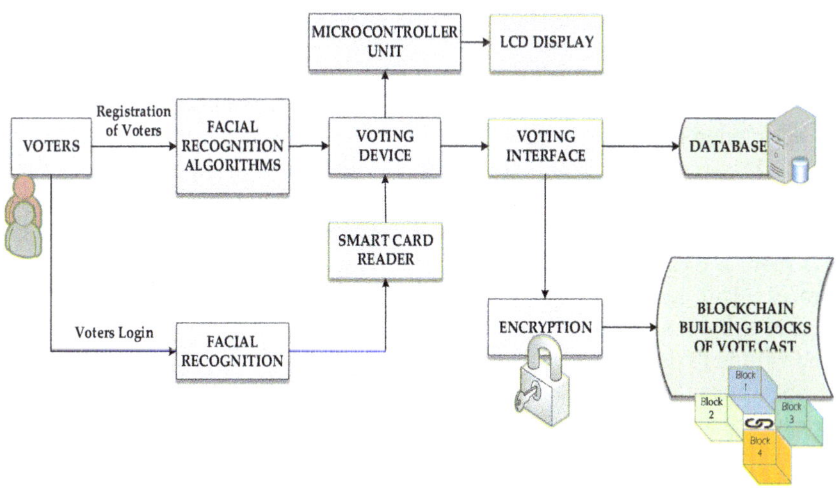

Fig. 2 Block diagram of the secured e-voting system

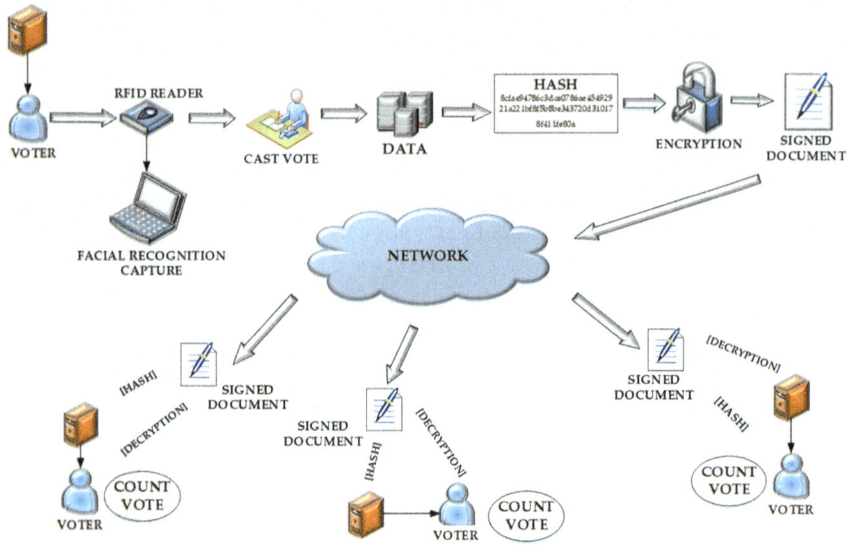

Fig. 3 Proposed secured e-voting architecture

5.1 System Hardware Design Consideration

This section presents the integration and design process of the system hardware components. More specifically, it describes the authentication module, the microcontroller unit as a whole, and the interaction of the various components in the process of authenticating valid eligible voters. The components include an Arduino ATMEGA, an LCD, a personal computer, an RFID reader, and an RFID card reader, as depicted in Fig. 4

RFID is a contactless auto-identification system similar to smart cards. It enables the electronic labeling and wireless identification of an object using frequency-shift keying (FSK) modulation [20]. Information exchange in an RFID system is done via radio waves where no contact or line of sight is needed for the identification process. This makes RFID relatively secure since readers can be designed to locate tags at a distance of several meters [21]. As a contactless auto-ID system, reading and writing of data in the RFID system are done through an RFID tag's nonvolatile memory using an RF signal by the reader. The reader emits an RF signal, and data is exchanged when the tag comes in proximity to the reader signal. Tags can be categorized as follows: a) active tags in which a battery supplies power and which are therefore costly, b) semi-passive tags that use batteries to power the tag IC but not for communication, and c) passive tags that have a battery. The absence of a power supply makes passive tags cheaper and more reliable than active tags..

Due to cost considerations, our e-voting authentication system is designed using a passive RFID reader (i.e., MF-522ED) that can only detect a passive RFID tag

Fig. 4 MIFARE 13.56Mhz RC522 RFID card reader

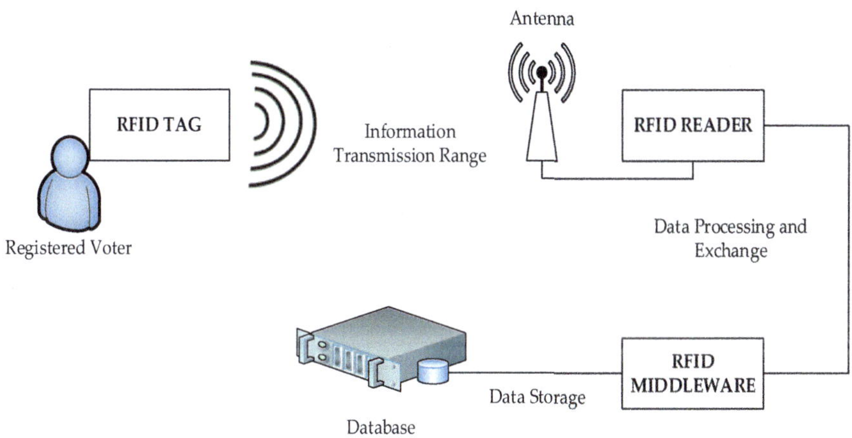

Fig. 5 Data transmission process between an RFID reader and an RFID tag

at a few centimeters away from the reader. The reader operates with contactless communication and uses MIFARE transfer speeds up to 10Mbit/s in both directions [22]. The specific RFID reader used in the system is a low-cost reader for reading passive RFID tags, as shown in Fig. 3. It operates at temperatures between 20 C and 80 C, humidity levels between 5% and 95%, at a frequency of 13.56 MHz, a working current of 13–26 mA/3.3 V DC, and a standby current of 10–13 mA/3.3 V DC power supply [22]. The effective detection range of the MF-522ED reader is around 5–8 cm. Each RFID tag has a unique serial number or ID. In this design, each voter is identified through the passive RFID card/tag. Figure 5 illustrates how data transmission is performed between an MF-522ED RFID reader and a voter's card/tag.

The design of the second factor, namely, facial recognition, implements a face API library. Face API is a JavaScript module built with the TensorFlow open-source software library, which implements several convolutional neural networks (CNNs) to solve face detection, face recognition, and face landmark detection, optimized for the Web and mobile devices [23]. This system implements three face API models for facial recognition authentication: tiny face detector model, face recognition model, and face expression recognition model [23]. The tiny face detector is a real-time face detector, which is fast and consumes few resources. The face recognition model is an architecture implemented to compute a face descriptor for any given image. The face expression recognition is a lightweight, fast, and reasonably accurate approach to match the facial expressions of a given image. The face API at the point of registration detects the human face and draws a canvas around it. The library gets the image of the detected face in the canvas and converts it to a float array, which is then saved to the blockchain.

During authentication, a new image of the detected face is taken and is then converted to a float array by the face API. The library verifies the similarity between the image taken at the point of registration and the image taken during authentication by computing the mean distance between the float arrays. The distance threshold is 0.6 meters, and if the mean distance between the arrays is greater than 0.6 meters, then the face does not match. But if the mean distance between the arrays is less than 0.6 and the face matching is successful, then the users are granted access to vote. The facial recognition implemented in this system has a very high capacity and works efficiently on a Windows 10 HP, 6th generation Intel Core i5 (2.3–2.8GHz) processor, 8GB RAM, and 500GB Hybrid Hard Drive. The system might not work efficiently on systems with less capacity.

5.2 System Software Design Consideration

The system software structure comprises the client Web application and the facial recognition application (FaceAPI). The client Web application provides an interface for the user to interact with the hardware components and connects to both the private blockchain and FaceAPI to ensure vote security and authentication, respectively. It allows the voters to gain access to the voting interface after comparing the password and username, unique facial recognition ID of the voter, and verified RFID ID of the voter. The voting interface allows voters to cast votes for their preferred candidate. In this proposed design, the blockchain provides the required integrity, verifiability, and post-electoral auditing of ballots based on a tamper-resistant public ledger for assurance of security and reliability of the distributed stored data.

The proposed system implements a permissioned private blockchain in which only those who have permission can join the Ethereum blockchain network. The blockchain is based on hashing, encryption, and decentralization. A private key is issued to each voter during registration. The private key is used to generate signatures on the vote during the election. The encrypted data are shared across the nodes in the blockchain, which makes it a decentralized system. The design considerations of blockchain technology in our proposed secure and robust voting mechanism extend work from Singh and Chatterjee [16] and integrates MFA of voters.

Preelection Steps

1. The voters need to register with the voting system. In the first step, the voters are required to:
   (a) Obtain a unique ID through the RFID tag/card.
   (b) Pre-enroll the facial image of the voter and obtain a unique facial ID (computed mean distance between the floating array stored image and real-time captured image).
   (c) Choose a unique password for login.
2. After successful registration with the system, the voter receives a voter ID.

Main Voting Steps

1. During the election period, the voter approaches the kiosk at the poll site and is then authenticated using the RFID tag ID and the generated Face ID and can log in with their assigned password.
2. After the successful login, the voter is verified by the EA and auditors.
3. If the voter is eligible for voting through the successful verification in step 1, the client Web application allows the voter to vote for his/her preferred candidate from the list of contestants.
4. The preferred vote/ballot is hashed with SHA-256 to assert vote integrity by the client Web application.
5. The hashed vote is signed for each voter by the voter's private key.
6. The signed, fingerprinted, and encrypted vote is then stored in the voting server. This is the first block of the blockchain.
7. Steps 1 to 6 are repeated for each legitimate voter, with each vote forming a new block that is added to the existing chain for the duration of the election period.

Postelection Steps

1. After the election is over at each poll site level, the individual blockchains of each poll site within the districts are joined together for the preparation of the zone-level blockchain.
2. The zone-level blockchains are joined together for the preparation of the state-level blockchain.
3. Finally, the EA and auditors check all the votes from the blockchain and declare the final result of the election.

The pseudo-code of this procedure is detailed in Algorithm 1, and the system flowchart is shown in Fig. 6.

6 Performance Evaluation

The hardware component comprises the RFID module, liquid-crystal display, and an Arduino Uno microcontroller development board. The software component consists of the facial recognition program and blockchain solution, which implements SHA

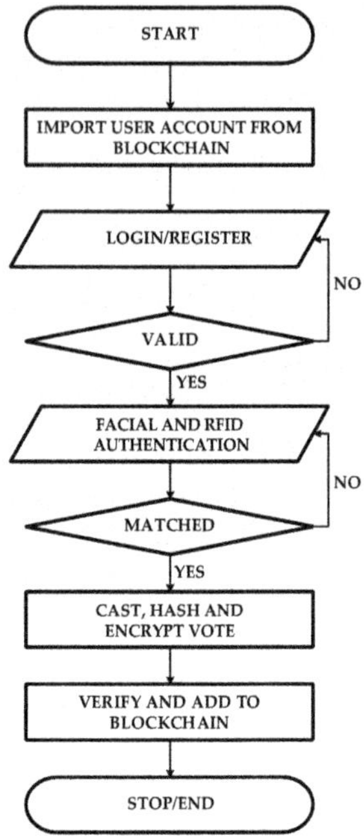

Fig. 6 System flowchart diagram

Algorithm 1: Voting Procedure

Input: voter unique id tag, voter face id

Output: Complete vote in the form of blockchain

Begin

1. The voter registered with the voting system.
2. Get the Voter Unique Tag Id and Voter Face ID and generate a private key.
3. If (Voter Unique Tag Id == registered voter Id) and (voter is eligible) and (Voter Facial Image == registered face ID), then go to step 4 else go to step 11.
4. Enter your password.
5. If Voter Unique Tag Id is not registered or he/she is not eligible or unregistered Face ID then deny voting and go to step 11, else go to step 6.
6. If (Password is correct) then go to step 7, else go to step 8.
7. Open the candidate choosing page, choose the candidate to vote, and go to step 9.
8. Enter the correct password and go to step 6.
9. Signing the encrypted data - SIGNV pricey (ENCRYPT(vote))
10. Generation of the block BLOCK (block header+ block data).
11. **End**

256 to encrypt votes. The RFID module validates the authentication in the electronic voting system. The Arduino Uno microcontroller receives a direct 5 V current through its USB connector, from which both the RFID module and the LCD are powered. When the RFID reader module is powered ON, it automatically detects and reads the data from an RFID tag/card data placed in the immediate vicinity of

(a) (b)

Fig. 7 (**a**) System authentication module and (**b**) module internal hardware integration

the module and transmits a signal to the microcontroller unit to decide on whether to grant access to vote or not.

The evaluation metrics used for the facial recognition process of the electronic voting system are the false acceptance rate (FAR) and false rejection rate (FRR). The FAR is the probability of cases where the system wrongly authorizes an unauthorized person; the equation for computing the FAR is given in (1). The FRR is the probability of cases where the system wrongly denies access to an authorized person; the formula for computing the FRR is given in (2). The permissioned private blockchain technique was evaluated based on the transaction time and transaction cost per voter. Meanwhile, the RFID auto-ID technique was evaluated based on the transmission distance between the tag and the reader. The overall system was evaluated based on the response time.

$$\text{False Acceptance Rate (FAR)} = \frac{Number\ \text{of False Acceptance}}{\text{Number of Identification Attempts}} \tag{1}$$

$$\text{False Rejection Rate (FRR)} = \frac{\text{Number of False Rejection}}{Number\ \text{of Identification Attempts}} \tag{2}$$

The prototype of the authentication system presented in the previous section is shown in Fig. 7a and Fig. 7b.

Figure 7b shows the hardware integration of the Arduino microcontroller, LCD, and an RFID module. The system is powered by an Arduino USB cable connected to the computer system that hosts the Web application, as shown in Fig. 7a. The RFID card of the voter is placed on the system module in Fig. 7a. The RFID reads the information on the card and compares it with the data stored inside the blockchain to grant voters access to vote or register. The performance of the RFID was evaluated by examining the read rate of voters' tags against the reader, as shown in Table 3. The read rate is the degree to which an RFID module reads tags with varying distances during voter authentication. Table 3 shows that the RFID module detected all tags up to 3.5 cm.

The software prototype for the system includes the client Web application, which contains different interfaces for registration, login and vote casting, and result viewing. During registration, the voters need to obtain a private key required to import an account from the blockchain to the Web browser and to encrypt the message sent to the blockchain. This is shown in Fig. 8.

After obtaining the private key for the voters, the account address is obtained from the blockchain using the MetaMask software. The process is as shown in Fig. 9.

After proper prior registration, the interface in Fig. 10 provides a platform for voters to provide all means of authentication of the system before being granted access to cast a vote in the election. Figure 11 shows the platform voters can use to express their vote after being successfully authenticated.

Table 3 The read rate of the voter's card against distance

S/N	Distance (cm)	No. of tags (N)	Read rate (R)	Read rate $\left(\frac{R}{N} * 100\right)$, %
1	0.5	15	15	100
2	1.0	15	15	100
3	1.5	15	15	100
4	2.0	15	15	100
5	2.5	15	15	100
6	3.0	15	15	100
7	3.5	15	15	100
8	4.0	15	0	0
9	4.5	15	0	0
10	5.0	15	0	0

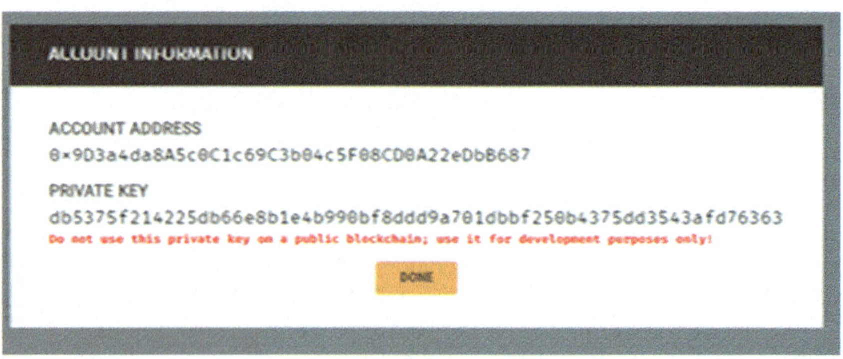

Fig. 8 The private key of an account

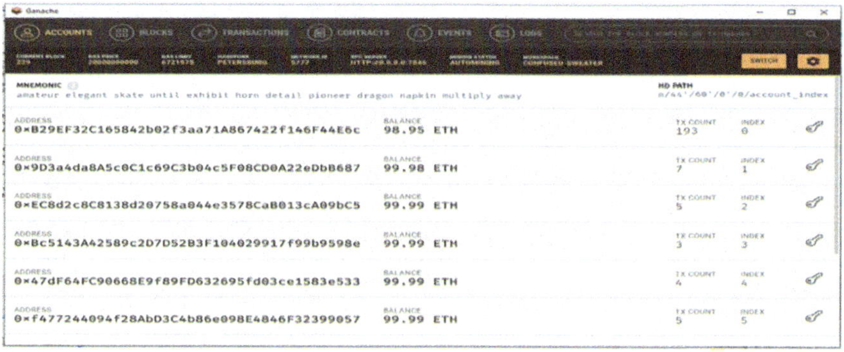

Fig. 9 Accounts in the blockchain

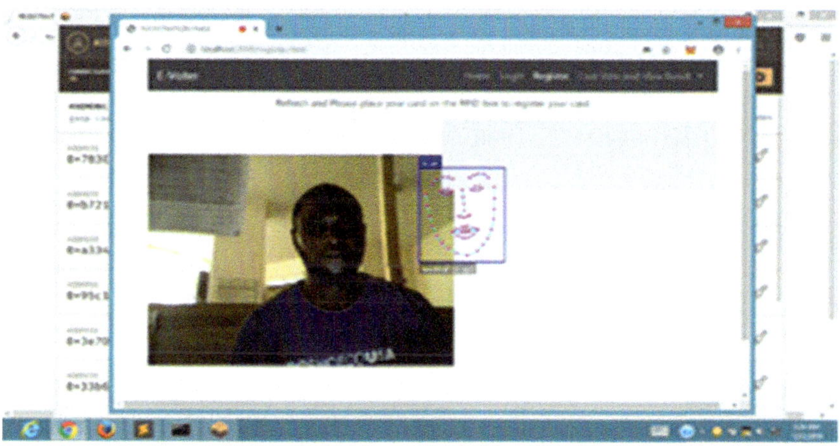

Fig. 10 Authentication after a successful registration

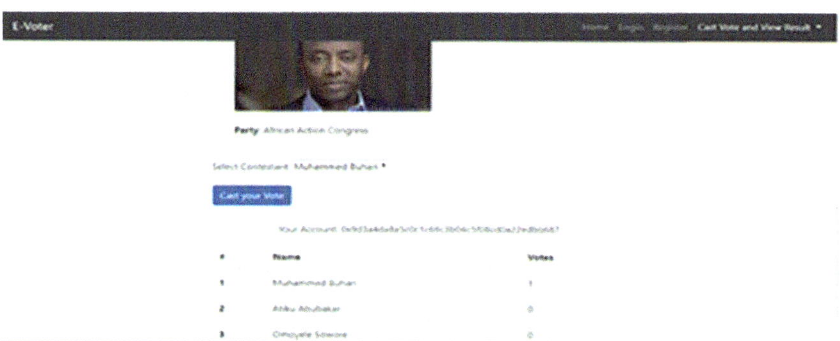

Fig. 11 Voting after successful MFA

Table 4 FAR of the developed system

Matching Tries	Accepted	Rejected	FAR
10	1	9	10%

Table 5 FRR of the developed system

User	Number of attempts	Number of times accepted	Number of times rejected	FRR (%)
1	12	11	1	8.3
2	12	11	1	8.3
3	15	13	2	13.3
4	12	11	1	8.3
5	15	13	2	13.3

Table 6 Transaction execution time and transaction fees for ten nodes

	Transaction execution time (Minutes)			Transaction fees (Ether)		
S/N	Slow	Avg.	Fast	Slow	Avg.	Fast
1	11.54	1.12	0.30	0.0012	0.0029	0.0097
2	8.30	1.24	0.30	0.0014	0.0023	0.0096
3	10.12	1.48	0.36	0.0014	0.0025	0.0096
4	1.24	1.24	0.36	0.0035	0.0035	0.0074
5	24.18	3.24	0.30	0.0010	0.0030	0.0074
6	13.24	3.48	0.36	0.0014	0.0048	0.0096
7	23.48	3.48	0.30	0.0010	0.0027	0.0074
8	1.30	1.30	0.36	0.0018	0.0018	0.0074
9	1.30	1.30	0.30	0.0018	0.0018	0.0074
10	13.24	1.12	0.36	0.0014	0.0029	0.0074

Table 4 shows the result of ten different voters' trials at facial recognition authentication to ascertain the module's efficiency. The FAR of the system was calculated using (1). It can be observed from Table 4 that out of 10 false face match attempts made, only one was granted access by the system. The false acceptance rate of 10% in the system is low, especially considering that this is just one part of the MFA system, and the single failed trial can likely be explained by the rotated and exaggerated skin distortion of the participating subject [24].

The system false rejection rate (FRR) was subsequently investigated. From Table 5, it could be observed that in all the attempts made to match a valid voter face with the one saved in the database, the rate of voter rejection is low. Table 5 shows that, while each valid voter was rejected at least once, these rejections comprise only a small percentage of the attempts made by each voter. From this low FRR, it can be deduced that the facial recognition authentication system is sufficiently reliable for authenticating voters in an election.

Similarly, the effectiveness of blockchain was investigated by examining the transaction execution time and the transaction fees for ten nodes. Table 6 shows the transaction execution time and the required transaction fees during election registration and casting of the vote to evaluate blockchain speed when 10 nodes are connected to the blockchain network. Transaction speed is the time taken to add a new voter and add a casting vote to the blockchain. A transaction fee is a monetary cost required to register a voter and to cast a vote in the blockchain.

The transaction execution time of slow, average, and fast, with the corresponding transaction fees of ten nodes in Table 6. Slow and average transaction execution times are determined by the network when an attempt is made to reduce registration and voting costs, while the fast transaction execution time is used when trying to increase the speed of adding registration and voting transactions to the blockchain network, albeit at a higher transaction cost. It can be observed from Table 6 that execution times differ greatly between slow and average, as opposed to differences in transaction fees. Thus, it can be inferred that slow and average transaction execution times should be avoided to increase the speed of the election process. Since the cost difference between the slow, average, and fast execution time is not much, the fast transaction execution time should be preferred during an election process. This also makes the system faster and more secure [39].

7 Security Analysis

This section provides the security analysis of the proposed system and highlights solutions to the threat model analysis mentioned in Sect. 3.

7.1 *Vote Consistency and Integrity*

The proposed model provides vote consistency since all nodes in the network maintain the same copy of the voting results using the blockchain time stamp. Furthermore, at any time of any update, the newly generated data blocks are subsequently updated. In the case of new voting requests, old votes in the blocks have to be committed in the chain before any new blocks can be inserted. Our model groups votes into blocks, and, anytime a vote is being cast, the voting authority adds the votes with other unverified votes to be accepted by other nodes after proper verification. The block also contains the hash of the previous block. We assume that the hash function is collision-resistant.

7.2 *Cast-as-Intended Transparency and Verifiability*

Our proposed scheme provides cast-as-intended voting by first providing integrity through a consensus protocol as defined in the previous section. Also, it uses double authentication to make sure that each voter is directly cast. Each voter is assigned a private key that is used as the nonce for hashing blocks into the blockchain. In the case of a corrupted system or a malware malfunction in the operating system, the vote cast into the blockchain will eventually be dropped since the consistency of the blocks is not maintained. The final polling outcome of all tallied votes is a summation of all the individual blockchains of each poll site within the districts, combined with the zone-level blockchain and state-level blockchain.

7.3 Vote Coercion Resistance

Our definition of resistance in this context is defined as our proposed system being able to resist modification by an adversary or a malicious entity after votes are being cast. Let's assume an adversary *A* tries to change a voter's option or an attempt to tamper with the votes stored in the blockchain. In the first case, this is not possible in our proposed scheme since each vote is secured through a collision resistance hash function such as SHA-256; afterward, the voters sign the vote using the private key. Additionally, each vote cast is sent and distributed on the entire decentralized network for approval and verification, meaning that a change in one node will invalidate the vote since the initially generated signature will be different on the other nodes using the voter's public key.

Our proposed scheme is secure against blockchain modification in the second case because each block has a hash pointer to the next block, creating a Merkle tree. For instance, if *A* makes an attempt to modify the vote on some blocks, the adversary will encounter a mismatch problem because the modified block will have an inconsistent hash value compared to the hash of the preceding blocks contained in the blockchain. In the worst-case scenario, if the adversary successfully breaks the hash of the previous block, the adversary will eventually fail when the head of the list is reached. Besides, every node in the network has a copy of the blockchain, making it very hard for an adversary to modify all the blocks in the entire network.

7.4 Double Voting

Our proposed system can thwart the instances of double voting through the blockchain's consensus protocol since each vote's authenticity is verified through time stamps and logs for each vote on the blockchain. Also, all nodes in the network can publicly verify votes in every block before committing it to the blockchain, ensuring that each voter votes for an option. Furthermore, each vote is signed by each voter using the private key, ensuring that the verifier can easily detect any falsification.

8 Conclusion

This chapter has presented an effective approach to solving the authentication, integrity, and verifiability issues of electronic voting using MFA and a private blockchain solution. The suggested procedure uses MFA and smart contracts to enable secure and cost-efficient election processes while guaranteeing voter privacy. The proposed blockchain approach provides high speed and scalability for casting votes as intended without incurring high transaction cost during slow, average, and fast transaction execution speed times. The proposed approach incurred a cost difference of 0.0085 Ether, 0.0068 Ether and 0.0017 Ether between fast, average, and slow transaction times. The strength of the system is in its synergistic application of MFA of facial recognition and RFID authentication with blockchain-based distributed ledger data storage. The proposed mechanism has shown that decentralized distributed electronic voting through blockchain technology offers a better possibility for countries to conduct a credible election without compromising critical attributes of integrity, confidentiality, and verifiability of voter's choice while being able to view the result of the election in real time. Adopting the proposed technique in future electronic democratic decision-making will help make vote casting easy, secure, and fast, which may encourage more citizens' participation in the electioneering process.

In the future, the authors would like to pay detailed attention to the communication complexity of the network of distributed computers [25] and to improve the overall system's performance, which is critical for a large-scale e-voting scenario.

Acknowledgments We want to thank the Federal University of Technology Minna, Nigeria, and Modul University Vienna, Vienna, Austria, for making the resources available to complete this work.

References

1. Mpekoa, N., & van Greunen, D. (2016). m-Voting: Understanding the complexities of its implementation. *International Journal for Digital Society, 7(4)*. https://doi.org/10.20533/ijds.2040.2570.2016.0149
2. Ashok, N., Teja, B., & Balakrishna, A. (2014). RFID and fingerprint recognition based electronic voting system for real-time application. *International Journal of Engineering Development and Research, 2*(4), 3850–3854.
3. Dogo, E. M., Nwulu, N. I., Olaniyi, O. M., Aigbavboa, C. O., & Nkonyana, T. (2018). Blockchain 3.0: Towards a secure Ballotcoin democracy through a digitized public ledger in developing countries. *I-Manager's Journal on Digital Signal Processing, 6, (2)*, 24. https://doi.org/10.26634/jdp.6.2.15593
4. Abayomi-Zannu, T. P., Odun-Ayo, I. A., & Barka, T. F. (2019). A proposed Mobile voting framework utilizing Blockchain technology and multi-factor authentication. *Journal of Physics. Conference Series, 1378*, 32104. https://doi.org/10.1088/17426596/1378/3/032104
5. Iwuoha, V. C. (2018). ICT and elections in Nigeria: Rural dynamics of biometric voting technology adoption. *Africa Spectrum, 53*(3), 89–113. https://doi.org/10.1177/000203971805300304

6. Fusco, F., Lunesu, M.I., Pani, F.E., Pinna, A. (2018). Crypto-voting, A Blockchain based e-voting system, *10th International Conference on Knowledge Management and Information Sharing (KMIS)*, Seville, Spain, pp. 221–225.
7. Anil, K. D., Rakshith, D., Manoj, C. V., & Poornachandra, N. U. (2017). RFID based voting machine. *International Journal of Current Engineering and Scientific Research, 4*(6), 23–25.
8. Okediran, O. O., Sijuade, A. A., & Wahab, W. B. (2019). Secure electronic voting using a hybrid cryptosystem and steganography. *Journal of Advances in Mathematics and Computer Science*, 1–26. https://doi.org/10.9734/jamcs/2019/v34i1-230201
9. Umar, B., Olaniyi, O. M., Ajao, L., Maliki, D., & Okeke, I. (2019). Development of a fingerprint biometric authentication system for secure electronic voting machines. *Kinetik (Malang), 4*(2), 115–126. https://doi.org/10.22219/kinetik.v4i2.734
10. Oke, B.A., Olaniyi, O.M., Aboaba, A.A., & Arulogun, O.T. (2017). Developing multi-factor authentication technique for secure electronic voting systems", *Proceedings of IEEE International Conference on Computing, Networking and Informatics (ICCNI 2017)*, pp. 48–53 https://doi.org/10.1109/ICCNI.2017.8123773.2017.
11. Gabriel, A. J., Alese, B. K., Adetunmbi, A. O., Adewale, O. S., & Sarumi, O. A. (2019). Post-quantum crystography system for secure electronic voting. *Open Computer Science, 9*(1), 292–298. https://doi.org/10.1515/comp-2019-0018
12. Olaniyi, O. M., Arulogun, O. T., Omidiora, E. O., & Adeoye, O. (2013). Design of secure electronic voting system using multifactor authentication and cryptographic hash functions. *International Journal of Computer and Information Technology (IJCIT), 2*(6), 122–1130.
13. Li, M., et al. (2019). CrowdBC: A Blockchain-based decentralized framework for crowdsourcing. *IEEE Transactions on Parallel and Distributed Systems, 30*(6), 1251–1266. https://doi.org/10.1109/tpds.2018.2881735
14. Jiao, Y., Wang, P., Niyato, D., & Suankaewmanee, K. (2019). Auction mechanisms in cloud/fog computing resource allocation for public Blockchain networks. *IEEE Transactions on Parallel and Distributed Systems, 30*(9), 1975–1989. https://doi.org/10.1109/tpds.2019.2900238
15. Habibu, T., Sharif, K., & Nicholas, S. (2017). Design and implementation of electronic voting system. *International Journal of Computer & Organization Trends, 7*(4), 1–6. https://doi.org/10.14445/22492593/IJCOT-V45P301
16. Singh, A., & Chatterjee, K. (2018). SecEVS : Secure electronic voting system using Blockchain technology. *International Conference on Computing, Power and Communication Technologies (GUCON)*, pp. 863–867.
17. Oluwatobi, A. N., Ayeni, T. P., Arulogun, O. T., & A.A., Ariyo, and K.A. Aderonke. (2020). Exploring the use of biometric smart cards for voters' accreditation: A case study of Nigeria electoral process. *International Journal on Advanced Science, Engineering and Information Technology, 10(1)*, 80. https://doi.org/10.18517/ijaseit.10.1.8459
18. Olaniyi, O. M., Arulogun, O. T., Omidiora, E. O., & Okediran, O. O. (2015). Enhanced Stegano-cryptographic model for secure electronic voting. *Journal of Information Engineering and Applications (JIEA), 5*(4), 1–15.
19. Olaniyi, O. M., Folorunso, T. A., Ahmed, A., & Joseph, O. (2016). Design of Secure Electronic Voting System Using Fingerprint Biometrics and Crypto-Watermarking Approach. *International Journal of Information Engineering and Electronic Business, 8*(5), 9.
20. Zaid, A., Firas, A.A., & Hussein, A. (2008). Design and implementation of RFID system, *Proceedings of 5th International Multi-Conference on Systems, Signals and Devices*, https://doi.org/10.1109/SSD.2008.4632787.
21. Kassim, M., Mazlan, H., Zaini, N., & Salleh, M.K. (2012). Web-based student attendance system using RFID technology, *Proceedings of 2012 IEEE Control and System Graduate Research Colloquium* (ICSGRC 2012), 213–218.
22. Allelectronics. (2020). *RFID Read and Write Module*. Available: https://www.nxp.com/docs/en/datasheet/MFRC522. 22nd August 2020.
23. ITNEXT. (2020). *face-api.js — JavaScript API for Face Recognition in the Browser with tensor-flow.js*. Available: https://itnext.io/face-api-js-javascript-api-for-face-recognition-in-the-browser-with-tensorflow-js-bcc2a6c4cf07. 31st August 2020.

24. Faridah, Y., Haidawati Nasir, A. K., Kushsairy, S. I., Safie, S. K., & Gunawan, T. S. (2016). Fingerprint biometric systems. *Trends in Bioinformatics, 9*, 52–58. https://doi.org/10.3923/tb.2016.52.58
25. Xu, L., & Bruck, J. (1998). Deterministic voting in distributed systems using error-correcting codes. *IEEE Transactions on Parallel and Distributed Systems, 9*(8), 813–824. https://doi.org/10.1109/71.706052
26. Hardwick, F.S., Gioulis, A., Akram, R.N., Markantonakis, K. (2018). E-voting with blockchain: An e-voting protocol with decentralization and voter privacy. In *2018 IEEE International Conference on Internet of Things (iThings) and IEEE Green Computing and Communications (GreenCom) and IEEE Cyber, Physical and Social Computing (CPSCom) and IEEE Smart Data (SmartData)*, pp. 1561–1567. IEEE, 2018.
27. N. Kshetri, J. Voas Blockchain-enabled e-voting. *IEEE Software*. 2018 Jul 6;35(4):95–99.
28. Bartolucci, S, Bernat, P., & Joseph, D. (2018). SHARVOT: Secret SHARe-based VOTing on the blockchain. In *Proceedings of the 1st International Workshop on Emerging Trends in Software Engineering for Blockchain-WETSEB'18,* Gothenburg, Sweden, 27 May–3 June 2018; pp. 30–34.
29. Thuy, L.V.-C., Cao-Minh, K., Dang-Le-Bao, C., Nguyen, T.A. (2019). Votereum: An Ethereum-based E-voting system. In *Proceedings of the 2019 IEEE-RIVF International Conference on Computing and Communication Technologies (RIVF),* Danang, Vietnam, 20–22 March 2019; pp. 1–6.
30. Yavuz, E., Koc, A.K., Cabuk, U.C., & Dalkilic, G. (2018). Towards secure e-voting using Ethereum blockchain. In *Proceedings of the 2018 6th International Symposium on Digital Forensic and Security (ISDFS)*, Antalya, Turkey, 22–25 March 2018; pp. 1–7.
31. Zhang, W., Yuan, Y., Hu, Y., Huang, S., Cao, S., Chopra, A., & Huang, S.(2018). A privacy-preserving voting protocol on Blockchain. In *Proceedings of the 2018 IEEE 11th International Conference on Cloud Computing (CLOUD)*, San Francisco, CA, 2–7 July 2018; pp. 401–408.
32. Sathya, V., Sarkar, A., Paul, A., & Mishra, S. (2019). Blockchain based cloud computing model on EVM transactions for secure voting. In *Proceedings of the 2019 3rd International Conference on Computing Methodologies and Communication (ICCMC),* Erode, India, 27–29 March 2019; pp. 1075–1079.
33. Qureshi, A., Megías, D., & Rifà-Pous, H. (2019). SeVEP: Secure and verifiable electronic polling system. *IEEE Access, 7*, 19266–19290.
34. Taş, R., & Tanrıöver, O. O. (2020, Aug). A systematic review of challenges and opportunities of blockchain for e-voting. *Symmetry., 12*(8), 1328.
35. Nakamoto, S. *Bitcoin: A peer-to-peer electronic cash system.* https://bitcoin.org/bitcoin.pdf.
36. Wood, G. (2014). Ethereum: A secure decentralized generalized transaction ledger. *Ethereum Project Yellow Paper, 151.*
37. Zhang, Y., Xu, C., Lin, X., & Shen, X. S. (2021). Blockchain-based public integrity verification for cloud storage against procrastinating auditors. In *IEEE Transactions on Cloud Computing. 9*, 923–937. https://www.semanticscholar.org/paper/Blockchain-Based-Public-Integrity-Verification-for-Zhang-Xu/ed0019791de620e7235d596107fabaea68b1ba85
38. Abayomi-Zannu, T. P., Odun-Ayo, I., Tatama, B. F., & Misra, S. (2020). Implementing a mobile voting system utilizing blockchain Ttechnology and two-factor authentication in Nigeria. In *Proceedings of first International Conference on Computing, Communications, and Cyber-Security (IC4S 2019)* (pp. 857–872). Springer.
39. Awotunde, J. B., Ogundokun, R. O., Jimoh, R. G., Misra, S., & Aro, T. O. (2020). Machine learning algorithm for cryptocurrencies price prediction. In S. Misra & A. Kumar Tyagi (Eds.), *Artificial Intelligence for Cyber Security: Methods, Issues and Possible Horizons or Opportunities* (Studies in computational intelligence) (Vol. 972). Springer. https://doi.org/10.1007/978-3-030-72236-4_17
40. Misra, S. (2021). A step by step guide for choosing project topics and writing research papers. In *ICT Related Disciplines, Communications in Computer and Information Science* (Vol. 1350, pp. 727–744). Springer.

Enhanced Security and Privacy Issue in Multi-Tenant Environment of Green Computing Using Blockchain Technology

Emmanuel Abidemi Adeniyi, Roseline Oluwaseun Ogundokun, Sanjay Misra, Joseph Bamidele Awotunde, and Kazeem Moses Abiodun

1 Introduction

Recently, the computer and information technology (IT) industry has recognized the value of being green, both in terms of environmental problems and costs reduction, which has resulted in a remarkable drift in IT industry strategies and policies. The motivation behind this transition began from the ever-growing demand for business computing, ever-rising energy costs, and a growing understanding of global warming issues. Therefore, green computing is a way of allowing an effective use of computing resources, intended to improve technologies by reducing energy and environmental waste [1, 2]. In an age of growing IT, resources need to be used efficiently without impacting the climate. The IT sector comprises many issues for promoting green technologies like reusable hardware devices, cloud services, paper consumption reduction, low energy management, and green production, among others.

Green technology is the technique of making the environment relate friendly with computing tools that require the deployment of central processing units (CPUs) applications, hardware, and servers for energy efficiency [3–5]. These technologies focus on reducing resource/energy usage and increasing the use of automated waste. Multi-tenancy in the cloud climate facilitates the full use of infrastructure as a

E. A. Adeniyi · R. O. Ogundokun (✉) · K. M. Abiodun
Department of Computer Science, Landmark University, Omu-Aran, Nigeria
e-mail: ogundokun.roseline@lmu.edu.ng

S. Misra
Department of Computer Science and Communication, Østfold University College (HIOF), Halden, Norway

J. B. Awotunde
Department of Computer Science, University of Ilorin, Ilorin, Nigeria

S. Misra, A. Kumar Tyagi (eds.), *Blockchain Applications in the Smart Era*, EAI/Springer Innovations in Communication and Computing,
https://doi.org/10.1007/978-3-030-89546-4_4

service hardware tool which renders servicing much simpler in a minimized cost-effective way in the organization. The shared multi-tenancy role encourages cloud services to have full flexibility for resource use. However, there are protection and confidentiality apprehensions that extremely need to be concentrated on in a multi-tenant cloud computing setting [6, 7]. There are four green approaches to green information technology computing, they are green designing, green manufacturing, green disposal, and green use [6]. The four complementary methods (green design, manufacturing, use, and disposal) are employed to foster green computing principles at all possible levels.

Blockchain does have a stronger reliability comparison to the collection of all data in a single repository [8]. In terms of data collection and maintenance, the harm incurred by attacks on the servers should be avoided. Besides that, BC has an open feature, and it could offer clarity in the data once extended to an environment engaging in the revelation of information. As a consequence, concerning these capabilities, BC can be used in several fields, such as the multi-tenant green computing segment, and its implementations are supposed to improve multi-tenant protection and privacy problems [9, 10]. The blockchain completes the transaction history via the job authorization process, where an individual who borrows digital currency creates a link by merging cryptocurrencies via the Internet [11]. The hash function is thereby created by inspecting and attaching the preceding block. This block is modernized regularly and mirrored in the specific segment of the virtual cash transaction to display the most current transaction information block. This method ensures protection for the exchange of online funds and facilitates the utilization of a secure device [12]. The BC is a hierarchical archive that deposits records in a manner equivalent to the disseminated archive and is programmed to make it unbearable to achieve individually as the system operators save and check the BC. Each block comprises a header and a body. The header comprises the hash values of the preceding, existing, and nonce blocks. The block data is checked in the database by employing the index procedure, whereas the block does not hold the next block hash value, it is introduced as a procedure [13].

Multi-tenancy is the process where multiple customers have access to the computing hardware/software resources, which covers the IaaS, PaaS, and SaaS [14]. When a massive number of tenants share the same resources in the cloud environment, a variety of issues exist concerning the privacy and security of each tenant. These issues must be sensibly meticulous and manage because they can lead to crucial susceptibility and privacy breaches via probable data seepage [15]. Hence, this paper proposed the application of BC technology to increase the safety and privacy issues of multiple tenants in a green computing environment.

Therefore, in this paper, various forms of green computing, the need for green computing, benefits, and security concerns of multi-tenancy in cloud environments are discussed. Finally, this paper proposed BC technology as a method to deal with the security and privacy dispute in a multi-tenant environment. The study implements blockchain in a cloud environment with the use of Ganache and MetaMask to create a dummy secure account for each cloud tenant.

The remaining part of the manuscript is organized as follows: Section 2 discussed the review of previous pieces of literature. The related works in the field of study are similarly conferred in this section. Section 3 conferred the materials and methods used for the implementation of the projected scheme. Similarly, the projected scheme architecture was discussed in this section. Section 4 discussed the results deduced from the system implementation, and the interpretation of the finding was discussed in the section as well. The article was concluded in Sect. 5, and also future research was suggested in the section.

2 Literature Review

Cloud users make requests to cloud service providers (CSP) for services. CSPs are third-party companies that offer cloud depository to their customers. Third-party accountant (TPA) and quality expert (QE) are two more third-party service providers that are expected to deliver security functionality in the cloud [16]. As it is identified, security and trust are the further most substantial and vital challenges when it comes to cloud benefits for enterprises and institutions. There are several causes for security and trust in the cloud system, including [17, 18]. Cloud customers' data is in great danger of being lost, disclosed, or attacked, and they have no remedy to get out of this bad predicament. Users of the cloud are unaware of who they are working with or with whom they are exchanging data. Transparency is also a big issue; that is, cloud customers do not know who is using their data or how it is moving around inside the cloud. Blockchain is a new and growing technology that cloud users may utilize to increase data security and trust while outsourcing and obtaining services from the cloud [19–21]. In comparison to centralized database security, blockchain (BC) can provide enhanced security. Blockchain (BC) continually checks the list of records that are connected and secured to the preceding block using a cryptographic hash function [22, 23]. A BC is a dispersed record that can document transactions and protects them from being tampered with. Blockchain is often run on a peer-to-peer link and is meant to prevent manipulation [24, 25]. BC has the potential to provide data security comparable to that of a central database [26, 27]. Data storage damages and threats may be avoided from a management standpoint. Furthermore, because the BC has an honesty quality, it may give data pellucidity when used to an area that requires data disclosure. It may be used in a variety of fields, for instance, the fiscal subdivision and the Internet of Things (IoT) setting, with its utilizations are likely to grow as a result of its capabilities [28–32]. Because of its efficiency and availability, cloud computing (CC) has been accepted by many data processing settings. Furthermore, critical security elements of cloud protection and confidentiality have been explored [33].

If consumers' confidential information is exposed in the CC system, this might result in financial and psychological damages. In the cloud computing context, we primarily investigate data security when sending and saving information, such as integrity and privacy. If blockchain is upgraded to a suitable service level and integrated with the CC set, it could assure security [34]. A secure electronic card is mounted when the BC system is used. If the electronic card is not correctly erased, the information handler may be abused. The operator's leftover data could be utilized to extract the consumer's statistics. Double transactions on the blockchain and faking the ledger of Bitcoin pose a significant difficulty. To deal with such security concerns, you'll need a secure and dependable electronic card. Normally, electronic cards installed on PCs are used, but as mobile devices become more widespread, it is more important than ever to rigorously test the security of the electronic cards on mobile devices. As a result, a transaction is only complete when the accuracy and integrity of the time stamp produced on a mobile device assure the transaction's security [35]. A protected electronic card must be built by lessening, verifying, and validating glitches that may arise throughout the design, requirements analysis, implementation, and testing stages, as well as ongoing maintenance.

A safe and dependable restoration of the electronic card must be implemented if the security is broken or hacked by the attackers. It must maintain the security of user transaction data deposited in the electronic card, together with the sceneries required to maintain and run the electronic card. When the electronic card is not in use, it must provide a way for effectively and securely removing the remaining user data, and it must then trash the remaining data.

In a multi-tenancy CC, several companies could access and employ an application utilizing the same infrastructure [36–38]. Multi-tenancy at the data-center levels may be a facility vendor that leases data center space and offers servers, routers, and so forth [39]. Multi-tenancy can be accomplished at the network layer via software heaps, where a single heap belongs to a different client. This multi-tenancy network layer saves costs compared to the multi-tenancy data center layer as such stacks are configured according to real customer accounts [39, 40].

Multi-tenancy deployment at the CC submission service level necessitates architectural changes at equally the software and substructure layers. Existing multi-instance software architectures must be modified, and diversity of multi-tenant designs must be employed across the utilization layer [39]. There are three popular approaches employed in cloud applications to accomplish multi-tenancy:

Database: This uses the configuration of the database with data separation supported on the application layer. The development of various aspects of the application is used to continuously change their dissimilar activities at runtime for dissimilar tenants.

Virtualization: It employs software to build application hosting ecosystems that can deliver rational barriers, particularly for IaaS, between the tenants themselves. It can also run several copies of server operating systems in a single physical machine and share physical hardware.

Physical separation: Tt relies on the entire tenant having their unique hardware properties to deploy, for instance, to allocate distinct physical servers to specific tenants or to give a large client a notable segment of a data center [41–44].

Several researchers have described multi-tenancy as a security problem in the cloud world. Researchers [36] suggested eliminating the virtualization layer to enhance device stability. However, for existing systems (especially large clouds), the cost of such a change will be high. Under such a case, the useful feature of virtual machine (VM) reallocation will also not be feasible, which will result in underperformance depravity (for instance, the reduced extent of resource usage).

Anthony and Syed [45] considered the protection dangers and apprehensions in CC. Enlightened phases that an organization could be engaged in to lessen the protection threats as well as defend their resources were similarly demonstrated. The paper also explained various benefits, weaknesses, and application areas of cloud computing to information risk management. The work aimed at providing more insight on enterprise cloud security and ways to resolve security issues and concerns before implementing any technology.

Xu et al. [46] implemented a multi-tenancy approval scheme using shibboleth for cloud-based applications with federated identity. This approach used a shibboleth-known method to promote the authentication, authorization, and identity federation implementation process. The goal was to deliver a further efficient method of connecting a consumer with a facility vendor.

Odun-Ayo, Misra, Abayomi-Alli, and Ajayi [47] examined the current inclinations in the domain of cloud multi-tenancy and provided a direction for imminent investigation. The aim was to examine a comprehensive study connecting to a current investigation in multi-tenancy, current trends, and the advancement in cloud multi-tenancy; prevailing research articles; and various benefits to prospective cloud customers and similarly cloud vendors.

Kamaran and Ahmed [48] discussed the impact and benefits of multi-tenancy in a software as a service (SaaS) layer of cloud computing which provides a flexible data model. They reviewed their architecture, approaches, and performance. The aim was to provide various benefits and drawbacks of a multi-tenancy SaaS database and suggested the need to develop an ideal database system for SaaS.

Archana and Rekha [49] discussed the former green computing achievements, how green computing protects our environments from cloud negative impacts, and green characteristics like power management, virtualization, green data center, recyclability, and so on. The work gave a comprehensive report on green cloud analysis to help the naïve GC investigation individuals acquire knowledge about GC subjects and comprehend the GC imminent investigation difficulties.

Kumar and Bhatt [50] developed elliptic curve cryptography (ECC) for tenant certification, data encoding, and decoding. The suggested ECC-based certification solution allows an authorized user to access confidential data while also successfully protecting the data against other associated threats. The investigators integrated a nature-enthused optimization, for instance, a moth search algorithm (MSA), with ECC to determine the proper and best value of the elliptic curve to build a more secure data encoding system. DNA encoding was employed with the ECC encryption technique in the projected encryption and decryption technique. The DNA scrambled ECC approach provided multiple layers of protection while requiring little processing effort. The suggested method's security analysis was presented to demonstrate its efficacy contrary to invasions, for instance, denial-of-service, impersonation, reply, readable text, and chosen-scrambled text assaults. The security model's scrambled period, decryption period takes 83.153 and 86.076 sec to complete correspondingly. The assessment shows that the suggested methodology delivers dual-level protection with the smallest key size and the least amount of storage space.

Suresh-Kumar and Jagadeesh-Kannan [51] described a learning-centered scheduler for multi-tenant cloud computing that caters to diverse software and hardware resources. The experiment was conducted using a GC emulator, and the outcomes were likened to existing techniques such as minimal execution period; it followed the process of first-in, first-served, and backfilling. The findings showed that the proposed technique is an excellent way of leveraging cloud resources while also drastically lowering the cost to 0.

The deductions from the surveyed state of the arts discovered that most of the researchers did not consider the protection and confidentiality of multi-tenancy in cloud green computing environments. The related works considered and discussed in this study were summarized in Table 1 as follows:

3 Materials and Method

This section discussed the necessity for green computing, multi-tenancy involved in green computing, security and privacy challenges in the multi-tenancy environment of GC, application of blockchain technology to secure multi-tenancy in GC environment, and the architecture for blockchain technology and multi-tenants in the GC environment.

3.1 The Need for Green Computing (GC)

GC is the systematic way of how computers, servers, and related subsystems, including displays, printers, computing units, networking, and communication systems, are built, manufactured, used, and disposed of efficiently and effectively environmental impacted. This required techniques to reduce the use of not so environmentally friendly equipment, optimizing power efficiency, and reuse mobile equipment and IT garbage. Green computing offers plans for the future. Green IT, therefore, covers the spectrum of eco-friendly sustainability, the economy of energy efficiency, and the overall cost of owning it, including the cost of efficiently disposing of and reusing it with some major characteristics which include consolidation, and cloud computing [52]. Initially designed to cater to the IT industry, cloud computing is now also being used for optimum use in many sectors (retail, supply chain, healthcare, etc.). Another role of cloud computing is to reduce an organization's paper and power consumption and move into a greener or greener cloud computing environment [53]. The best way to save money and cut carbon (IV)

Table 1 Summary of related works

Authors	Problem considered	Contribution	Gap
Anthony and Syed [45]	The protection threats and apprehensions in cloud computing were considered in this research	Various benefits, weaknesses, and application areas of cloud computing to information risk management were considered	Trust and confidentiality were not considered
Xu et al. [46]	The authors implemented a multi-tenancy approval scheme	This approach used a shibboleth-known method to promote the authentication, authorization, and identity federation implementation process	The authors didn't consider trust and confidentiality in their research
Odun-Ayo, Misra, Abayomi-Alli, and Ajayi [47]	The authors examined the current inclinations in the domain of cloud multi-tenancy	The aim was to examine a comprehensive study connecting to a current investigation in multi-tenancy, current trends, and the advancement in cloud multi-tenancy; prevailing research articles; and various benefits to prospective cloud customers	The authors didn't consider multi-tenancy safety, confidentiality, and trust
Kamaran and Ahmed [48]	The authors discussed the impact and benefits of multi-tenancy in a software as a service (SaaS) layer of cloud computing	The aim was to provide various benefits and drawbacks of a multi-tenancy SaaS database and suggested the need to develop an ideal database system for SaaS	The authors only proposed a prototype architecture for an ideal database system for SaaS
Archana and Rekha [49]	The study discussed the former green computing achievements, how green computing protects our environments from cloud negative impacts, and green characteristics like power management, virtualization, green data center, recyclability, and so on	The work gave a comprehensive report on green cloud analysis to help the naïve GC investigation individuals acquire knowledge about GC subjects and comprehend the GC imminent investigation difficulties	The authors only conducted a literature review. They did implement any system or even suggested a secured green cloud computing system for multi-tenants

(continued)

Table 1 (continued)

Authors	Problem considered	Contribution	Gap
Kumar and Bhatt [50]	The study proposed an enhanced multi-tenancy security in the CC using hybrid data encoding techniques	The researcher employed hybrid ECC-based data encryption techniques	The average execution time for both the proposed encoding and decoding of the system was too high
Suresh-Kumar and Jagadeesh-Kannan [51]	The study described a learning-centered scheduler for multi-tenant cloud computing that caters to diverse software and hardware resources	GC emulator was employed to experiment and the findings showed that the proposed technique is an excellent way of leveraging cloud resources while also drastically lowering the cost to 0	The study time complexity was not evaluated

oxide CO_2 emissions was going green. It means greening is completely necessary to reduce IT operational costs which will be useful to solve environmental issues.

3.2 *Multi-Tenancy in GC*

A tenant is a community of users who hold the same opinion of an application that they are using. This view includes access information, configuration, user management, particular features, and related nonfunctional properties. The parties are usually representatives of different legal bodies. This implies limitations (e.g., data protection and privacy). Multi-tenancy is a method for sharing an application instance among multiple tenants by supplying each tenant with a designated "share" of the instance, separated from other functions and data security shares [54]. A widely used illustration for clarification is a living complex, whereby various parties exchange some of their resources such as heating to reduce costs but also love to appreciate their confidentiality and thus require some degree of separation (especially when it comes to noise). There is also the notion of rental space beyond multi-tenancy. A tenant space refers to the condition where clients rent a specified resource space in which they can operate various instances of applications. One example is an IaaS service where a client purchases resources where he installs his chosen applications.

Multi-tenancy is an inevitable consequence of seeking to achieve an economic advantage in cloud computing through the use of virtualization and the sharing of resources [46]. Multi-tenancy involves the distribution of services in the cloud setting, although, in the sense of GC, where multi-tenancy is used separately from multiple service models, such a term is still common. As a service (SaaS) in applications, cloud service provider (CSP) offers applications as a service where the user is unable to track or manage the underlying infrastructure; here, multi-tenancy implies that two or even additional users employ the equivalent system or software offered by the CSP regardless of the actual services [54]. Multi-tenancy arises when the same physical machine (PM) is shared by virtual machines (VMs) distributed to different clients in infrastructure as a service (IaaS), in which the user can offer computing, processing, and networking services and can manage but could not handle the underlying infrastructure.

Multi-tenancy is regarded as one of cloud computing's important consequences for security and privacy. Davida, Well, and Kam [55] described multi-tenancy as the main cloud functionality and a major cloud security challenge aspect that involves a systemic solution from SaaS to infrastructure as a service (IaaS). Multi-tenancy can allow exposure to information and expand the potential of the attack that influences cloud security. VM mobility is one of those advantages where providers can quickly redeploy VMs to make better use and save energy usage [56–58]. Over-provisioning is regarded as one of cloud computing's main advantages, as it provides the CSP with the ability to seal more than its network capacity [56–58]. These functions are essential to cloud users as well as any possible solution that must add or at least seek to maintain them and not remove any of them.

3.3 Security and Privacy Challenges in the Multi-Tenancy Environment of GC

In green computing, multi-tenancy employs methods where both the intruder and the victim use a similar server (i.e., physical machine (PM)). Such a configuration could not be reduced by conventional security strategies and interventions basically because it is not intended to infiltrate servers and its control procedures are restricted to the system layer [59]. Three situations differentiate between multi-tenancy and traditional attacks. Situation 1 describes the Internet attacks where the client and intruder are both frequent users of the Internet. Conventional network protection techniques and tools are produced to protect against these attacks. Situation 2 describes the attacks within cloud provider; the victim and attacker are a client of the same service provider with each on a different host. This situation is related to virtualization in which cloud providers incorporate virtual network protection devices and technologies to protect the clients [59]. In situation 3, the client and the attacker share the same server. This situation is known as multi-tenancy. Protection of clients in this situation requires more tasks since both the client and attacker network contact in the physical machine are restricted. There is more traffic to the physical machine, and this makes it difficult to target the intruder.

An attacker should start a network probing for VMs in a PM to demonstrate how this can be done. If a target VM has been defined, it can then launch a brute force attack. This is one of the potential drawbacks of multi-tenancy when hosted in the same PM, that is, an intruder will collect data from the victims using a side-channel attack. Perhaps such attacks could not be detected by the hypervisor or resident operating system of the PM. Multi-tenancy techniques cannot be eliminated in green computing. To keep its benefits, the application of blockchain technology could minimize the risk of multi-tenancy for customers and restrict the number of potential attackers [59].

3.4 Application of BC Technology to Secure Multi-Tenancy in GC Environment

Blockchain is a digital ledger that requires all participants to manage a dataset that holds the entire contract data and to change their archives to uphold transparency anytime an innovative contract takes place [59, 61]. Since the advent of the Internet and cryptographic technologies has made it conceivable for the complete participants to check the security of the contract, a solitary event of catastrophe resulting from reliance on an approved third party has been overcome; thus, this paper proposed a blockchain technology to enhance security and privacy issues in a multi-tenancy environment of green computing. The blockchain is a groundbreaking technology for trustworthy measurement and tracking of energy-related properties, anticipated to be further transparent as self-generation and micro-grid market situations are distributed across both the public and private sectors [60, 61]. A BC is a collective and disseminated database that consists of related transaction blocks. Unlike other database methods, blockchain ensures that authorized transactions will be stored in a tamper-proof way. Because of its distributed and decentralized structure, IoT uses blockchain, for instance, for managing system setup, storing sensor data, and allowing micro-payments. The Bitcoin network represents the first successful use of blockchain technology. To manage the network, the system uses democratic consensus, which means it is not centrally governed by a bank, company, or government. The larger the network, the more decentralized it becomes, and the more efficient it becomes [60, 61].

The blockchain development promise isn't limited to Bitcoin. As such, it has gained significant demand in several companies, like financial institutions, hospitals, and nonprofit organizations, entertainment, and e-commerce. With a blockchain, many individuals can submit inputs to the information record, and a user group can monitor how the information record is changed and updated. Blockchain-created distributed database has a fundamentally different digital backbone. This is also the most distinct and important feature of blockchain technology [60–62]. The combination of proven technologies applied in a new way (the Internet, private key cryptography, and a protocol governing incentivization) made the idea so useful in the multi-tenancy application. With blockchain technology configuration, user authentication, and authorization in a multi-tenancy platform of the cloud environment will be established. Figure 1 depicts the proposed architecture of blockchain in a multi-tenancy environment activity for green computing.

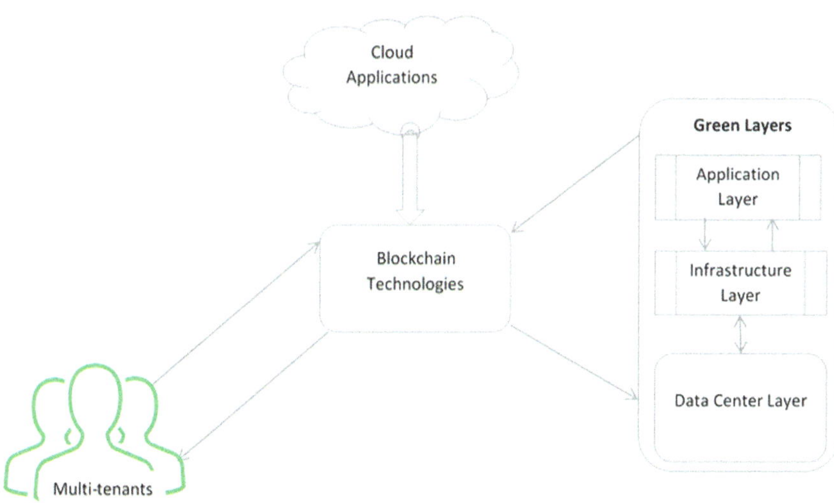

Fig. 1 Proposed architecture for blockchain technology and multi-tenants in the green computing environment

Blockchain is a digital infrastructure, related to the extremely secured and centralized databases operated today by governments or banks, or insurance companies. Control of consolidated databases remains with their operators, including maintenance of updates, entry, and cyber threat protection [60, 61].

Security Solution in Multi-Tenant Environment of Green Computing: Multi-tenancy has various potential implementation models for all multiple tenants, from multiple instances to a specific instance for each client. It does, however, require that the SaaS application manage multi-tenancy and occupant data and process isolation itself or hand over processing and storage tasks to third parties. This raises new protection and confidentiality problems that aren't addressed in multi-instance single-tenant applications. There is a necessity for building proper security measures for every element of the SaaS utilization, and for the entire virtual IaaS facility. These safety subjects depend on four essential trends:

Filtering: Using an intermediate surface amid an occupant and a data basis acting as a strainer leaves an occupant unaware of the presence of other occupants and makes the occupant look as if the occupant's data is the solitary data in the collective database.

Permissions: Using access-control lists (ACL) to assess data rights (who may access data in the application) and data processing models (what data activity is permitted).

Encryption and Obfuscation: Secret sensitive data and/or processing of each tenant to prevent unauthorized parties from accessing it.

Blockchain cryptosystem: this is a digital signature through the use of public-key cryptographic techniques to protect user privacy and information. Figure 2 shows the proposed BC technology to secure the multiple tenants in the CC environment.

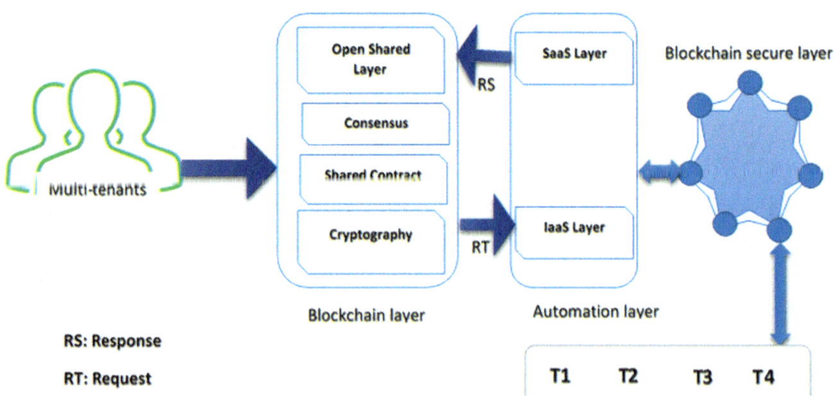

Fig. 2 Proposed blockchain technology to secure multiple tenants in cloud environment

4 Results and Discussion

This section discussed the execution of the system. Ganache blockchain was used to create a secure contract account for each tenant in the cloud green computing environment. The interfaces gotten from the execution of the proposed system were also demonstrated and discussed in this section.

4.1 Implementation

The proposed technique consists of several layers such as the multi-tenants, the blockchain layer, the automation layer, the BC secure layer, and the cloud storage where the client's information is store. This technique allows each client to use the cloud facilities as if there is the only one present in the cloud, thereby enhancing the privacy and security of multiple tenants in a green computing environment, because if client data is leaked as a result of multiple tenants' characteristic of green computing platform, monetary and psychological harm can affect the tenant due to leakage of user confidential information. The second layer of the proposed method, which is the blockchain layer, will help to guarantee the privacy of each tenant. The blockchain secure layer that was paired with a cloud storage environment will provide a seamless service that offers improved security to each tenant's records. User confidentiality will be guaranteed if the blockchain approach is used to store user knowledge in the green computing setting.

4.2 Discussion

The execution of blockchain for security in a green computing environment was proposed and implemented in this study. Figures 3 to 5 describe the implantation of blockchain in a cloud environment using Ganache and MetaMask to create a dummy secure account for each cloud tenant. Figure 3 shows the Ganache blockchain interface, and this is where the truffle-shuffle interface was used to create a secure contract account key for each tenant in the cloud GC environment. Figure 4 shows the various tenant accounts with secured contract identity. Figure 5 also shows each of the tenants on the cloud GC environment with their secured blockchain keys, which were employed to carry out all secured and privacy activities that are the interface which displays the profile page view, where each tenant in the cloud environment can view and export their private keys. Table 2 shows the comparative analysis with the state of the art, and it was deduced that the proposed system performed better than the existing ones with an execution time of 52.548 sec over that of Kumar and Bhatt [50] having 83.153 sec execution time.

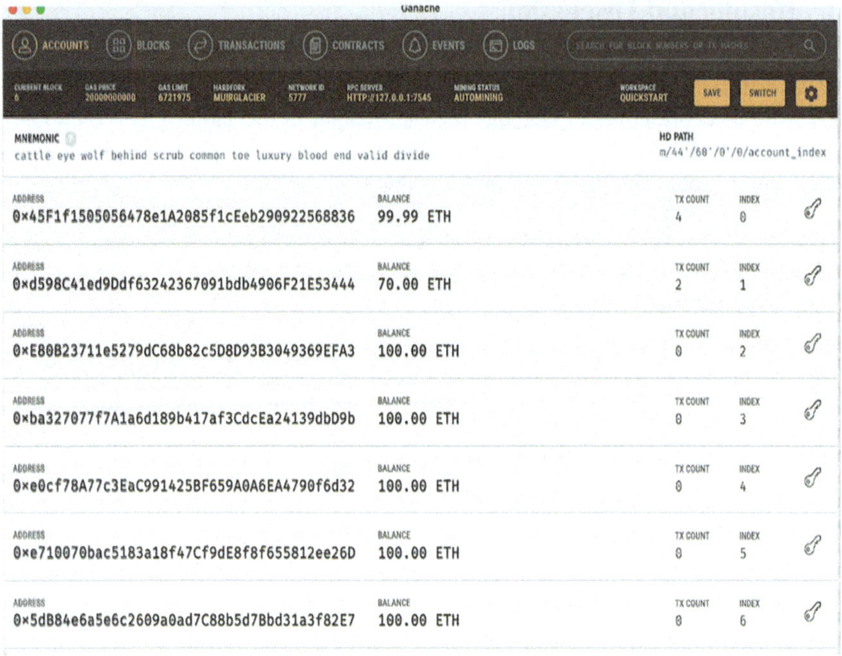

Fig. 3 Ganache blockchain interface

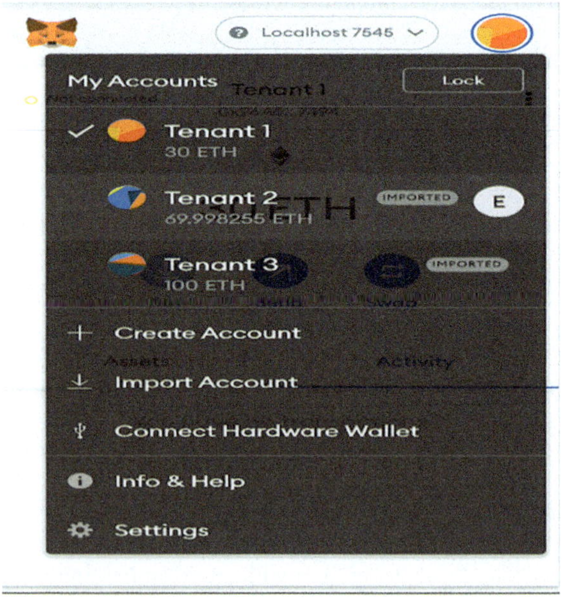

Fig. 4 Various tenant accounts with secured contract identity

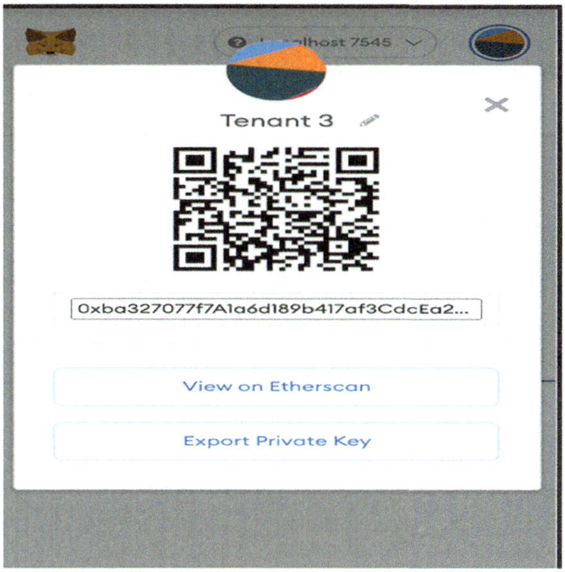

Fig. 5 Profile view of each tenant

Table 2 Comparative study of existing research

Authors	Methods	Platform	Execution time
Kumar and Bhatt [50]	Elliptic curve cryptography (ECC)	Cloud-based environment	83. 153 secs
Reantongcome, Visoottiviseth, Sawangphol, Khurat, Kashihara, and Fall [63]	Smart contract	Cloud-based environment	The system was not evaluated
Proposed system	**Ganache blockchain and meta-mask**	**Green computing environment**	**52.548 secs**

5 Conclusion

A blockchain technology eliminated the presence of the central authority from the registry and allowed purchases by individuals who collectively store electronic documents and, eventually, accepted contracts using peer-to-peer network technologies. This technique is proposed to be suitable for a multi-tenancy environment in green computing by enhancing the current solutions, and more advanced and innovative approaches to make sure green computing environment benefits are fully realized as its adoption accelerates. Green computing is still very new; its wide adoption will depend on the effectiveness of multi-tenancy security and privacy pattern. Blockchain technology is sometimes identified as a foundation for the cloud's transaction layer; the idea that cryptographic keys and shared ledgers will allow users to protect and formalize digital relationships has wild-running imaginations. It was deduced from the implementation of the proposed system that it performed better than the existing ones with an execution time of 52.548 sec over that of Kumar and Bhatt [50] having 83.153 sec execution time. Thus, the application of blockchain cryptosystem in a multi-tenancy environment of green computing will enhance and address the security issues since everybody seeks to develop a trusted transaction in the cloud environment.

References

1. Kaur, A., & Kaur, S. (2019). Green computing: Emerging issues in IT. *International Journal of Trend in Scientific Research and Development (IJTSRD), 3*(5) ISSN: 2456 – 6470. www.ijtsrd.com
2. Jaiswal, A., Kumar, S., Kaiwartya, O., Prasad, M., Kumar, N., & Song, H. (2021). Green computing in IoT: Time slotted simultaneous wireless information and power transfer. *Computer Communications, 168*, 155–169.
3. Rani, R., Kumar, S., Kaiwartya, O., Khasawneh, A. M., Lloret, J., Al-Khasawneh, M. A., … Alarood, A. A. (2021). Towards green computing oriented security: A lightweight postquantum signature for IoE. *Sensors, 21*(5), 1883.
4. Naji, H. Z., Zbakh, M., & Munir, K. (2017, October). *A review of green cloud computing techniques*. In International Conference of Cloud Computing Technologies and Applications (pp. 264–283). Springer.
5. Raza, K., Patle, V. K., & Arya, S. (2012). A review on green computing for eco-friendly and sustainable it. *Journal of Computational Intelligence and Electronic Systems, 1*(1), 3–16.
6. Raza, K., Patle, V. K., & Arya, S. (2014). A review on green computing for eco-friendly and sustainable IT. *Journal of Computational Intelligences and Electronic System, 1*, 1–14.
7. Tuttle, J., Chen, Y., Jiang, T., Hunter, L., Waldren, A., Ghosh, S., & Ingram, W. A. (2020). *Multi-tenancy cloud access and preservation.*
8. Le Nguyen, B., Lydia, E. L., Elhoseny, M., Pustokhina, I., Pustokhin, D. A., Selim, M. M., … Shankar, K. (2020). Privacy-preserving blockchain technique to achieve secure and reliable sharing of IoT data. *Computers, Materials & Continua, 65*(1), 87–107.
9. Beikverdi, A., & Song, J. (2015). *The trend of centralization in Bitcoin's distributed network.* In 2015 IEEE/ACIS 16th International Conference on Software Engineering, Artificial Intelligence, Networking, and Parallel/Distributed Computing (SNPD) (pp. 1–6). IEEE.
10. Huh, S., Cho, S., & Kim, S. (2017). *Managing IoT devices using a blockchain platform.* In 2017 19th international conference on advanced communication technology (ICACT) (pp. 464–467). IEEE.
11. Jain, S., & Simha, R. (2018, July). *Blockchain for the common good: A digital currency for citizen philanthropy and social entrepreneurship.* In 2018 IEEE International Conference on Internet of Things (iThings) and IEEE Green Computing and Communications (GreenCom) and IEEE Cyber, Physical and Social Computing (CPSCom), and IEEE Smart Data (SmartData) (pp. 1387–1394). IEEE.
12. Zhang, J., Xue, N., & Huang, X. (2016). A secure system for pervasive social network-based healthcare. *IEEE Access, 4*, 9239–9250.
13. Ziegeldorf, J. H., Matzutt, R., Henze, M., Grossmann, F., & Wehrle, K. (2018). Secure and anonymous decentralized bitcoin mixing. *Future Generation Computer Systems, 80*, 448–466.
14. Mahmood, Z. (2011). Cloud computing for enterprise architectures: Concepts, principles, and approaches. In *Cloud computing for Enterprise architectures* (pp. 3–19). Springer.

15. Cheung, A. S., Weber, R. H., & (Eds.). (2015). *Privacy and legal issues in cloud computing*. Edward Elgar Publishing.
16. Li, J., Jia, C., Li, J., & Chen, X. (2012). Outsourcing encryption of attribute-based encryption with MapReduce. In T. W. Chim & H. Yuen (Eds.), *ICICS 2012. LNCS* (Vol. 7618, pp. 191–201).
17. Hur, J., & Noh, D. K. (2011). Attribute-based access control with efficient revocation in data outsourcing systems. *IEEE Transactions on Parallel and Distributed Systems (TPDS), 22*(7), 1214–1221.
18. Verizon 2015. (2015). *2015 Data Breach Investigations Report*. http://www.verizonenterprise.com/DBIR/2015/. Accessed 20 Sept 2017.
19. Awotunde, J. B., Ogundokun, R. O., Misra, S., Adeniyi, E. A., & Sharma, M. M. (2021). Blockchain-based framework for secure transaction in Mobile banking platform. *Advances in Intelligent Systems and Computing, 1375*, 525–534.
20. Gupta, A., Siddiqui, S. T., Alam, S., & Shuaib, M. (2019). Cloud computing security using blockchain. *Journal of Emerging Technologies and Innovative Research, 6*(6).
21. Polina, M., Lucy, O., Yury, Y., Alex, O., Alex, B., Pavel, P., et al. (2018). Converging blockchain and next-generation artificial intelligence technologies to decentralize and accelerate biomedical research and healthcare. *Oncotarget, 9*(5), 5665–5681.
22. Khan, F. A., Asif, M., Ahmad, A., Alharbi, M., & Aljuaid, H. (2020). Blockchain technology, improvement suggestions, security challenges on the smart grid, and its application in healthcare for sustainable development. *Sustainable Cities and Society, 102018*.
23. Bibri, S. E. (2018). The IoT for smart sustainable cities of the future: An analytical framework for sensor-based big data applications for environmental sustainability. *Sustainable Cities and Society, 38*, 230–253.
24. Awotunde, J. B., Ogundokun, R. O., Jimoh, R. G., Misra, S., & Aro, T. O. (2021). Machine learning algorithm for cryptocurrencies Price prediction. *Studies in Computational Intelligence, 972*, 421–447.
25. Halim, N. S. A., Rahman, M. A., Azad, S., & Kabir, M. N. (2017). *Blockchain security hole: Issues and solutions*. In: Proceedings of the International Conference of Reliable Information and Communication Technology, 739–746.
26. Ferrag, M. A., Derdour, M., Mukherjee, M., Derhab, A., Maglaras, L., & Janicke, H. (2018). Blockchain technologies for the internet of things: Research issues and challenges. *IEEE Internet of Things Journal, 6*(2), 2188–2204.
27. Abayomi-Zannu, T. P., Odun-Ayo, I., Tatama, B. F., & Misra, S. (2020). *Implementing a mobile voting system utilizing Blockchain technology and two-factor authentication in Nigeria*. In Proceedings of first international conference on computing, communications, and cybersecurity (IC4S 2019) (pp. 857–872). Springer, Singapore.
28. Beikverdi, A., & JooSeok, S. (2015). *The trend of centralization in Bitcoin's distributed network*. In Proceedings of the 2015 16th IEEE/ACIS international conference on software engineering, Artificial Intelligence, Networking and Parallel/Distributed Computing (SNPD), Takamatsu, Japan.
29. Bonneau, J., Miller, A., Clark, J., Narayanan, A., Kroll, J.A. and Sok, F.E.W. (2015). *Research perspectives and challenges for bitcoin and cryptocurrencies*. In Proceedings of the 2015 IEEE symposium on security and privacy (SP), San Jose, CA.
30. Christidis, K., & Michael, D. (2016). Blockchains and smart contracts for the internet of things. *IEEE Access, 2016*(4), 2292–2303.
31. Huang, H. Chen, X. Wu, Q. Huang, X., & Shen. J. 2016. Bitcoin-based fair payments for outsourcing computations of fog devices. *Future Generation Computer Systems*
32. Huh, S., Sangrae, C. and Soohyung, K. (2017). *Managing IoT devices using a blockchain platform*. In Proceedings of the 2017 19th international conference on advanced communication technology (ICACT), Bongpyeong, Korea.
33. Singh, S., Jeong, Y.-S., & Park, J. H. (2016). A survey on cloud computing security: Issues, threats, and solutions. *Journal of Network and Computer Applications, 75*, 200–222.
34. Haber, S., & Stornetta, W.S. (1990). *How to time-stamp a digital document*. In Proceedings of the Conference on the Theory and Application of Cryptography, Sydney, NSW, Australia.

35. Alam, S. Siddiqui, S. T. Masoodi, F. and Shuaib M. 2018. *Threats to information security on cloud: Implementing Blockchain*. 3rd international conference on SMART computing and informatics (SCI), 21–22 December 2018, Kalinga Institute of Industrial Technology, Odisha. Springer. SPRINGER-SIST series.
36. Keller, E., Szefer, J., Rexford, J., & Lee, R.B. (2010). *NoHype: virtualized cloud infrastructure without the virtualization*. The 27th Annual International Symposium on Computer Architecture (June 19–22, 2010).
37. Shao, Q. (2011). *Towards effective and intelligent multi-tenancy SaaS*. Arizona State University.
38. Shaikh, F., & Patil, D. (2014, August). *Multi-tenant e-commerce based on saas model to minimize its cost*. In 2014 international conference on advances in Engineering & Technology Research (ICAETR-2014) (pp. 1–4). IEEE.
39. Petersson, J. (2011). *Best practices for cloud computing multi-tenancy*.
40. Meiers, J. (2011). *Best practices for cloud computing multi-tenancy*. White paper, IBM, 6.
41. Fiaidhi, J., Bojanova, I., Zhang, J., & Zhang, L. J. (2012). Enforcing multitenancy for cloud computing environments. *IT Professional Magazine, 14*(1), 16.
42. Bojanova, I., Zhang, J., & Zhang, L. J. (2012). Enforcing multitenancy for cloud computing environments. *IT Professional, 14*(1).
43. Aljahdali, H., Townend, P., & Xu, J. (2013, March). *Enhancing multi-tenancy security in the cloud IaaS model over public deployment*. In 2013 IEEE seventh international symposium on service-oriented system engineering (pp. 385–390). IEEE.
44. Cai, H., Wang, N., & Zhou, M. J. (2010, July). *A transparent approach to enabling SaaS multi-tenancy in the cloud*. In 2010 6th World Congress on Services (pp. 40–47). IEEE.
45. Anthony, B., & Syed, M. R. (2011). An overview of the security concerns in enterprise cloud computing. *International Journal of Network Security & its applications (IJNSA), 2*(1).
46. Xu, Y., Musgrave, Z., Nobel, B., and Bailey, M. (2014). *Workload-aware provisioning in public clouds*. IEEE internet computing, 18(4), 15–21, IEEE.
47. Odun-Ayo, I., Misra, S., Abayomi-Alli, O., & Ajayi, O. (2017, December). *Cloud multi-tenancy: Issues and developments*. In Companion Proceedings of the the10th International Conference on Utility and Cloud Computing (pp. 209–214).
48. Kamaran F. & Ahmed Y. (2018). *The impact and benefits of multi-tenancy database in a cloud computing environment*. Copublished by the IEEE Computer and Reliability Societies.
49. Patil, A. & Patil, R. (2019). *An analysis report on green cloud computing current trends and future research challenges*. International conference on sustainable computing in science, technology & management (SUSCOM-2019). Amity University Rajasthan, Jaipur, India. pp. 813–820.
50. Kumar, P., & Bhatt, A. K. (2020). Enhancing multi-tenancy security in cloud computing using hybrid ECC-based data encryption approach. *IET Communications, 14*(18), 3212–3222.
51. Suresh Kumar, D., & Jagadeesh Kannan, R. (2020). Reinforcement learning-based controller for adaptive workflow scheduling in multi-tenant cloud computing. *The International Journal of Electrical Engineering & Education, 0020720919894199*.
52. Sato, H. (2011). *Eco-Labelling and Green Procurement Schemes for IT Products: The Japanese Approach*. http://enviroscope.iges.or.jp/modules/envirolib/upload/1511/attach/Paper%209.pdf. Retrieved 31 Dec 2011.
53. Erenben, C. (2009). *Cloud computing: The economic imperative, school news*, Vol. 13, March, available at www.eschoolnews.com/emails/esntoday061509.html.
54. Rouven, K, Christof, M, & Samuel, K. (2012). *Architectural concerns in multi-tenant SaaS applications*. In Proceedings of the 2ndInternational conference on cloud computing and services science (CLOSER-2012) (pp. 426–431). ISBN: 978-989-8565-05-1.
55. Davida, G. I., Wells, D. L., & Kam, J. B. (1978, November). *Security and privacy*. In IEEE computer society's second international computer software and applications conference, 1978. COMPSAC'78. (pp. 194–203). IEEE.
56. Fernando, N., Loke, S. W., & Rahayu, W. (2013). Mobile cloud computing: A survey. *Future Generation Computer Systems, 29*(1), 84–106.

57. Qin, J., Wu, Y., Chen, Y., Xue, K., & Wei, D. S. (2019). Online user distribution-aware virtual machine re-deployment and live migration in SDN-based data centers. *IEEE Access, 7*, 11152–11164.
58. Xiao, H., Hu, Z., & Li, K. (2019). Multi-objective VM consolidation based on thresholds and ant colony system in cloud computing. *IEEE Access, 7*, 53441–53453.
59. Jansen, W.A. (2011). *Cloud hooks: Security and privacy issues in cloud computing*. Proceedings of the 44th Hawaii International Conference on System Sciences.
60. Augusto Ciuffoletti. (2010). *Monitoring a virtual network infrastructure*, (October 2010).
61. Imbault, F., Swiatek, M., de Beaufort, R., & Plana, R. (2017). *The green blockchain: Managing decentralized energy production and consumption*. 2017 IEEE International Conference on Environment and Electrical Engineering and 2017 IEEE Industrial and Commercial Power Systems Europe (EEEIC/I&CPS Europe). https://doi.org/10.1109/eeeic.2017.7977613.
62. Almorsy M., Grundy J., & Ibrahim A. S. (2012). *Smurf: Supporting multi-tenancy using a re-aspects framework*. In Engineering of complex computer systems (ICECCS) (pp. 361–370). IEEE, 2012.
63. Reantongcome, V., Visoottiviseth, V., Sawangphol, W., Khurat, A., Kashihara, S., & Fall, D. (2020, April). *Securing and trustworthy Blockchain-based multi-tenant cloud computing*. In 2020 IEEE 10th Symposium on Computer Applications & Industrial Electronics (ISCAIE) (pp. 256–261). IEEE.

Application of Crypto-Blockchain Technology for Securing Electronic Voting Systems

Lukman Adewale Ajao, Buhari Ugbede Umar, Daniel Oluwaseun Olajide, and Sanjay Misra

1 Introduction

The irregularities and malpractices in democratic positions (election) are becoming normal practices among African countries. This is due to low technology of voting system materials, poor governance, and oligarchy tendencies. For instance, Nigeria is a case study where election malpractices are domicile during the democratic process of voting, resulting in wrong selection of leadership, chaotic, unfortunate free, fair, and credible election processes and many other challenges. Voting is a democratic process by which a democratic society or citizens approved to determine their choice of leadership selection [1]. E-Voting is an acronym for electronic voting system that encompasses the integration of electronics and information technology to support the counting of electorate votes [2]. This e-voting system has been improved with recent technology in the design architecture but is still found susceptible to voters' privacy, tampering, manipulation, and frauds by the individual electoral parties' agent or authority [3]. Some of the misconduct includes ballot snatching, multiple voting, failure of smart card readers' (SCR) performance, and biometric authentication mischief. However, electronic voting systems' poor performance results from system error performance, network security challenges, and data insecurity [4].

Biometric authentication is an identification technique and control of access based on physiological or behavioral characteristics [5]. Biometrics authentication

L. A. Ajao (✉) · B. U. Umar · D. O. Olajide
Department of Computer Engineering, Federal University of Technology, Minna, Nigeria
e-mail: ajao.wale@futminna.edu.ng

S. Misra
Department of Computer Science and Communication, Østfold University College (HIOF), Halden, Norway

S. Misra, A. Kumar Tyagi (eds.), *Blockchain Applications in the Smart Era*, EAI/Springer Innovations in Communication and Computing,
https://doi.org/10.1007/978-3-030-89546-4_5

relies on individual authenticity, unlike password authentication, which depends on the generation of a key to which brute force attacks or man-in-the-middle attacks can compromise. Biometric authentication methods involve comparing a biometric sample (biometric template or identifier) registered or enrolled with a newly captured biometric sample.

A unimodal fingerprint biometrics electronic voting system using Advanced Encryption Standard (AES)-based wavelet and cryptographic watermarking was developed to improve the disorder in the election system. The unimodal biometrics authentication is implemented to solve the problem of voting irregularities in a situation where the voter's fingerprint cannot be authenticated by the system [6, 7]. This system ensured the integrity and credibility of votes. A system that allowed people to vote on a website using voter ID and PIN code has been developed, which builds confidence in e-voting by improving the verification process and auditing of election results voting [8]. But the central system could be exposed to denial-of-service (DoS) attacks due to the inadequacy of security countermeasure.

An e-voting system using RSA and MD5 algorithms for encryption and securing votes was developed to maintain a high level of security with the use of two encryption algorithms. Still, it is computational complexity that slows the performance of the system [9]. However, the encryption schemes required some procedure for the decryption algorithm to achieve original votes before tallying ballots, which slow the system's performance and make security authentication inefficient. Cetinkaya and Doganaksoy implemented an e-voting system using dynamic balloting and Pseudo-Voter Identity (PVID) scheme to provide adequate security. The system considered recasting of votes as a solution for coercibility problems in uncontrolled environments [10]. But it cannot prevent the manipulation of electoral results by the central authority.

The development of a smart system using cryptography and hashing methods has been a way forward to reduce the security threats in electronics or embedded system applications. But this security authentication on the electronic system is still inefficient in some application areas to prevent unauthorized third party from accessing or tampering with legitimate records, as in the financial transaction, electronic businesses, online marketing, e-health records, e-voting system, and many other areas of smart technology. Therefore, blockchain (with immutability, transparency, and decentralization advantages) has been popularly proposed as a recent security countermeasure to reduce third-party mischievous, fraudulent, stealing of legitimate information, and data privacy breaches. Therefore, this study proposed developing a bi-factor crypto-blockchain technology for securing electronic voting systems as our contribution to the existing works. The bi-factor authentication of a biometric system using fingerprint and iris identity is proposed to reduce the false acceptance rate (FAR) of the counterfeit candidates and the false rejection rate (FRR) of authentic candidates during the election voting processes. Also, private blockchain technology is proposed as a decentralized system with immutability and transparency records management system using the Paillier homomorphic encryption algorithm (PHEA) to improve the efficient performance of the chain network.

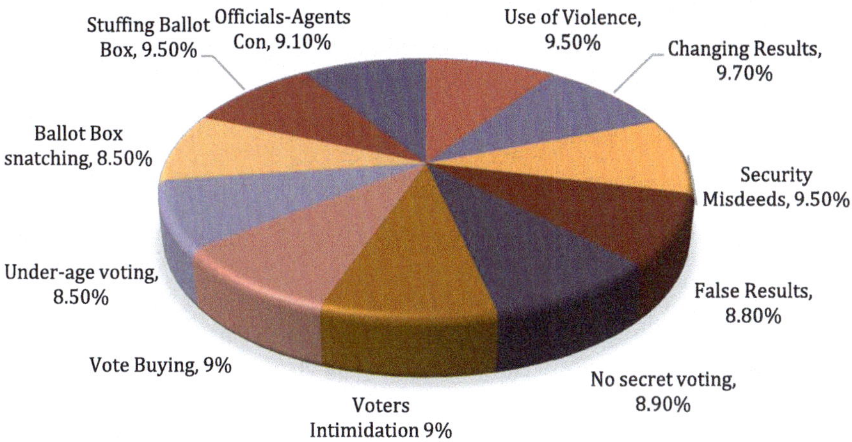

Fig. 1 Analysis of Nigeria polls' misconduct during the general election

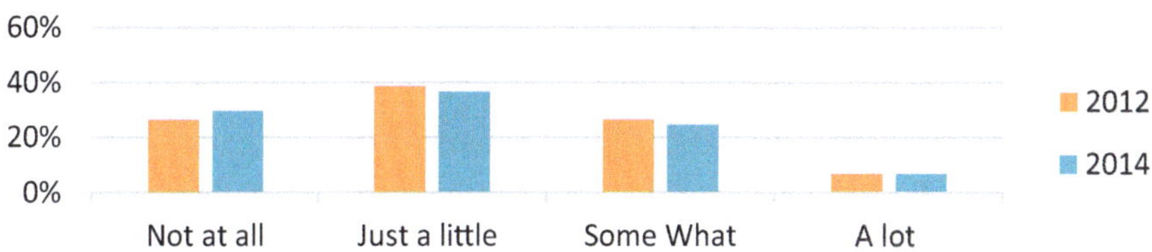

Fig. 2 Independent electoral commission evaluation

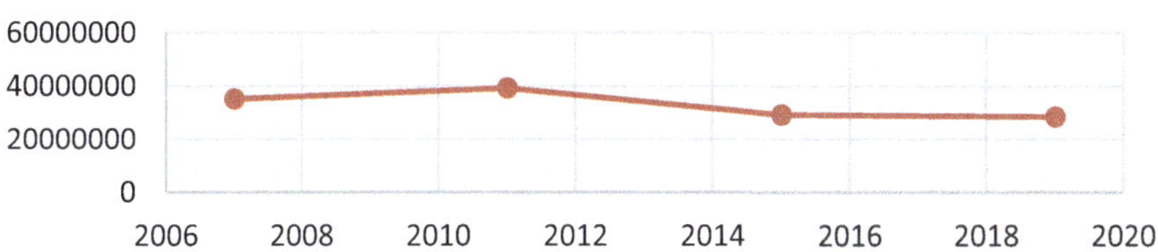

Fig. 3 Nigerian voter participation in general elections

The statistical analysis of a misconduct during a general election is presented (see Fig. 1), while the irregularities from independent electoral commission (INEC) during the 2011 and 2015 general election were analyzed (see Fig. 2). Therefore, due to the brutal experience of electorates during election processes, gross misconduct of electoral body, inefficient electronic voting system, and voters' privacy breaches have drastically reduced the citizen's participation in the democratic process. See Fig. 3 for the detailed analysis of the voters' participation in the election between 2007 and 2019.

This research is organized as follows: Section 1 introduces the general research background of the study, electronic voting system, and biometric classifications. Section 2 introduces blockchain technology and its consensus algorithm in Sect. 2.1, blockchain classification in Sect. 2.2, and the area of related blockchain applications in Sect. 2.3. The proposed system methodology with flowchart and algorithm is presented in Sect. 3. Section 4 discussed results and system performance evaluation, while Sect. 5 concludes the research investigation.

2 Blockchain Technology

Blockchain is a distributed public ledger that contains a set of blocks that are interconnected to embrace a digital signature of the preceding block with the next hash block to make it irrefutable, immune to tampering, and transparent [11]. Blockchain technology does not agree with third-party interactions in the chain participant and does not support a singular authority system [12]. A connected blockchain in the networks is known as a ledger which is shared between the chain participants in a public distributed ledger [13]. This complex cryptographic algorithm of blockchain makes it difficult for hackers or any third party to tamper, delete, and modify records [14]. So, the blockchain network participant jointly verifies the transactions and archives transaction information to ensure the integrity and reliability of records transactions. The blockchain transaction process begins with the creation of a block to store the transaction; each of the blocks is unified with a timestamp and linked to the previous blocks. The node of the blockchain participant examines the transaction before added to the hashing chain in the consensus algorithm network. Finally, some created blocks and connected chains are encrypted, and each encoding block takes a reference from the previous hashing blocks.

2.1 The Consensus of Blockchain Algorithm and its Classification

A consensus blockchain algorithm is a mechanism that issues certificate agreement to the blockchain network for mutual interrelation and verifies an agreement for the record validation. A blockchain network is a decentralized and distributed ledger in nature that exists among several nodes in the chain participant to avoid autonomous centrality, control, and validity of a transaction [15]. The consensus algorithm is the backbone of blockchain technology that can be described and classified as follows (see Fig. 4).

Proof of Work (PoW) This cryptographic technique ensures the authenticity of transactions (proof) between the parties (prover and verifier) in the chain networks using Diffie-Hellman-based puzzle, which can be subsequently verified or confirmed with little effort. For instance, PoW is popularly adopted as a prover and verifier for the consensus of Bitcoin transactions through cryptocurrencies.

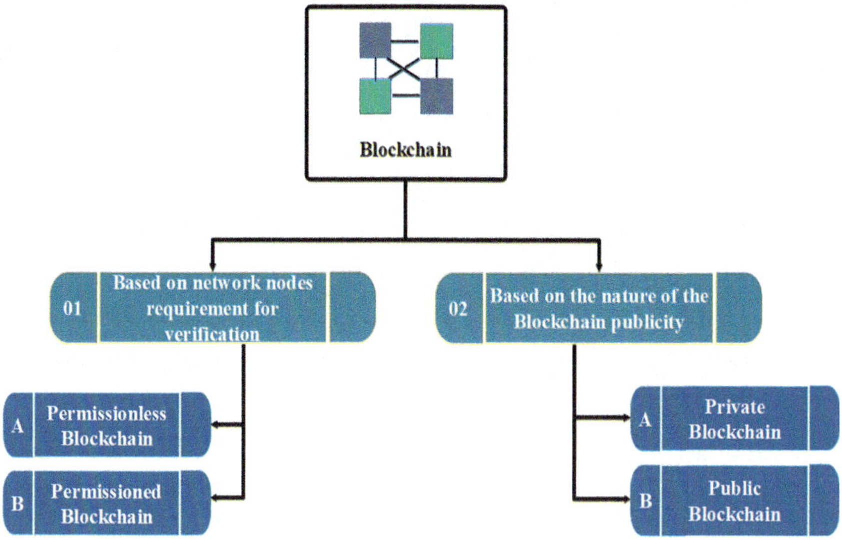

Fig. 4 Classification of blockchain

The disadvantage of this POW (consensus algorithm) is high computing power requirements [16].

Proof of Stake (PoS) This protocol was adopted for securing cryptocurrency and Ethereum operation through the selection of validators, appended on the transaction which is proportional to the number of linked cryptocurrencies in the consensus mechanism. This prevents a malicious actor and groups of users in the transactions chain networks. The PoS is efficient for fast transaction of the blockchain verification with low energy consumption and minimum hardware requirements [17].

Proof of Authority (PoA) This type of consensus algorithm is automated with the use of software program to secure the transactions in the blockchain network and validated by permitted authority or accounts called validators [18].

The Blockchain-Based Network Nodes for Certificate Verification

Permissionless Blockchain Network This type of blockchain credential verification is called public blockchain, and it does not require any permission (participants' consensus) to become or join the chain network for participation and interaction [19]. This model is suitable for operating digital currencies with effective management. The permissionless approach allows individual users to create and address the issues in network for the interaction by either validating the transaction or direct the transactions to another participant on the networks. The use-cases environment are

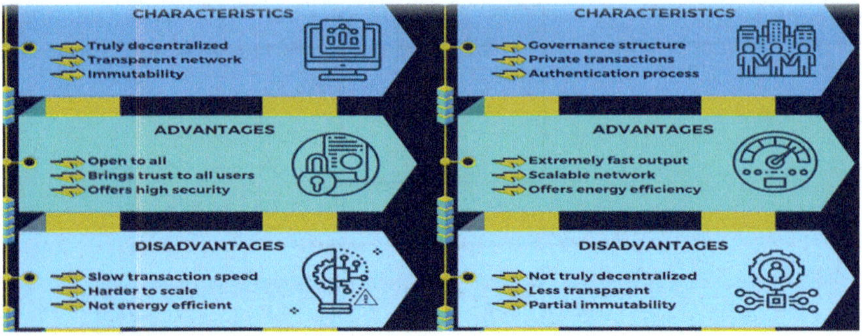

Fig. 5 Permissionless (left) and permissioned (right) blockchain characteristics

balloting systems, digital identity, and fundraising. The example includes Bitcoin and Ethereum.

Permissioned Blockchain This type of blockchain credential verification does not allow third-party participants to join the chain network for transactions without the mutual agreement of the authentic network administrator [20]. The permissioned blockchain adopts a partial decentralization technique for recording and information storage authentication. It is commonly used in the banking sectors, research, internal voting, supply chain management, and many institutions for data security and policy regulations. Ripple (XPR) is a popular example. The characteristics, advantages, and disadvantages of both permissionless and permissioned blockchain are shown in Fig. 5.

Based on the Nature of the Blockchain Publicity

Private Blockchain This type of blockchain profile-raising is a sole autonomy network and partial decentralization that allows a single organization to have permission and authority over the control of the network chain. It allows read and write for single organization access, fast transaction speed, permissioned consensus, and high efficiency and supports partial immutability [21].

Public Blockchain This type of blockchain network allows an individual participant to join the chain network without restrictions to access the ledger (decentralized records) for consensus processes. But the transaction speed is slow and supports full immutability with low efficiency [22].

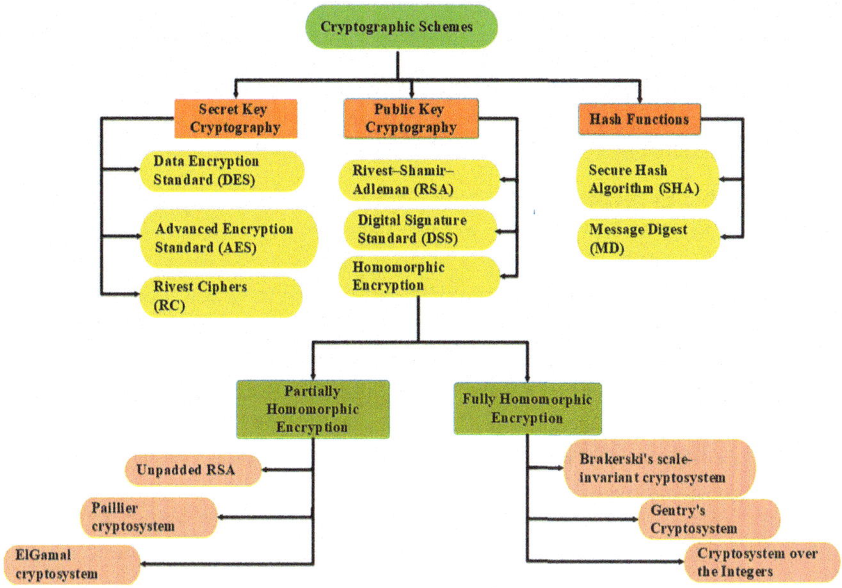

Fig. 6 Types of cryptography schemes

2.2 The Cryptography Techniques for Securing Blockchain-Based Electronic Voting System

This technology (blockchain) is a decentralized digital ledger that helps in securing the transaction processes in the blockchain network using steganography or hashing techniques to achieve optimal security (see Fig. 6). These techniques are widely used to improve the applicability of blockchain as a security countermeasure in most systems. It provides transparent, immutable, and fraud-resistant platform and subsists the confidentiality, integrity, and user's data privacy breach [23].

The Symmetric or Secret-Key Cryptography (SKC) This type of cryptography key refers to the balanced encryption schemes that adopt a single key for the implementation of both encoding and decipherment of messages. This symmetric key approach is mainly utilized to ensure data privacies (confidentiality), integrity, and authenticity [24]. The subcategories include data encryption standards, advanced encryption standards, and the Rivest ciphers.

The Asymmetric or Public-Key Cryptography (PKC) This encryption type of cryptography scheme utilizes pairs of different and encoded keys in nature with secrecy. It includes an irregular process of encrypting data (plaintext) using a public key and a private key for lopsided ciphertext called decipher. This cryptographic method is mainly focused to achieve data authentication and non-repudiation of messages. Some of these asymmetric cryptography types include digital signature standard and Rivest-Shamir-Adleman homomorphic encryption schemes.

Hash Functions This cryptography technique uses a mathematical transformation on the ciphertext to make it irrevocable and to provide trust, privacy, confidentiality, integrity, and authenticity.

Homomorphic Encryption Algorithm (HEA) This is an encryption method that allows users to carry out computations over the encrypted message without decryption reasoning. This computation process over the encrypted message generates an identical form of result when decrypted to provide privacy and protective storage outsourced. The homomorphic encryption algorithm is suitable for securing automated systems like electronic voting systems using blockchain technology for safety services improvement and removing barriers that prevent data distribution of consolidating ballots. The homomorphic encryption algorithms can be classified into partial or full types. The partial homomorphic steganography includes unpadded RSA, Paillier, ElGamal, Goldwasser, Benaloh, Boneh-Goh-Nissim, Ishai-Parkin, and Sander-Young-Yung cryptosystem. While the fully homomorphic includes Gentry's cryptosystem, cryptosystem over the integers, and Brakerski's scale-invariant cryptosystem.

Fully Homomorphic Encryption/Cryptosystem This is a class of homomorphic encryption that supports both additive and multiplicative homomorphism operations. They are versatile but require more computational power than the partially homomorphic class.

Partially Homomorphic Encryption/Cryptosystem This is a class of homomorphic encryption that is capable of executing the additive or multiplicative homomorphism processes. They achieved high level of performance but only support one type of computation, either multiplication or addition operations [25]. Examples of partially homomorphic cryptosystems are:

(i) Unpadded RSA cryptosystem: this steganography method supports multiplicative operations and is expressed as given in Eq. 1, 2, and 3.

$$\Theta\left(s_1\right)\cdot\Theta\left(s_2\right)=s_1^e s_2^e \bmod n \tag{1}$$

$$=\left(s_1 s_2\right)^e \bmod n \tag{2}$$

$$=\Theta\left(s_1\cdot s_2\right) \tag{3}$$

where.

$\Theta(s)=s^e \; mod\; n$ is the encryption of a message s and e is the encryption exponent.

(ii) The ElGamal homomorphic cryptosystem has a cyclic group P of order q, with generator p as expressed in Eq. 4, 5, and 6.

$$\Theta\left(s_1\right)\cdot\Theta\left(s_2\right)=\left(p^{r_1},s_1\cdot n^{r_1}\right)\left(p^{r_2},s_2\cdot n^{r_2}\right) \tag{4}$$

$$=\left(p^{r_1+r_2},\left(s_1\cdot s_2\right)\ p^{r_1+r_2}\right) \tag{5}$$

$$=\Theta\left(s_1\cdot s_2\right) \tag{6}$$

where.

(P, q, p, n) is the public key, $n = p^s$ is the secret key, and $\Theta(s) = (p^r, s \cdot n^r)$ is the encryption of the message *s*.

(iii) The Goldwasser-Micali cryptosystem is expressed as in Eq. 7, 8, and 9.

$$\Theta\left(d_1\right)\cdot\Theta\left(d_2\right)=s^{d_1}r_1^2s^{d_2}r_2^2 \bmod n \tag{7}$$

$$=s^{d_1+d_2}\left(r_1r_2\right)^2 \bmod \tag{8}$$

$$=\Theta\left(d_1\oplus d_2\right) \tag{9}$$

where.

$\oplus$ is the addition modulo 2 (exclusive-or), $\Theta(d) = s^d\ r^2\ mod\ n$ is the encryption of a bit *d,* and *s* is the quadratic non-residue.

(iv) Benaloh cryptosystem and Okamoto-Uchiyama cryptosystem are expressed as in Eq. 10, 11, and 12.

$$\Theta\left(s_1\right)\cdot\Theta\left(s_2\right)=\left(p^{s_1}r_1^c\right)\left(p^{s_2}r_2^c\right) \bmod n \tag{10}$$

$$=p^{m_1+m_2}\left(r_1r_2\right)^c \bmod n \tag{11}$$

$$=\Theta\left(s_1+s_2\right). \tag{12}$$

where.

$\Theta(s) = g^m r^c\ mod\ n$ is the public key.

2.3 Application of Blockchain Technology Survey

The blockchain has popularly gained interest in several application areas to enhance the security challenges that have become a threat to the normal operation of the smart system, online transaction, and e-business. Adopting this technology (blockchain) with the integration of cryptographic and hashing techniques has relaxed the security pressure against hackers, social network threats, malicious actors, and many others. Especially in the financial institution which gave respect to the functionality and advantage provided by this blockchain application in terms of adequate security measures against the third-party assailant in the network. Also, it is widely used in online or electronic business transactions to prevent a third-party participant or intruder from tampering, altering, or illegally modifying the records without the chain participant consolidations. Blockchain with hashing techniques has been widely adopted to secure smart devices, sensor networks, software-defined networks (SDN), and Internet of Things (IoT) infrastructures.

Several researchers have contributed with different methods and approaches to resolve the problems affecting the voting system using secure authentication Schemes [26]. However, this research area is still open for further investigation and development of an efficient security countermeasure for a smart technology like an e-voting system that may concern the security of voter privacy and protection of ballot tallying and auditing. The new trend in the blockchain application and steganography schemes for securing e-voting systems and other emergent technology of smart systems are investigated.

A hyperledger blockchain is implemented on an e-voting system to ensured immutability and record tampering as a token-free system [27]. But the system could not resist some security issues such as confidentiality and ensure voter privacy. The Ethereum private blockchain was developed and implemented on an electronic voting system to achieve a higher degree of transaction processing and to guarantee the electorate's privacy [28]. The open source-based blockchain technique was implemented on the automated voting system using cryptographic techniques based on ElGamal to ensure the privacy of the electorate. The system guarantees the immutability of the ballots and could not be interfered with by the third party in the chain network [29]. But the encoding method required the decryption of a secret ballot before it could be tallied which makes the system to be time appealing and does not guarantee privacy.

Gupta et al. developed a telesurgery scheme with multilevel user authentication to ensure a privacy-oriented and interoperable telesurgery system using a blockchain [30]. It further experiments an Ethereum blockchain system with a fifth generation-enabled tactile Internet (5GTI) to evaluate the system efficiency and privacy orientation and secure real-time delivery of health-care services through unmanned aerial vehicles (UAV). Rahman et al. developed a decentralized blockchain-based framework with mobile edge computing (MEC) to allow for the support of a large user pool with low latency. The system achieved a good level of security and user anonymity over a centralized database for the storage of actual health-care multimedia data [6]. The concept of a scalable, secured, and user-centric collection of health-care data from personal wearable devices is developed using a blockchain technology [31]. This system ensured adequate protection of personal health records but failed to consider the variety of datasets that could exist from a wide variety of wearable devices. Vora et al. proposed a blockchain framework with varying contract classification to allow for a balanced privacy-oriented and readily accessible storage of electronic health records. The proposed scheme ensured ease of use and the complete encryption of the records. It however did not put in place the consideration for the time and computational resources that would be required to decrypt the records whenever they were needed [32].

Bodkhe et al. presented a token-based blockchain with proof of collaboration and zero-knowledge proofs (ZKP) with deep learning to allow for the management and protection of tourists' data from identity theft and payment clearance cycle attacks [33]. The system however failed to put in place modalities for the security of stakeholders. A blockchain system with energy trading validation and stability was developed for the management of a smart grid system using lightweight security [34]. Wei et al. projected a scheme for a blockchain-based framework with a proxy model and Merkle hash tree for monitoring data changes and to ensure the integrity of storage data [35]. A self-sovereign identity management system based on Ethereum smart contracts was proposed for securing and managing digital assets, reputation, and personal identity. But the system did not offer privacy protection for the user recovery delegates, which could be exploited for security attacks [36].

Blockchain technology using RSA and SHA-256 cryptography algorithms was proposed to secure the process of digital banking. During testing, the system achieved high accuracy of 88% with good reliability but failed to guarantee the integrity of data in transit [37]. The bio-cryptographic systems using Gabor filter images (GFI) and lifting wavelet transforms (LWT) were proposed to safeguard an automated voting system. The results of 0.0001% and 0.1% are obtained for FAR and FRR, respectively. The system ensures security measures like authentication, confidentiality, and integrity, but computational complexity reduces the response [38, 39].

An ElGamal homomorphic encryption algorithm was developed to ensure user privacy and integrity of electorates records through the automated voting system [40]. The system encrypted votes with the matched format of the decipher to ensure privacy and verifiability of the ballots. But the tallying process of ballots was inactive due to the nature of the homomorphic encoding method used. The Paillier homomorphic cryptography was adopted to secure an electronic voting system. The system evaluation proved efficient with a high level of voter privacy but vulnerable to the manipulation of the ballot through conspiring adversaries [41]. Other security areas of blockchain technology applications are discussed with methods, strengths, and limitations (such as health care, smart home, vehicular area network (VANET), electronic voting system, financial institution, business management, and gas exploration industry. A detailed investigation of the existing works in literature is presented (see Table 1).

3 The Proposed Crypto-Blockchain Scheme and Bi-factor Authentication for Securing E-Voting Systems

An electronic voting system is proposed using a private blockchain with the Paillier homomorphic steganography and bi-factor biometrics certification of iris fingerprint systems. The biometric-based automated voting system is a reliable node with integration of a private blockchain technique for immutable and data privacy breach

Table 1 Summary of the existing research investigation

Method	Strength	Limitation
The fingerprint biometric authentication was used [39]	It solves the problems of authentication	It is not efficient enough as unimodal biometric authentication was used
The automated voting system using BEVS blockchain [42]	It enhanced security and transparency in the voting system	The system is complex to interact with and takes a longer time to operate
Electronic voting system-based blockchain using test-driven approach [43]	It ensured the ballots are safe and difficult for tampering	The confidentiality of voters can be compromised
Blockchain-based e-voting with proof-of-work consensus algorithm [44]	It ensured that ballots were safe from third-party tampering	The privacy of the electorate is not failsafe
The electronic voting system using ElGamal steganography and open source-based blockchain [45]	The ballots storage privacy is guaranteed	The decryption of ballots is essential before votes can be matched which results in excess time
A private key and digital signature-based blockchain is used [46]	It secures the ballots from third-party tampering and is openly verifiable	But it is abortive to address the privacy concerns of ballots and openly verifiable blockchain
The tokens transfer on the Ethereum blockchain network [28]	It safeguarded the openness and immutability of ballots	Not user-friendly with a high start-up cost approach
The automated voting system based on a hyperledger blockchain network [27]	It ensured the immutability of ballots	It does not adopt the cryptosystem technique to render voter privacy advantages

security. The implementation of this blockchain helps to confiscate the autonomous central authority control in the elective system by executing a decentralization system for the transactions among the participants (nodes) on the networks. The use of the Paillier homomorphic steganography is to safeguard the privacy of the voters and eliminate the requirement for decryption of the encrypting data before tallying ballots can be successfully performed. The bi-factor method of using biometrics verification of iris fingerprint (IF) helps to resolve variances that occur during voters verification. Fig. 7 shows the implementation of a crypto-blockchain-based electronic voting system model. The secure smart e-voting system architecture using crypto-blockchain technology is illustrated (see Fig. 8).

The smart e-voting system process begins with the voters that cast their votes at the polling unit. The votes get encrypted using the Paillier homomorphic cryptosystem, after which a block containing the encrypted vote, the voter blockchain id, hash of the previous blocks, and hash of the current block's transaction is created and mined unto the private blockchain ledger. This process repeats itself continuously until the admin (electoral authority) decides to tally the encrypted ballots to obtain

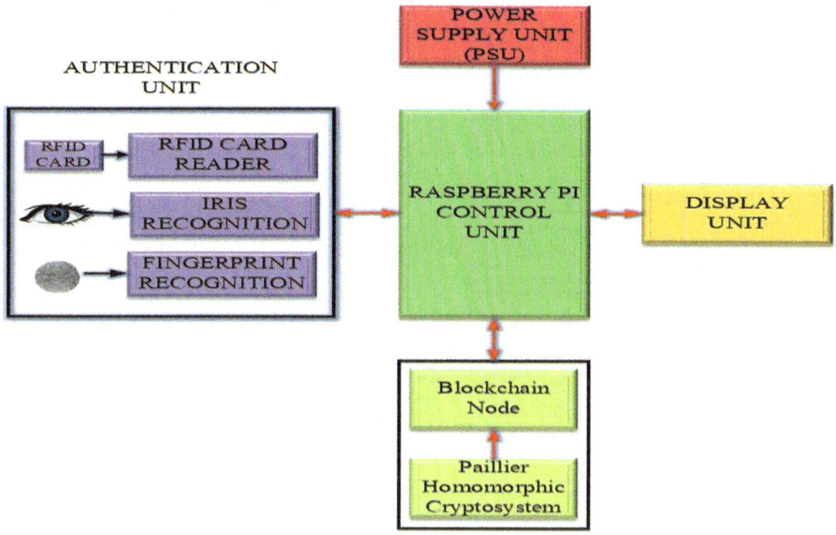

Fig. 7 Block diagram of proposed system architecture

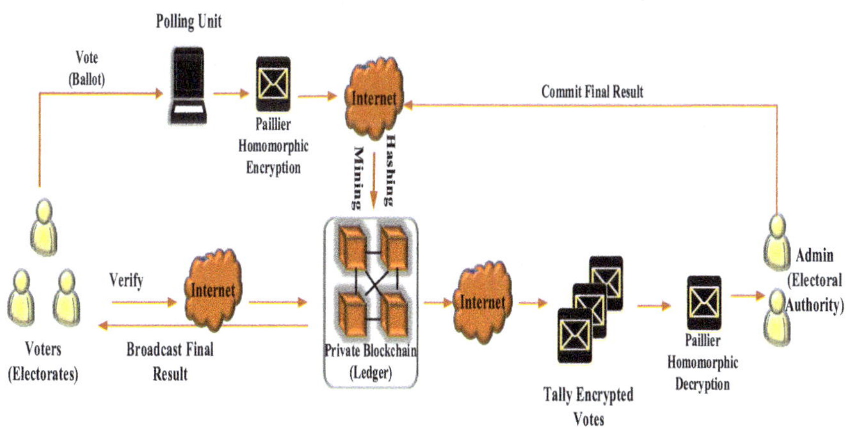

Fig. 8 The secure smart e-voting system architecture using crypto-blockchain technology

a single final encryption sum value that becomes decrypted homomorphically (see Figs. 9, 10, and 11). Therefore, the final result gets announced and broadcasted to the voters over a secure and privacy-oriented network. The Paillier homomorphic encryption algorithm-based blockchain is presented in Table 2.

The Paillier homomorphic encryption cryptosystem property is expressed in Eq. 13, 14, and 15. The Paillier encryption algorithm for the key generation, encryption, and decryption process are contained in Table I.

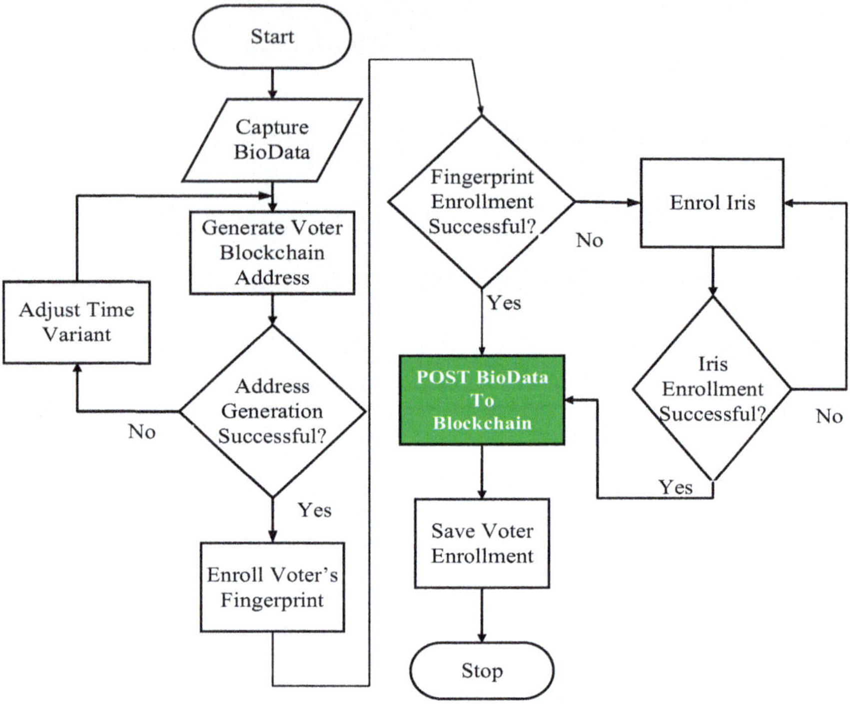

Fig. 9 Voter enrollment process

$$\Theta(s_1)\cdot\Theta(s_2) = \left(p^{m_1}r_1^n\right)\left(p^{m_2}r_2^n\right) \bmod n^2 \tag{13}$$

$$= p^{m_1+m_2}(r_1r_2)^n \bmod n^2 \tag{14}$$

$$= \Theta(s_1 + s_2) \tag{15}$$

where.

$\Theta(s) = p^m\ r^n\ mod\ n^2$ is the encryption of a message s.

4 Results and Discussions

The smart electronic voting system was implemented using Raspberry Pi 3B+ as a control system with the integration of a fingerprint sensor (ZFM-60) and camera to accept the voter's registration and enrollment during voting. The data interface touchscreen with a resolution of 800x480px was used as both input and output systems. This smart system was programmed using a python programming

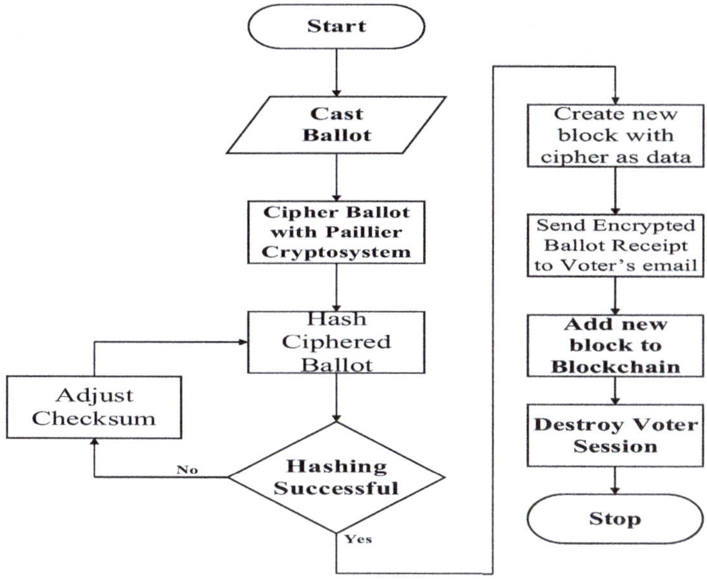

Fig. 10 Blockchain ballot casting process

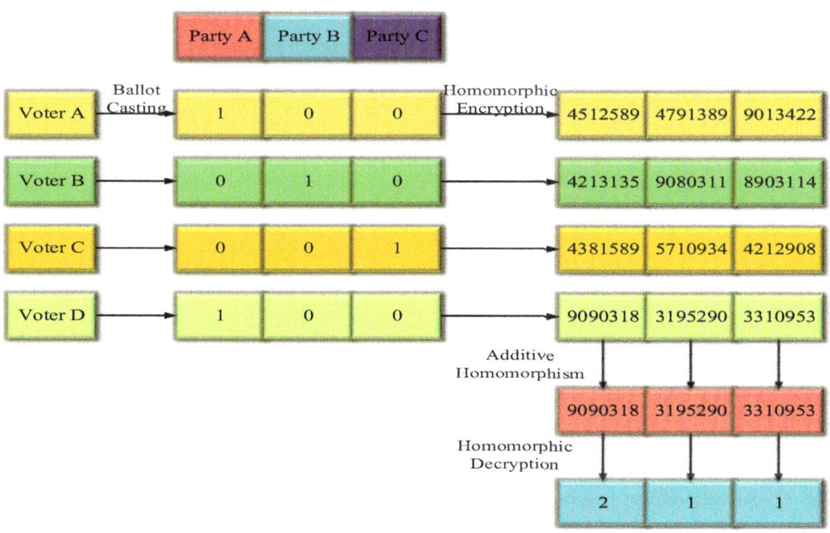

Fig. 11 Homomorphic encryption and decryption of ballots

language as a console that eases the interaction, control, and coordination of the system response and performance. The graphic user interface (GUI) was designed using MySQL, PHP, CSS, HTML, and JavaScript to sectionalize the web view page and manage the flows of data that includes admin log in page, polling unit name,

Table 2 Paillier encryption cryptosystem algorithm

1. Key generation algorithm	
Step 1	Select large, random prime numbers m and p, which is independent of each other. $Gcd(mp, (m-1)(p-1)) = 1$
Step 2	Calculate the value of n and λ using, $n = mp$ as well as $\lambda = lcm(m-1, p-1)$, where; *lcm is the least common multiple*
Step 3	Choose an integer, k at random, Where $k \in Z^*_{n^2}$
Step 4	Confirm, by using the function J, *to* check for the modular multiplicative inverse $\mu = (J(k^{\lambda} \ mod\ n^2))^{-1} \ mod\ n$, to make sure that n can divide the order of k $J(s) = \frac{s-1}{n}$
Step 5	The public key, therefore, is (n,g) while the private key is (λ, μ)
2. Encryption algorithm	
Step 1	Given message, s to be encrypted, Where $0 \leq s \leq n$
Step 2	Choose a number, y at random, Such that $0 \leq y \leq n$ and $y \in Z^*_n$
Step 3	Calculate the ciphertext of the message as: $c = k^m \cdot y^n \ mod\ n^2$
3. Decryption algorithm	
Step 1	A ciphertext, c to be decrypted, such that $c \in Z^*_{n^2}$
Step 2	Obtain the plaintext, s as $s = L(c^{\lambda} \ mod\ n^2) \cdot \mu \ mod\ n$

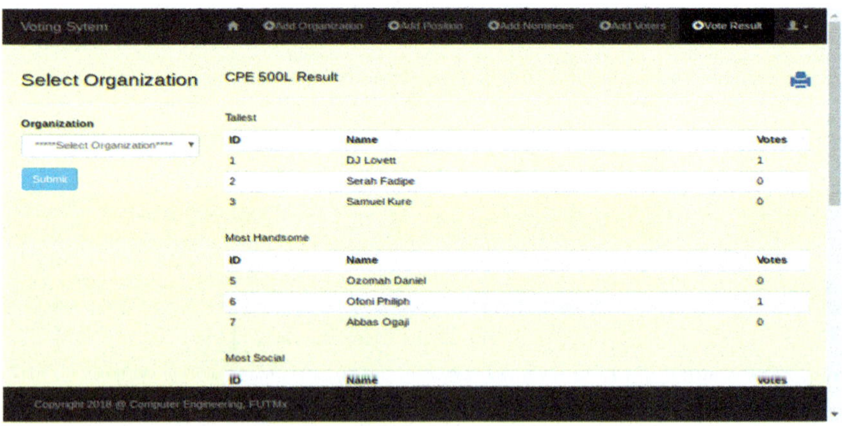

Fig. 12 Secure e-voting system log in page

positions, voters' log in, and votes casting log in page and many others (see Fig. 12).

Performance Evaluation The secure smart electronic voting system is subjected to testing, and it was evaluated using false acceptance rate (FAR) and false rejection rate (FRR). The metrics were selected for the performance evaluation to cross-check and validate the correct voters' enrolment and nonregistered candidates (see Tables

Table 3 The result of FAR during testing

Matching trials	Rejected	Accepted	FAR
9	9	0	0%
21	21	0	0%
36	36	0	0%
45	44	1	2%
111	110	1	0.02

Table 4 False rejection rate FRR

Matching trial	Accept	Reject	FRR
9	9	0	0%
21	21	0	0%
36	34	0	0%
50	45	5	10%
116	109	5	0.1

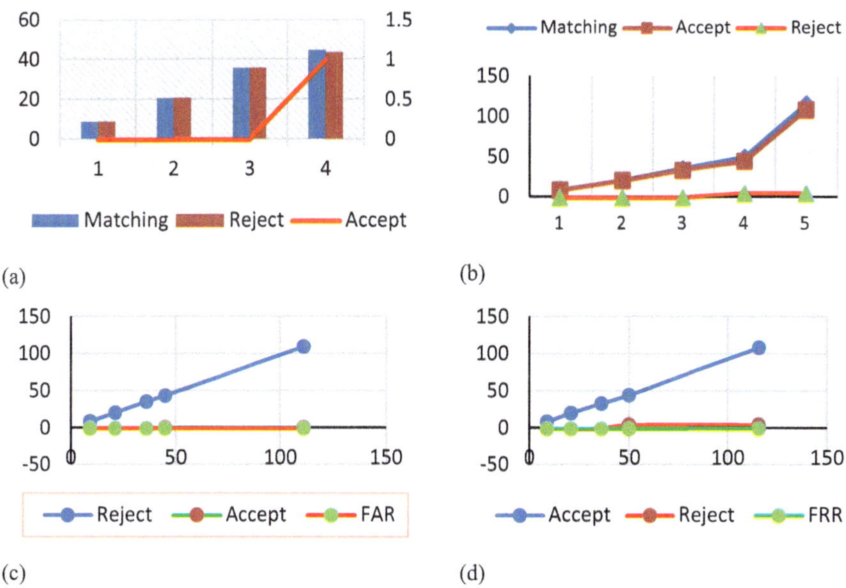

Fig. 13 The system evaluation performance during testing was presented, (**a**) the number of trials for rejection to acceptance, (**b**) the number of trials of acceptance to rejection, (**c**) the rejection to the accepted rate FAR, and (**d**) the acceptance to the rejection rate (FRR)

3 and 4). The FRR is used to evaluate and benchmark the correctly authenticated registered voters and enrolment in the system. The electronic voting system was evaluated based on the number of FAR, FRR, and latency (see Figs. 13 and 14). The blockchain certification performance using the Paillier homomorphic method was evaluated based on the retrieval time (latency), certificate size, and execution (see Fig. 15).

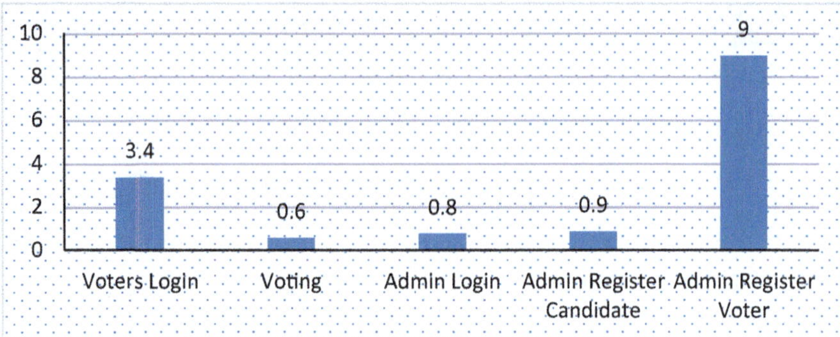

Fig. 14 System response time during testing

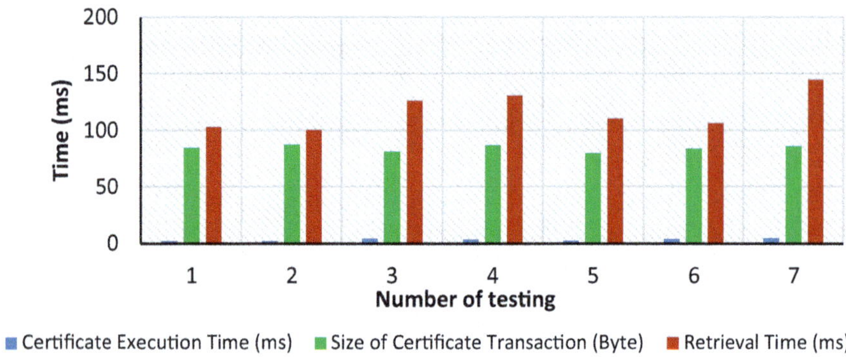

Fig. 15 Crypto-blockchain performance testing

5 Conclusions

This research study investigates the areas of an application using blockchain technology with cryptography algorithm as a recent efficient security countermeasure to avert the data privacy, confidentiality, and trust breaches. It further proposed efficient security measures against vulnerability to data tampering, fraud, and hacking of electronic voting systems. A private blockchain technology for securing decentralized ledger or database was adopted to prevent the central authority from fraud and data tampering. A Paillier homomorphic encryption algorithm was implemented with blockchain to make the system immutable, transparent in the chain transaction, and secured. Also, a bi-factor authentication technique (iris and fingerprint) was used for the voter's registration, verification, and genuine authenticity, which performs efficiently with 0.02% FAR and 0.1% FRR. The system response time was measured, and it was relatively fast as it takes less than 1 min to accept registered candidates during the verification process. The credential execution time, retrieval time, and size of the certificate transaction were measured in milliseconds and give better performance response time.

References

1. Banerjee, S. P., & Woodard, D. L. (2012). Biometric authentication and identification using keystroke dynamics: A survey. *Journal of Pattern Recognition Research, 7*(1), 116–139.
2. Aluaigba, M. T. (2016). Democracy deferred: The effects of electoral malpractice on Nigeria's path to democratic consolidation. *Journal of African Elections, 15*(2), 136–158.
3. Ayinde, A. F., & Idowu, A. O. (2016). Nigeria's 2015 elections: Permanent voter's cards, smart card readers and security challenges. *Journal of African Elections, 15*(2), 50–68.
4. Udu, L. E. (2015). INEC and 2015 general elections in Nigeria: Matter arising. *Democracy, 5*(12), 96–108.
5. Inalegwu, O. C., Dogo, E. M., Kolo, J. G., Bima, M. E., Ajao, L. A., & Inechioma, J. (2018). Development of a biometric-based car park access control and billing system. In *The second international Engineering Conference (IEC)* (pp. 421–425). Nigeria.
6. Okokpujie, K., Etinosa, N. O., John, S., & Joy, E. (2018). Comparative analysis of fingerprint preprocessing algorithms for electronic voting processes. In *IT Convergence and Security 2017* (pp. 212–219). Springer.
7. Abo-Rizka, M., & Ghounam, H. R. (2007). A novel e-voting in Egypt. *International Journal of Computer Science and Network Security, 7*(11), 226–234.
8. Rahman, M. A., Hossain, M. S., Loukas, G., Hassanain, E., Rahman, S. S., Alhamid, M. F., & Guizani, M. (2018). Blockchain-based mobile edge computing framework for secure therapy applications. *IEEE Access, 6*, 72469–72478.
9. Aditya, S. N., Kishore, M. V., & Suresh, C. (2018). A secure e-voting system using RSA and md5 algorithms using random number generators. *International Journal of Applied Engineering Research, 18*(11), 9468–9473.
10. Abayomi-Zannu, T. P., Odun-Ayo, I., Tatama, B. F., & Misra, S. (2019). Implementing a mobile voting system utilizing blockchain technology and two-factor authentication in Nigeria. In *Proceedings of First International Conference on Computing, Communications, and Cyber-Security (IC4S 2019)* (pp. 857–872).
11. Ajao, L. A., Agajo, J., Olaniyi, O. M., Jibril, I. Z., & Sebiotimo, A. E. (2019). A secure tracking automobile system for oil and gas distribution using telematics and blockchain techniques. *Journal of Electrical and Computer Engineering, 7*(3), 257–268.
12. Hjálmarsson, F. P., Hreiðarsson, G. K., Hamdaqa, M., & Hjálmtýsson, G. (2018). Blockchain-based e-voting system. In *2018 IEEE 11th international conference on cloud computing (CLOUD)* (pp. 983–986).
13. Ajao, L. A., Agajo, J., Adedokun, E. A., & Kargong, L. (2019). Crypto-hash algorithm-based blockchain technology for managing decentralized ledger database in oil and gas industry. *International Journal of Molecular Sciences, 2*(3), 300–325.
14. Awotunde, J. B., Ogundokun, R. O., Jimoh, R. G., & Misra, S. (2021). Machine learning algorithm for cryptocurrencies price prediction. In S. Mistra & T. A. Kumar (Eds.), *Artificial intelligence for cybersecurity: Methods, issues, and possible horizons or opportunities* (Studies in computational intelligence) (Vol. 972). Springer.
15. Velmurugadass, P., Dhanasekaran, S., Anand, S. S., & Vasudevan, V. (2021). Enhancing blockchain security in cloud computing with IoT environment using ECIES and cryptography hash algorithm. *Materials Today: Proceedings, 37*, 2653–2659.
16. Shanaev, S., Shuraeva, A., Vasenin, M., & Kuznetsov, M. (2019). Cryptocurrency value and 51% attacks: Evidence from event studies. *Journal of Alternative Investments, 22*(3), 65–77.
17. Saleh, F. (2021). Blockchain without waste: Proof-of-stake. *The Review of Financial Studies, 34*(3), 1156–1190.
18. Zhang, R., & Chan, W.-K. (2020). Evaluation of energy consumption in block-chains with proof of work and proof of stake. *Journal of Physics: Conference Series, 1584*(1), 12–23.
19. Shaikh, M. Z. (2021). A review on cryptocurrency with distributed ledger technology for blockchain technology. *Turkish Journal of Computer and Mathematics Education, 12*(9), 143–151.

20. Singhal, B., Dhameja, G., & Panda, P. S. (2018). *Introduction to the blockchain. In beginning blockchain* (pp. 1–29). Apress.
21. Li, X., Wang, Y., Vijayakumar, P., He, D., Kumar, N., & Ma, J. (2019). Blockchain-based mutual-healing group key distribution scheme in unmanned aerial vehicles ad-hoc network. *IEEE Transactions on Vehicular Technology, 68*(11), 11309–11322.
22. Jindal, A., Aujla, G. S., & Kumar, N. (2019). SURVIVOR: A blockchain-based edge-as-a-service framework for secure energy trading in SDN-enabled vehicle-to-grid environment. *Computer Networks, 153*, 36–48.
23. Aziz, N., Ridiah, R., & Susanto, H. (2021). Encryption of digital banking transaction records: A blockchain cryptography security approach. *International Journal of Computers and Applications, 975*, 8887.
24. Belej, O., Staniec, K., & Wieckowski, T. (2020). The need to use a hash function to build a crypto algorithm for blockchain. In *International conference on dependability of computer systems* (pp. 51–60). Springer.
25. Mondal, A. H., Ranjan, M., & Saikia, M. (2015). A brief overview of homomorphic cryptosystem and their applications. *International Journal of Computers and Applications, 975*, 8887.
26. Mistra, S. (2021). *A step by step guide for choosing project topics and writing research papers in ICT related disciplines, communications in computer and information science* (Vol. 1350, pp. 727–744). Springer.
27. Sadia, K., Masuduzzuaman, M., Paul, R. K., & Islam, A. (2020). Blockchain-based secure e-voting with the assistance of smart contract. In *IC-BCT 2019* (pp. 161–176). Springer.
28. Dhulavvagol, P. M., Bhajantri, V. H., & Totad, S. G. (2020). Blockchain ethereum clients performance analysis considering e-voting application. *Procedia Computer Science, 167*, 2506–2515.
29. Kshetri, N., & Voas, J. (2018). Blockchain-enabled e-voting. *IEEE Softwares, 35*(4), 95–99.
30. Ch, R., Srivastava, G., Gadekallu, T. R., Maddikunta, P. K. R., & Bhattacharya, S. (2020). Security and privacy of UAV data using blockchain technology. *Journal of Information Security and Applications, 55*, 102670.
31. Hobil, M., Kompara, M., Kamisalic, A., & Nemec Zlatolas, L. (2018). A systematic review of the use of blockchain in healthcare. *Symmetry, 10*(10), 470.
32. McGhin, T., Choo, K. K. R., Liu, C. Z., & He, D. (2019). Blockchain in healthcare applications: Research challenges and opportunities. *Journal of Network and Computer Applications, 135*, 62–75.
33. Sharma, M., Sehrawat, R., Daim, T., & Shaygan, A. (2021). Technology assessment enabling blockchain in hospitality and tourism sectors. *Technological Forecasting and Social Change, 169*, 120810.
34. Jindal, A., Aujla, G. S., Kumar, N., & Villari, M. (2020). GUARDIAN: Blockchain-based secure demand response management in smart grid system. *IEEE Transactions on Services Computing, 13*(4), 613–624.
35. Wei, P. C., Wang, D., Zhao, Y., Tyagi, S. K. S., & Kumar, N. (2020). Blockchain data-based cloud data integrity protection mechanism. *Future Generation Computer Systems, 102*, 902–911.
36. Houtan, B., Hafid, A. S., & Makraksi, D. (2020). A survey on blockchain-based self-sovereign patient identity in healthcare. *IEEE Access, 8*, 90478–90494.
37. Khan, K. M., Arshad, J., & Khan, M. M. (2018). Secure digital voting system based on blockchain technology. *International Journal of Electronic Government Research, 14*(1), 53–62.
38. Yi, H. (2019). Securing e-voting based on blockchain in a P2P network. *EURASIP Journal on Wireless Communications and Networking, 2019*(1), 1–9.
39. Umar, B. U., Olaniyi, O. M., Ajao, L. A., Maliki, D., & Okeke, I. C. (2019). Development of a fingerprint biometric authentication system for secure electronic voting machines. *KINETIK Journal, 4*(2), 115–126.

40. Jabbar, I., & Alsaad, S. N. (2017). Design and implementation of secure remote e-voting system using homomorphic encryption. *International Journal of Network Security, 19*(5), 694–703.
41. Al-Anie, H. K., Alia, M. A., & Hnaif, A. A. (2011). E-voting protocol based on public-key cryptography. *International Journal of Network Security & Its Applications, 3*(4), 87–98.
42. Lalam, N., Nithinn, M. S., & Jebakumar, D. R. (2020). BEVS-blockchain based e-voting system. *International Journal of Advanced Science and Technology, 29*, 6241–6249.
43. Hsiaso, J., Tso, R., Chen, C. M., & Wu, M. E. (2017). Decentralized e-voting systems based on the blockchain technology. In *Advances in computer science and ubiquitous computing* (pp. 305–309). Springer.
44. Panja, S., & Roy, B. (2021). A secure end-end verifiable e-voting system using blockchain and cloud server. *Journal of Information Security and Applications, 59*, 102815.
45. Bulut, R., Kantarci, A., Kesskin, S., & Bahtiyar, S. (2019). Blockchain-based electronic voting system for election in Turkey. In *2019 4th international conference on computer science and application engineering* (pp. 183–188). IEEE.
46. Yellamma, P., Anupama, P., Lakshmibhavani, K., Priya, U. J. S., & Ch, K. (2020). Implementation of e-voting system using Blockchain technology. *Journal of Critical Reviews, 7*(6), 865–870.

Enhanced Hash Value and Public Key Infrastructure Generations for Blockchains Using Sooner Lightweight Cryptography

Abraham Ayegba Alfa, John Kolo Alhassan, Olayemi Mikail Olaniyi, and Morufu Olalere

1 Introduction

Blockchain technology (BT) assembles diverse computer technologies such as consensus, encryption, distributed storage, smart contracts, and peer-to-peer (P2P) network. The typical attributes of BT including security, immutability, openness, trustworthiness, and smartness became possible because of these technologies [1]. Also, diverse transactions can be unceasingly linked or tied to blockchain. With blockchain, historical data and transaction records are mutually running and untampered on the central database system [2]. The ability of Internet users to establish contracts without face-to-face meetings or central trust became feasible through the point-to-point ledger, smart contracts, or digital encryption. Recently, these several uniqueness of blockchain technology continues to draw further probing and research on fields other than the original cryptocurrency (Bitcoin technology) [3, 4].

Blockchain provides fresh prospects in evolving new kinds of digital services to tackle problems faced by businesses and enhance business practices by changing the status of transaction information into a public resource [5]. According to Henry et al. [6], privacy can be best attained by combining the concept of cryptography and decentralization as utilized for digital currencies (Bitcoin). The central idea behind blockchain is the decentralization attribute that enables the storage of data across

A. A. Alfa (✉)
Kogi State College of Education, Ankpa, Nigeria
e-mail: abraham.alfa@kscoeankpa.edu.ng

J. K. Alhassan · O. M. Olaniyi · M. Olalere
Federal University of Technology, Minna, Nigeria
e-mail: jkalhassan@futminna.edu.ng; mikail.olaniyi@futminna.edu.ng; lerejide@futminna.edu.ng

S. Misra, A. Kumar Tyagi (eds.), *Blockchain Applications in the Smart Era*, EAI/Springer Innovations in Communication and Computing,
https://doi.org/10.1007/978-3-030-89546-4_6

the distributed network, thereby minimizing the risks of holding data centrally or complete system not centrally operated by a third party [1, 3].

BT provides privacy protection for applications through data encryption and data verification process [8, 9]. Nevertheless, it is anticipated that the nodes are able of performing all tasks of verifying encrypted (or protected) data. However, this approach increases blockchain networks' overheads especially at the point of computation and communication in the case of IoT systems and networks. Quite a lot of leakages have been reported all through the mathematical generations of ciphers causing cases of breaches of personally identifiable information (PII) and data [9]. However, there are other cases of long public keys accumulating more problems to BT-based IoT applications and systems.

Another study by Huh et al. [10] identified certain shortfalls of blockchain including: (a) longer time for generating block usually 10 minutes or more, which is rather unsupportive of faster transactions. (b) there is no use of loops. (c) it is Turing incomplete. It is feasible to adapt blockchain to suit favorite applications such as authentication and verifications in private cloud computing [10, 11]. With Ethereum, there is the chance to configuring IoT devices and manage authentication procedures for PKIs [10].

Unlike the traditional blockchains, there is a need to evolve lightweight and computationally efficient methods before incorporating them into IoT-based systems. The reason for this option is to cope with the overheads of traditional blockchains but providing the same level of data privacy and security in the aftermath [12]. Blockchains are capable of overcoming cloud server availability, faulty architecture, constraints of costs and capability, vulnerability to data manipulation, and entire security framework [13]. This paper proposes enhanced hash value and PKI generations based on Sooner lightweight cryptography for blockchains to overcome these drawbacks. Subsequent sections are arranged as follows [14]: section two is the literature review, section three is the research methodology, section four is the results and discussion, and the conclusion is in section five.

2 Literature Review

2.1 Overview of Blockchain Technology

Blockchain is a well-organized back-linked list of transactions boxed inside of blocks. In practice, blockchain is stored in the form of a flat file inside of a conventional database. The chain describes each block as back-linked to the preceding block. It suffices to explain blockchain as vertical transactions pile with the initial block serving as the base of the stack. In other words, the block is composed of several transactions, while the blockchain is built using these sets of blocks. Aside from transaction details, the block warehouses diverse information for uniquely recognizing transactions in the chain [2, 7]. The structure of the block is made up of an 80 bytes header and transaction capacity of close to 250 bytes capacity. A common transaction has the originator, the recipient, and other information concerning messages in blocks, which are secured with encryption codes [5].

The rise of blockchain technology (BT) followed the unprecedented increases in Internet technology and interconnected devices to information transmission in a more secured manner. According to Yang et al. [15], blockchain is a dispersed database system having unfalsified and decentralized attributes capable of spearheading technological innovations. It was first effectively implemented in digital currency, Bitcoin. The distributed and anti-attack properties are well-suited for IoT systems. Technical features make it possible to actualize distributed privacy and security in IoT. Notable scholars have explained the concept of BT as discussed as follows.

The study by Brandão et al. [7] provided a background of blockchain as beginning from the idea of the ledger for cryptocurrency applications (such as Bitcoin) according to Nakamoto in 2018. The BigchainDB concept of traceability system was applied to food supply chain monitoring on the real-time situation based on the analysis of hazard and critical point controls. BTs provide an information communication platform for neutrality, transparency, reliability, and security. In smart homes, blockchain security offers more confidentiality, availability, integrity, processing time, traffic consumption, and energy dissipation.

More so, the decentralized structure of BTs reduces widespread frauds and manipulations by participants because of identity and access systems management of blocks. In IoT networks, BTs use DistBlockNet to reroute attacks in real-time scenarios because of its high performance [16]. In terms of security, privacy, and costs of BTs, deployments make BTs successful in enhancing financial services provisions. However, there are still open issues on the effectiveness of BTs in privacy preservations in IoT-based systems.

The basic framework of BTs was highlighted by Tasatanattakool and Techapanupreeda [17]. BT is a kind of database storage, which is decentralized, reliable, and highly resistant to attacks. In fact, Bitcoin uses blockchain (BC) public ledger functionality to carry out transactions on peer-to-peer networks. BT is the backbone of complex security applications such as hyperledger and smart contracts. There is the possibility of customizing BC for diverse applications such as IoTs other than Bitcoin. Typically, BTs are a form of database warehoused in a decentralized system. BT can be described under the following points [15].

Decentralization. The accounting, verification, and transmission of BTs rely on distributed systems. The mode of accounting and storage is distributed. There is no well-defined centralized pattern of processing. The participating nodes have fair and equal status. The technique of node verification and information broadcast minimizes the chances of malicious node propagating attacks across the network. There is a high prospect of load management from IoT devices.

Credit Mechanism. This includes transaction transparency, trustworthiness, mutual trust of nodes, and Things.

Timestamp. It is used to identify, trace, and record data transactions which makes the BTs high traceability.

Data Encryption. It makes use of an asymmetric cryptography system for data encryption. The distinct encryption and decryption keys are generated distinctively. There is no need for security channels. The proof-of-work mechanism is mostly deployed for transaction verification. The parties require no exchanges and any agreement to be established for data confidentiality. Essentially, data privacy is achieved through cryptography in which encryption of IoT devices' data and users' privacy is enhanced.

It was noted by Ben and Belhajji [5] that BTs have been considered for integration in IoTs for the reasons as follows: IoT is the most favored technology for the consumer and business sector with BT as most innovative for IoT evolution. Traditionally, IoT systems operate in a federal model in which a hub or broker regulates communications among diverse devices. It is impracticable in cases where devices have to broadcast data with each other in autonomous forms.

Consequently, decentralized IoT platforms come to the fore as a new opportunity [18, 19]. Blockchain gives a decentralized IoT platform for trusted and secured data exchanges among devices. It serves as a regular ledger to keep a reliable log for all information and data exchanges among various things, objects, or devices. Modern IoTs having decentralized controls require the following: peer-to-peer messaging (such as Telehash), distributed file sharing (such as BitTorrent), and autonomous device communication (such as Ethereum blockchain protocol).

This concept has been adopted by manufacturers (such as IBM and Samsung) to implement Autonomous Decentralized Peer-to-Peer Telemetry (ADEPT) using features of Bitcoin design to build IoT devices [18, 20].

2.2 Types of Blockchain Technology

Two categories have been identified for three types of blockchain according to Atlam and Wills [21] including permissioned (public blockchain) and permissionless (private blockchain and federated blockchain). The differences in the three types of blockchains are presented in Table 1 [21, 22].

Table 1 Blockchain technology types

Parameter	Private blockchain	Public blockchain	Federated blockchain
Mode of access	Specific organizations can read/write	Anyone can read/write	Several selected organizations can read/write
Efficiency	High	Low	High
Speed of operation	Faster and lighter	Slower	Faster and lighter
Type of security	Authorized participants, multiparty consensus, and voting	Proof of work, proof of state, and consensus mechanisms	Authorized participants, multiparty consensus, and voting
Immutability	Susceptible to tampering	Highly resistant to tampering	Susceptible to tampering
Asset	Any asset	Native asset	Any asset
Network architecture	Partially decentralized	Totally decentralized	Partially decentralized
Applications	Ripple, Everledger, and Eris	Ethereum, bitcoin, and NXT	R3, EWF, and B3i

2.3 *Characteristics of Blockchains*

The four major characteristics of blockchain technology on the basis of the different types available whether private or public blockchain were identified by Dinh et al. [22]:

Distributed Ledger This is used to describe a kind of data structure composed of transactions in a well-defined arrangement or ordered list. Distributed ledger is analogous to conventional ledger systems used for keeping records of monetary transactions or item exchanged within parties. In the case of a blockchain-based system, all nodes retain a ledger copy in the network. Again, the chains are used to bind groups of blocks together. Hence, the distributed ledger enables append-only iterative data layout. The benefit of this method is that it is possible to keep records and updates the operation history of the states of blockchains. The consensus protocol is required for nodes undertaking updates activities on the ledger as in the case of Bitcoin. The parity schemes create the write to ledger privileges for a certain group of owners using blocks signing.

Cryptography The success of blockchain systems relies heavily on cryptographic approaches to enforce ledgers' integrity (i.e., tampering detection of the blockchain data). The attribute overcomes trust issues of public infrastructure as in the case of blockchain-based applications such as cryptocurrency applications. Bitcoin is used to provide currency valuations utilizing the integrity of the ledger that prevents double-spending. Similarly, private infrastructure-based blockchains require integrity through efficient authentication of nodes to guard against malicious activities.

BT offers two major integrity schemes, namely, the global status is secured using a Merkle hash tree having its source hash saved to a block. A fresh root hash is created for a corresponding change of state. The states are depicted in leaves of the Merkle tree, while the children's hashes are stored in the internal nodes. The hyperledger v0.6 system makes use of a bucket hash tree that group states by hashing property into a specified quantity of buckets. Ethereum system takes advantage of a Patricia-Merkle tree whose leaves represent states with key values.

The protection of the block history is immutable or nearly permanent upon appending to the blockchains. The basic idea is to interconnect the blocks in form of a chain made up of cryptographic hash pointers: the space for the block, $Q + 1$, holds the block hash, Q. The security mechanism of blockchain combines hash pointers and Merkle tree to safeguard data in a more efficient model which trails all past alterations to the global status.

The model of security for blockchain holds the assumption of PKI cryptography availability. The entire users' and transaction details/identities are obtained through public key certificates. In effect, protected key administration is an open issue in blockchain applications. Currently, several key and identity management approaches exist. However, there is a need to modify the traditional blockchain architecture design using new and less complex cryptographic protocols for IoT.

Consensus Blockchain ledger content indicates the historical and present status of states maintained. The replicative nature of blockchain enforces the consensus of several parties for updates to be carried out on ledger content data. Blockchain consensus protocols enable untrusted nodes to perform transactions transparently to deal with the Byzantine behaviors of certain nodes. At present, there are many distributed consensus protocols for blockchains. One type of distributed consensus protocols is the purely computational protocols for controlling and regulating adjoining operations as in the case of proof-of-work (PoW) for Bitcoin.

The truly communication-related protocols are another type of consensus protocol in which consensus of operations is reached utilizing equal votes and numerous rounds of exchanges. These protocols are most suited for private scenarios because of the authentication of nodes. Later, hybrid protocols came about to enhance the effectiveness of PoW and PBFT.

The inadequacies of PoW in public blockchains are overcome by new trusted hardware approaches such as proof of elapsed time (PoET) for Intel SGX. Other consensus mechanisms used interchangeably with PoW include Elastico and Algorand. Recently, hybrid protocols have evolved for private blockchains including proof of authority (PoA), Stellar, and Ripple, as improvements over PBFT in federated network settings.

2.4 Applications of Blockchain Technology

Three main applications of blockchain technology were reported in Ejaz and Anpalagan [13], Ben and Belhajji [5], Lin et al. [23], and Abayomi-Zannu et al. [24]. The main applications of BTs, domain, roles of BC, objectives, and strengths are shown in Table 2.

2.5 Limitations of Blockchain Technology

With the rapid advancements of blockchain applications, its data security comes into focus because it plays a crucial role and capability of numerous blockchain-based applications. One major drawback is the exploitation of the attributes of the blockchain itself to propagate data within the blockchain. This continues to pose a serious challenge to the success and effectiveness of BT and its subsequent adoption as mentioned by Zhu et al. [2]. These include:

(i) The privacy of users is susceptible to compromises due to the openness of blockchain data. Oftentimes, transactions can be used to analyze and establish interrelationships between addresses by experienced attackers.
(ii) The blockchain data become available whenever attacks on the blockchain network cause the authorized or abnormal right of use. Practically, an internal protocol (IP) address can be easily linked to a Bitcoin address and the reverse is the case. Consequent to this, attackers can monitor the one-to-one relationships obtainable in users, addresses, and genuine identity.
(iii) The blockchain data become vulnerable to manipulation or tampering once an attacker is able to access the consensus mechanism of the blockchain. There is the prevalence of selfish mining threats on blockchain, which further rubbishes the blockchain data integrity at large.
(iv) The weaknesses of the smart contracts have severe consequences on the uncontrollability of blockchain data by owners. Other security risks include miner and mining pool attacks, which impact on the privacy of blockchain data applications.
(v) BC-based applications (such as cryptocurrency) have no firm regulatory structure and high unpredictability [4].

The threats and attacks associated with blockchains are still developing due to the relative newness of the technology. Blockchains are potent as long as it is mathematical impracticable for a single party to crack or modify the keys. But, it is unrealizable due to the growth of quantum computing that enables public and private keys to crack within a relatively shorter time. Thus, whenever blockchain keys are broken, the entire system becomes vulnerable.

The entire model of operation of Bitcoin or blockchain is anonymous whereby different user nodes are capable of sending and receiving money regardless of

Table 2 Application areas of BTs

Domain	Roles of blockchain	Objective(s)	Strength(s)
Smart homes	Inclusion of local BT on a resource-enabled node	Enable a miner to initiate message exchanges between inside and outside of the smart home Local storage maintains BT ledgers	IoT nodes performed better by reducing the energy dissipation and transaction times
Traceability system for food supply chain	BT and decentralized IoT (RFID tags)	Gather and broadcast data associated with food items in a distributed manner Members append, update, and search diverse information concerning food items Confidence boosts for food subsector players	Digital profile information of food supply chain members to store and access on BT database Public and private keys for real-time, transparent, and efficient access to food products
Public services	Property registration for prospective dealers in the republic of Georgia	Reliable records of property information and ownership and relevant information for property dealers Decentralization of information	Secure and validation of official business transactions of government such as land title administration Minimize costs and time required for searching records and consummating transactions of property
Voting system	Blockchain technology inclusion to traditional electoral systems	Improved security, reliability, trust, and anonymity of the entire election process Two-factor mobile-based authentication	Further deepened the trust of the people in their governments due to the transparent and decentralized nature
Health-care system	MedRec – a blockchain-based application in private peer-to-peer network	Sharing of medical data with improved trust, security, and reduced cost of infrastructure Block describing ownership of data and viewership permissions of information shared	Smart contract mechanism makes use of automation and traceability of state transactions for rights modification or creation of fresh records
Smart agriculture	BT and information communication technology (such as IoT) design considerations	Distributed and immutable ledger systems for managing records	Environmental and agricultural data security and privacy safeguards for users

unknown identities. There is no linkage of Bitcoin addresses to originating users' nodes, rather assigned to a pseudonym. There is the guarantee of linking the same pseudonym to other users which could erroneously reveal transaction details to unauthorized persons. User's identity can be divulged on the hosted wallet or the Internet service provider can track the Internet protocol (IP) addresses visited by a user. The hosted wallet and ISPs maintain all users' information and records which can be requested by relevant agencies of governments across the world.

The majority of connected devices can be controlled remotely using apps. Consequently, there are frequent cases of malevolent and inappropriate use or references for carrying out distributed denial of service (DDoS) attacks. More so, Bitcoin digital currency exchange markets have faced several threats of DDoS following the trading success experiences in 2011 and beyond. The value of Bitcoin diminished significantly after the successful launch of DDoS attacks.

Several fraudulent online services transact on Bitcoin wallets similar to those of legitimate ones, but the Bitcoin of their supposed victims is confiscated at a certain level of threshold. These include easy coin, Bitcoinwallet.in, and onion wallet.

Other BT challenges were reported in the study by Zhu et al. [2] as categorized into privacy, availability, controllability, and integrity.

Data privacy threat is associated with data breaches or data acquired employing analysis approach of attackers conducted on the privacy of identity and transactions. Data availability threats happen whenever there is an incorrect or abnormal right to use data stored on blockchains, which is attained through network eclipse and traceability attacks.

Data integrity threat is a situation in which data on blockchains are manipulated or tampered with. These kinds of threats are occasioned by mining pools, double-spending, and miner attacks. Nevertheless, data controllability threats occur when there is an accidental manipulation of blockchain data caused by weaknesses in smart contracts.

2.6 Data Retrieval and Manipulation on Blockchains

A blockchain-based application enables data-sharing platform for users in which privately stored data can be accessed upon authorizations. Anonymity is offered by blockchain through the allocation of changeable public key infrastructures (PKIs) for its users. Often, it is possible to deanonymize users through classification of transactions such as flow of the inputs/outputs to identify a user. Also, it is possible to reveal the identity of a user using real-time network traffic analysis in which the device network address is associated with one or more PKIs.

The public networks are susceptible to higher privacy risks using the transacted data to disclose the activity patterns of a user. Malicious attackers rely on a user's activities to identify owners of transactions across the networks. Concerns are raised about user deanonymization with users' node credentials thereby exposing transactions and identity of users [25].

The concept of multiparty private computation approaches including homomorphic encryption algorithms or Enigma is considered in preventing reconstruction of original data by enabling pieces of data processing to protect privacy or exploitation of data in big data applications. In effect, PBC or permissionless blockchains should increase anonymity, which is the basics of their designs because of the lack of accountable data processors [26]. There are some ways in preserving privacy in public blockchains including:

(i) Combination of diverse types of cryptographic operations.
(ii) Private computation schemes.
(iii) Decentralized validation of transactions while protecting private elements or contents.
(iv) Use of hash values to minimize redaction and support erasure.
(v) Designated data controller for decentralized verification by removing links and separating hash values of data from real data.

3 Research Methodology

3.1 Description of the Proposed Sooner-C Lightweight Cryptography

The recently renewed interests in lightweight cryptosystems can be attributable to less-memory usages, minimal computing resources, and lower energy requirements in the process of securing devices. Again, they are efficient, lesser key and message block sizes, faster, and simpler relative to traditional systems. However, they are less secure when compared to the traditional cryptosystems [27, 28].

The proposed lightweight encryption and hash algorithms support stream ciphers and shorter characters in the ciphertexts. The vulnerability to attacks by lightweight cryptosystems has motivated the use of hardening procedure [6, 28].

In addition, the enhanced hash value and PKI generation approach target blockchains that predominantly utilize less-secure and complex computation for deriving hash values and PKIs. Therefore, Sooner-C is composed of the lightweight versions of the traditional cryptosystems given in [12, 27, 29] based on Gentry et al.'s [30] homomorphism principle, AES-256, signing or hardening algorithms, and SHA-256. The main goal is to provide privacy, which disallows the acquisition of the information contained in a message without approval or authorization on the blockchain. Often, privacy is used to mean secrecy and confidentiality [31].

3.2 Mathematical Representation of the Sooner-C

Theorem 1 The blockchain's traditional cryptosystems are generally secured, but they are less applicable for real-time and protection of data for constrained devices due to complex structures, large keys formulation, and large memory capacity.

Proof 1 The lightweight cryptography reduced computation, ciphertext, encryption, and decryption keys. They are less secure for privacy preservations, which require hardening or signing encryption schemes to raise the level of security due to reduced sizes of ciphertexts and PKIs. The sizes of block ciphers, encryption keys, and decryption are represented for classical cryptosystems (CCs) and lightweight cryptosystems (LCs) in Eq. 1.

$$A_n + B_n + C_n = T\ (a_n + b_n + c_n) \tag{1}$$

where, A_n = block cipher size for CCs, B_n = encryption key size of CCs, C_n = decryption key size of CCs, a_n = block cipher size for LCs, b_n = encryption key size of LCs, c_n = decryption key size of LCs, n = instances of data, and T = lightweight transformation function.

Theorem 2 Block-based encryption schemes operate on a large chunk of bits representing data, which are faster, organized, and relatively easy to deploy against stream encryption that performs bit-by-bit encryption for data representation. Therefore, IoT privacy can be best preserved with its metadata and data during transmission from point of deployment at edge network shown in Eq. 2.

$$T\ (a_n + b_n + c_n) = p_i \varpi G \tag{2}$$

where, G = plaintext, k = public key, l = private key, p = block encryption function, ϖ = relationship function, and i = number of instances.

Definition 2 The strongest cryptosystems used traditionally for IoT and BT are AES-256, SHA-1, SHA-2, and SHA-3, which have been proven to be effective against attacks.

Theorem 3 The level of security offered by a cryptosystem or cryptographic algorithm is relative to the lengths of its encryption and decryption keys and block sizes. LCs propose shorter lengths of their encryption and decryption keys and message block sizes, which are vulnerable.

Definition 3 The security of LCs can be enhanced by increasing the number of cryptosystems used for derivations and fortifying with signing or hardening algorithms [6] as depicted in Eq. 3. The resultant ciphertexts and keys are secured and applicable in blockchains.

Table 3 Experimental parameters

Parameter	Attributes
Hardware	
Processor	AMD E1–1200 APU, Radeon™ HD graphics 1.40 GHz
RAM	4.00 GB
Hard disk drive (HDD)	282GB
System type	64-bit operating system, x64-based processor
Software	
Operating system	Windows 8 single language
Application programming interface	Visual studio code 16
Traditional cryptosystems	AES-256, base64code, cipher block chaining (CBC), Galois/counter mode (GCM), SHA 1–256
Evaluation metrics	Encryption/decryption time, length of messages, and public key infrastructure, and number of rounds of encryption operations.

$$p_i \varpi G = \sum_{j=2}^{m} \mathrm{Q} \oplus F \tag{3}$$

where, Q = encryption algorithms, F = hashing algorithms, $\bigoplus$ = random generator function, Q = initial numbers of cryptosystems for derivation exactly two, and m = last numbers of cryptosystems for derivation.

3.3 Experimental Setup

The minimal system configurations for validating the proposed Sooner-C lightweight cryptosystem are presented in Table 3.

4 Results and Discussion

Hash Value Generations Sooner lightweight cryptosystem (Sooner-C) implementation is composed of reduced classical AES256, CBC hardening algorithm, and SHA1–256. The computational performance on the basis of the length of plaintext and secret key (5 characters or 32-bits long) is presented in Table 4.

Similarly, the traditional cryptosystem implementation before the reduction operations on AES256, hardening algorithm (GCM), and SHA1–256 in which the

Table 4 Plaintexts and corresponding hash values using proposed Sooner-C

S/N.	Plaintext	Hash value
1.	howtodoinjava.com	BfNFPRgfKF8Ke9kpoNAagmcI4/Hya5o/rq9/fq97ZiA=
2.	23.33	viRSYFTq1J0rRDLwfrXg3w==
3.	1.98	BR8oAxAeV/cmXWRoYUzxjA==
4.	High	iNKG9N05FtGF6AEeJnwLhw==
5.	high1212	GzIH1TZ2Ft + TAz3unfbnaw==
6.	Withheld for PII reasons	JiLPe4nGJzOnBIKk/ghaDg==
7.	Withheld for PII reasons	dBbbtVUsFsH3/Z9hzSoH3TBMV04rxouwFQOuaOYMS9o=
8.	Withheld for PII reasons	Ciqj0SCaQh5Oh/KyLf8yBfb6jUXsL6COOrbI4w81wII=
9.	Withheld for PII reasons	0x9i8GrAA82N49n4mBWetq9FjWacMPfkC6GdVsdAFyw=
10.	Password	Gv/yIop0HyLy7AzjnrCBGA==

computational performance uses the lengths of plaintext and secret key is presented in Table 5.

Encryption Operations The differences between the Sooner-C and classical cryptosystem were determined using ciphertext sizes, encryption time, decryption time, and the number of rounds.

Ciphertext Sizes The comparisons of ciphertext sizes realized after the application of both cryptosystems are presented in Table 6.

From Table 6, the performance of both cryptosystems based on the resulting ciphertext sizes reveals better outcomes for the proposed cryptosystem due to its lesser sizes as shown in Fig. 1.

Encryption Time Compared The encryption time performance after applying both cryptosystems is shown in Table 7.

From Table 6, the performance of both cryptosystems based on the ensuing encryption times reveals that the proposed cryptosystem outclassed the classical cryptosystem due to its shorter time frame of plaintext encryption as shown in Fig. 2.

Decryption Time Compared The encryption time performance after applying both cryptosystems are shown in Table 8.

From Table 8, the performance of both cryptosystems based on the ensuing decryption times reveal that the proposed cryptosystem is better than the classical cryptosystem due to its shorter time frame of ciphertext decryption as shown in Fig. 3.

Number of Rounds The number of rounds per cycle performance after applying both cryptosystems is presented in Table 9.

From Table 9, the performance of both cryptosystems based on the ensuing number of rounds of encryption times shows that the new cryptosystem is more

Table 5 Plaintexts and corresponding hash values using proposed Sooner-C

S/N.	Plaintext	Hash value
1.	howtodoin-java.com	R52AMbobURxDY0vdrJf8PWZPBloPdrh8tMhfYqlHgvSj3GmaBy0JwulvBZJQxbFwVuBfdNzpL8VoNiGRjw==
2.	23.33	C5MmzQCh6IVHeL/REudAUNE50Kiv0sJkBBJLk/69ixEhq631V5q0Q4Hk + rBEzW0U6A==
3.	1.98	OnPQrVmYo6OIRa/WNgqlUP9X4mh5QrhMEdksQy6r28K1nerNp4f2a7kb + iBQidFF
4.	High	UroPDFvHGdJ5ks++00CAAy2CMWqz8AtJn7FXToseqfQN6g0JCG1JIJUxMphvooZr
5.	high1212	NhZ5NX6cCiseyNgFy8I1lObOPT5r9c8nique22emJw0rUQrAvgejAYWAz/iWEHtvLUJtPQ==
6.	Withheld for PII reasons	y2GjqOwM8CSTEc5IUm9b/8lMDapcDM + aa+pEI42HDP0izVlAeA4wiWClEOgGKzbHxeehTOLLNQ==
7.	Withheld for PII reasons	jnkhEIJOjcHSFe3pa/nb1v/d4OYDT3gbncmfg + QAjUVNubadXpk8PVvQn5lbgoJvn3IyvzjO6xAtZUYJyLnll2on
8.	Withheld for PII reasons	hIgwdUVrlCFq/nEe5rB + bgm0EOyNNXtRWNMQhgbCz5nEKcAKVjgINkW/HhocH4RI6Aw + 8HitwyxTicja33sfunjz95GJ8vjSRf8Y
9.	Withheld for PII reasons	zHIZY+3YF89EBZTwWfNsrJhR99WnbCgOPEP1gPQ7zrpFRYhk/iMuZeK3k7at3cfn4cZnuVJEdO8cYt2 + dH9Ii4wyDiWBgs/K6Q=
10.	Password	ttB3dXgWNU78dJ – ck + 4dTSDIMOEzud + 5SB + KgRoKXf/vLzQMhw4fQMt5gCdUNyRXwt3onw==

Table 6 Ciphertexts sizes compared

S/N.	Sooner-C	Classical cryptosystem
1.	44	84
2.	24	68
3.	24	64
4.	24	64
5.	24	72
6.	24	76
7.	44	88
8.	44	100
9.	44	100
10.	24	72

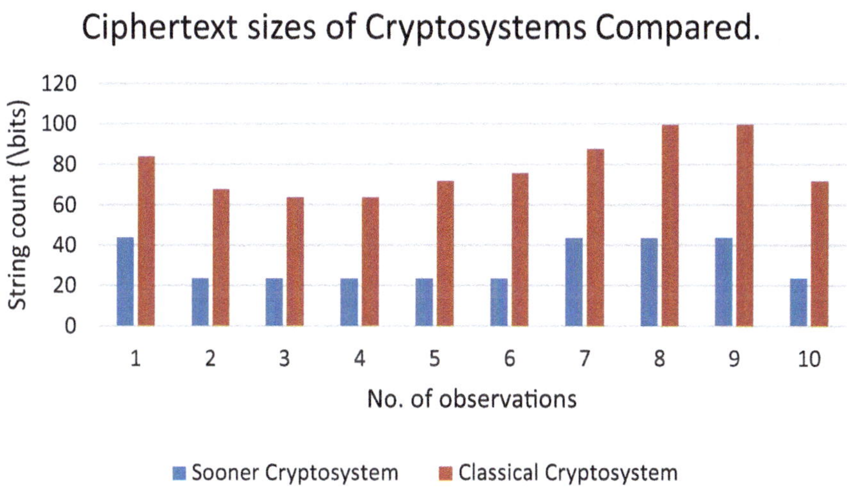

Fig. 1 Cryptosystem performances using ciphertext sizes

Table 7 Encryption time of cryptosystems compared (secs)

S/N.	Sooner-C	Classical cryptosystem
1.	1574	20,050
2.	2653	18,101
3.	1356	18,039
4.	1540	17,762
5.	1452	17,684
6.	1591	17,576
7.	1658	18,015
8.	1690	17,498
9.	1616	20,119
10.	1592	19,009

effective than the classical cryptosystem due to increased computation required to break the security system in order to obtain ciphertext as shown in Fig. 4.

Average Summary of Performance The average for both cryptosystems' performances using ciphertext size, encryption time, decryption time, and the number of rounds is presented in Table 10.

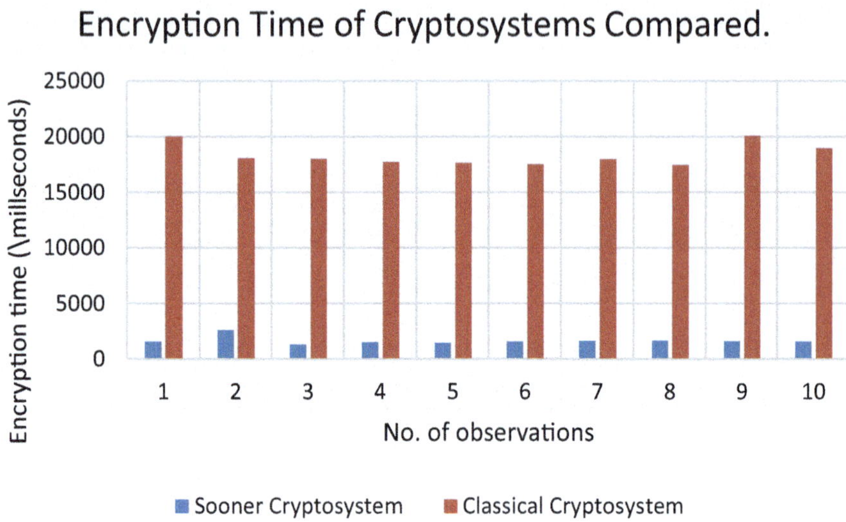

Fig. 2 Cryptosystem performances using encryption time

Table 8 Decryption time of cryptosystems compared (secs)

S/N.	Sooner-C	Classical cryptosystem
1.	584	1034
2.	756	872
3.	598	655
4.	585	722
5.	632	686
6.	643	807
7.	578	909
8.	693	855
9.	572	805
10.	547	720

From Table 10, the ciphertext size for the proposed cryptosystem is more effective than the classical cryptosystem by 28.88% to 71.12%, which underpinned the concept of lightweight cryptosystem investigated in this study. The encryption time has the smallest value (8.34%) for the proposed system indicating its suitability for data protection on blockchains.

Similarly, the decryption time for the proposed cryptosystem is slightly lower than the classical cryptosystem from 43.42% to 56.48%. In terms of security level, Sooner-C offered better security protection against classical cryptosystem by increasing the computation needed to break the ciphertext by an attacker from 61.21% to 38.79%.

The Sooner-C is capable of resisting the meet-in-the-middle attack on session keys due to multiple functions utilized in the formulating ciphertexts, hash values,

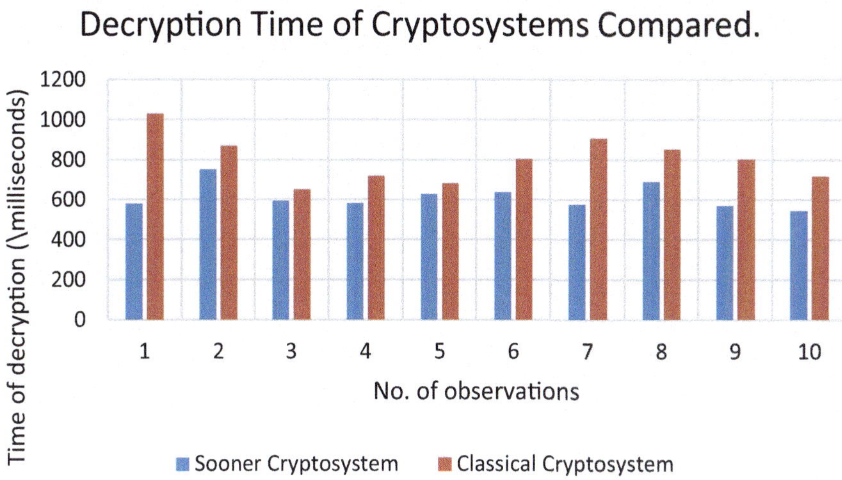

Fig. 3 Cryptosystem performances using decryption time

Table 9 Number of rounds of cryptosystems compared

S/N.	Sooner-C	Classical cryptosystem
1.	12	9
2.	17	10
3.	17	10
4.	17	10
5.	17	10
6.	17	9
7.	12	9
8.	12	9
9.	12	9
10.	17	10

and PKIs. Consequently, it satisfies good and strong lightweight encryption algorithm requirements in Bahrami and Naderi [32] and Janakiraman et al. [33].

5 Conclusion

The results obtained in this paper corroborated the arguments in literatures [2, 8, 12, 22, 34, 35] that LCs offer lesser memory consumption for shorter ciphertexts, shorter hash values, smaller time of encryption/decryption, and considerable low number of rounds required for encryption. The proposed Sooner-C holds promise for integrating BT in IoT-based networks and applications. In the case of encryption time, decryption time, ciphertext size, and number of rounds, Sooner-C offered

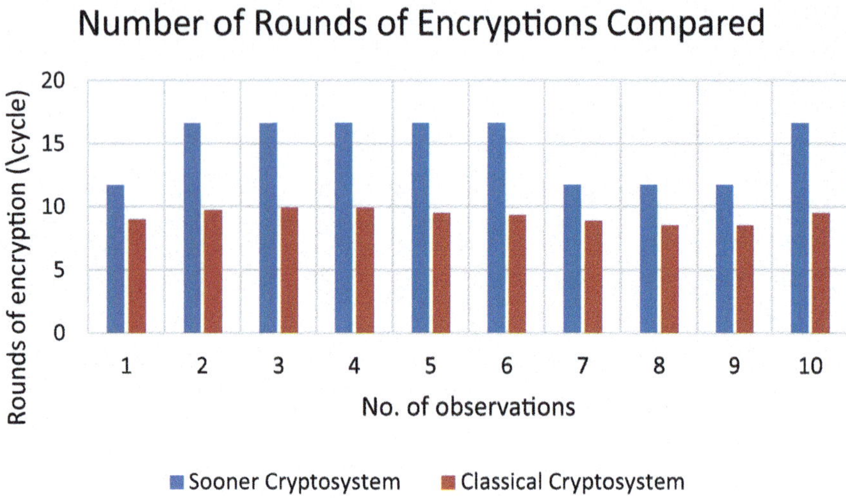

Fig. 4 Cryptosystem performances using number of rounds per cycle

Table 10 Average performances of cryptosystems compared

Evaluation metric	Sooner-C (proposed)	Classical cryptosystem
Ciphertext size (%)	28.88	71.12
Encryption time (%)	8.34	91.66
Decryption time (%)	43.42	56.48
Number of rounds (%)	61.21	38.79

better performance against classical cryptosystem by 1672.2 secs to 18385.3 secs, 618.8 secs to 806 secs, 32-bits to 79-bits, and 15 to 10, respectively.

Again, the level of security is high despite the reduced number of bits when compared to classical cryptosystems due to multiple encryptions and hashing schemes utilized in the formulation of the Sooner. The blockchain security is enhanced with Sooner-C than traditional cryptosystem by increasing the computation needed to break the ciphertext, hash values, and PKIs for an attacker from 61.21% to 38.79%. While only 5 characters or 32-bits long key sizes are required as against 8 characters or 256-bits for Sooner-C and traditional cryptosystem, respectively.

Future works need to consider the full implementation of the proposed lightweight cryptosystem (Sooner-C) in real-life situations especially in smart rice farming environments and other smart systems for improved security and privacy preservations.

References

1. Ji, Y., Zhang, J., Ma, J., Yang, C., & Yao, X. (2018). BMPLS: Blockchain-based multi-level privacy-preserving location sharing scheme for telecare medical information systems. *Journal of Medical Systems, 42*(8), 147. https://doi.org/10.1007/s10916-018-0998-2
2. Zhu, L., Zheng, B., Shen, M., Yu, S., Gao, F., Li, H., Shi, K., & Gai, K. (2018). Research on the security of blockchain data: A survey. *arXiv preprint.* http://arxiv.org/abs/1812.02009
3. Abayomi-Zannu, T. P., Odun-Ayo, I., Tatama, B. F., & Misra, S. (2020). Implementing a mobile voting system utilising blockchain technology and two-factor authentication in Nigeria. In *1st international conferenceon computing, communication, and cyber-security* (pp. 857–872).
4. Awotunde, J. B., Ogundokun, R. O., Jimoh, R. G., Misra, S., & Aro, T. O. (2021). Machine learning algorithm for cryptocurrencies price prediction. In S. Misra & A. Kumar Tyagi (Eds.), *Artificial intelligence for cyber security: Methods, issues, and possible horizons or opportunities* (Studies in computational intelligence) (Vol. 972, pp. 421–447). Springer. https://doi.org/10.1007/978-3-030-72236-4_17
5. Ben, A. A., & Belhajji, M. A. (2018). The Blockchain technology. *International Journal of Hyperconnectivity and the Internet of Things, 1*(2), 1–11.
6. Henry, R., Herzberg, A., & Kate, A. (2018). Blockchain access privacy: Challenges and directions. *IEEE Security and Privacy, 16*(4), 38–45. https://doi.org/10.1109/MSP.2018.3111245
7. Brandão, A., Mamede, H. S., & Gonçalves, R. (2018). Systematic review of the literature, research on blockchain technology as support to the trust model proposed applied to smart places. *Advances in Intelligent Systems and Computing, 745*, 1163–1174.
8. Alfa, A. A., Alhassan, J. K., Olaniyi, O. M., & Olalere, M. (2021). *Blockchain technology in IoT systems: current trends.*, methodology, problems, applications, and future directions, *7*(2), 115–143.
9. Ferrag, M. A., Derdour, M., Mukherjee, M., Derhab, A., Maglaras, L., & Janicke, H. (2018). Blockchain Technologies for the Internet of things: Research issues and challenges. *IEEE Internet of Things Journal, 6*(2), 2188–2204. https://doi.org/10.1109/JIOT.2018.2882794
10. Huh, S., Cho, S., Kim, S.: Managing IoT devices using blockchain platform. International conference on advanced communication technology, pp. 464–467 (2017). https://doi.org/10.23919/ICACT.2017.7890132
11. Olowu, M., Yinka-Banjo, C., Misra, S., & Florez, H. (2019). A secured private-cloud computing system. In H. Florez, M. Leon, J. M. Diaz-Nafria, & S. Belli (Eds.), *ICAI 2019, CCIS* (Vol. 1051, pp. 373–348). Springer. https://doi.org/10.1007/978-3-030-32475-9_27
12. Alfa, A. A., Alhassan, J. K., Olaniyi, O. M., & Olalere, M. (2021). Sooner lightweight cryptosystem: Towards privacy preservation of resource-constrained. In S. Misra & B. Muhammad-Bello (Eds.), *ICTA 2020, CCIS* (Vol. 1350, pp. 415–429). Springer Nature. https://doi.org/10.1007/978-3-030-69143-1
13. Ejaz, W., & Anpalagan, A. (2018). Blockchain Technology for Security and Privacy in internet of things. *Internet Things Smart Cities*. https://doi.org/10.1007/978-3-319-95037-2_5
14. Misra, S. A. (2021). Step by step guide for choosing project topics and writing research papers in ICT related disciplines. In *Communications in computer and information science* (Vol. 1350, pp. 727–744). Springer.
15. Yang, Y., Yang, Y., Chen, J., & Liu, M. (2018). Application of blockchain in internet of things. Lecture notes in computer science (including subseries lecture notes in artificial intelligence and lecture notes in bioinformatics). *LNCS, 11067*, 73–82. https://doi.org/10.1007/978-3-030-00018-9_7
16. Medhane, D. V., Sangaiah, A. K., Hossain, M. S., Muhammad, G., & Wang, J. (2020). Blockchain-enabled distributed secuity framework for next generation IoT: An edge-cloud and software defined network integrated approach. *IEEE Internet of Things Journal, 7*(7), 6143–6149. https://doi.org/10.1109/JIOT.2020.2977196
17. Tasatanattakool, P., & Techapanupreeda, C. (2018). Blockchain: Challenges and applications. In *2018 international conference on information networking* (pp. 473–475). IEEE. https://doi.org/10.1109/ICOIN.2018.8343163

18. Higgins, S. (2015). *IBM reveals proof of concept for blockchain-powered internet of things*. https://www.coindesk.com/ibm-reveals-proof-concept-Blockchain-powered-internet-things/.
19. Branco, F., Moreira, F., Martins, J., Au-Yong-Oliveira, M., & Goncalves, R. (2019). Conceptual approach for an extension to mushroom farm distributed process control system: IoT and Blockchain. In A. Rochas et al. (Eds.), *WorldCIST'19 2019, AISC* (Vol. 930, pp. 738–747). Springer Nature. https://doi.org/10.1007/978-3-030-16181-1_69
20. IBM. (2015). *Empowering the edge*. https://www-935.ibm.com/services/multimedia/GBE03662USEN.pdf.
21. Atlam, H. F. (2018). Wills, G. B. (2018). Technical aspects of blockchain and IoT. *Advances in computers, 115*, 1–39. https://doi.org/10.1016/bs.adcom.2018.10.006
22. Dinh, T. T. A., Liu, R., Zhang, M., Chen, G., Ooi, B. C., & Wang, J. (2018). Untangling Blockchain: A data processing view of Blockchain systems. *IEEE Transactions on Knowledge and Data Engineering, 30*(7). https://doi.org/10.1109/TKDE.2017.2781227
23. Lin, J., Shen, Z., Zhang, A., & Chai, Y. (2018). Blockchain and IoT based food traceability for smart agriculture. In *3rd international conference on crowd science and engineering* (pp. 1–6). ACM.
24. Abayomi-Zannu, T. P., Odun-Ayo, I., Tatama, B. F., & Misra, S. (2020). Implementing a mobile voting system utilizing blockchain technology and two-factor authentication in Nigeria. In *Proceedings of 1st International Conference on Computing, Communications, and Cyber-Security (IC4S 2019)* (pp. 857–872). Springer.
25. Dorri, A., Roulin, C., Jurdak, R., & Kanhere, S. (2018). On the activity privacy of blockchain for IoT. *arXiv preprint*. http://arxiv.org/abs/1812.08970
26. Ibáñez, L. D., Kieron, O., & Simperl, E. (2018). On Blockchains and the general data protection regulation. *EU Blockchain Forum and Observatory, 1–13*.
27. Sun, J., Zhong, Q., Kou, L., Wang, W., Da, Q., & Lin, Y. (2018). A lightweight multi-factor mobile user authentication scheme. In *2018 IEEE conference on computer communications workshops, INFOCOM WKSHPS* (pp. 831–836). IEEE. https://doi.org/10.1109/INFCOMW.2018.8406952
28. Buchanan, W. J., Li, S., & Asif, R. (2018). Lightweight cryptography methods. *Journal of Cyber Security Technology, 1*(3-4), 187–201. https://doi.org/10.1080/23742917.2017.1384917
29. Halabi, J., & Artail, H. (2019). A lightweight synchronous cryptographic hash chain solution to securing the vehicle CAN bus. In *2018 IEEE International Multidisciplinary Conference on Engineering Technology, IMCET-2018* (pp. 1–6). https://doi.org/10.1109/IMCET.2018.8603057
30. Gentry, C. (2009). *A fully homomorphic encryption scheme*. Ph.D. Dissertation. Stanford University.
31. Driscoll, K. (2018). Lightweight crypto for lightweight unmanned arial systems. In *2018 Integrated Communications, Navigation, Surveillance conference, ICNS 2018* (pp. 1–15). Honeywell. https://doi.org/10.1109/ICNSURV.2018.8384913
32. Bahrami, S., & Naderi, M. (2010). Image encryption using a lightweight stream encryption algorithm. *Advances in Multimedia, 2012*, 1–8.
33. Janakiraman, S., Sree, K. S., Manasa, V. L., Rajagopalan, S., Thenmozhi, K., & Amirtharajan, R. (2018). On the diffusion of lightweight image encryption in embedded hardware. 2018 international conference on computer communication and informatics. *ICCCI, 2018*, 1–6. https://doi.org/10.1109/ICCCI.2018.8441229
34. Conti, M., Sandeep, K. E., Lal, C., & Ruj, S. (2018). A survey on security and privacy issues of bitcoin. *IEEE Communications Surveys and Tutorials, 20*(4), 3416–3452.
35. Hassan, M. U., Rehmani, M. H., & Chen, J. (2019). Privacy preservation in blockchain based IoT systems: Integration issues, prospects, challenges, and future research directions. *Future Generation Computer Systems, 97*, 512–529. https://doi.org/10.1016/j.future.2019.02.060

A Prediction Model for Bitcoin Cryptocurrency Prices

Micheal Olaolu Arowolo, Peace Ayegba, Shakirat Ronke Yusuff, and Sanjay Misra

1 Background of the Study

Owing to its potential applications in finance, stock chain management, and cybersecurity, blockchain is quickly gaining popularity. Significant advancement of blockchain technology has the latent for users to directly exchange ownership of digital resources (such as cryptocurrencies) to one another inside a dispersed and distributed peer-to-peer network lacking the assistance of a third party (e.g., financial institution). Transactions change the global state of the blockchain network by transferring ownership of resources. Respective transaction published to the blockchain network necessitates digital autographs by the user, using a digital signature technique [1]. Bitcoin is a decentralized virtual integrated financial management information solution that allows peer-to-peer transactions without any centralized authority. Cryptography is used to secure network transactions and liquidity. Presently, the system has a vibrant open-source community and payment network. Bitcoin's ecosystem is attracting a lot of interest from enterprises, individuals, and financiers due to the distinct payment protocol and its rising popularity. For financial products and services that are presently accessible in our

M. O. Arowolo (✉) · P. Ayegba
Department of Computer Science, Landmark University, Omu-Aran, Nigeria
e-mail: arowolo.olaolu@lmu.edu.ng; peace.ayegba@lmu.edu.ng

S. R. Yusuff
Department of Computer Science, Kwara State University, Malete, Nigeria
e-mail: shakirat.yusuff@kwasu.edu.ng

S. Misra
Department of Computer Science and Communication, Østfold University College (HIOF), Halden, Norway
e-mail: Sanjay.misra@covenantuniversity.edu.ng

S. Misra, A. Kumar Tyagi (eds.), *Blockchain Applications in the Smart Era*, EAI/Springer Innovations in Communication and Computing,
https://doi.org/10.1007/978-3-030-89546-4_7

conventional to succeed, fiat money world must be replicated and customized to Bitcoin and other developing cryptocurrencies [2].

Among the most contentious and unclear breakthroughs in the current economic environment has been the rapid development of digital currencies during the last decade. The organization of the economy, capital industry, and payment systems are all changing as technology advances. Financial markets around the world have become more digitized than ever before, and a cashless society is on the horizon. People can now produce their own money (digital cryptocurrency), thanks to modern technology, and the role of central banks as lenders of the last resort is being debated [3]. Bitcoin, like some other cryptocurrencies, is a finance phenomenon in which data is viewed as money. Users (referred to as "miners") transmit and receive cryptocurrencies (data) electronically from their workstations in peer-to-peer information systems to purchase goods and services if other parties are willing to accept these transactions. In 2019, the market valuation of 2957 cryptocurrencies and the number of miners hit $221 billion (Bitcoin $147) and 42 million, respectively. From its inception in 2009 to late August 2020, the price of Bitcoin has risen dramatically, from $0.0008 to $10,168 per single coin [4].

Cryptocurrencies like Bitcoin are among the most contentious and difficult technological advancements in today's financial system. Given that Bitcoin is still an emerging technology with such a highly volatile market value, current valuation prediction models are scarce and ineffective in a production setting [5]. Investors in the cryptocurrency market establish trust connections by creating blockchains based on cryptography and hash algorithms. Bitcoin's inherent qualities, obtained from blockchain technology, have sparked a wide range of analytical concerns, with economics and cybersecurity and machine learning [6]. Machine learning techniques are being used since they do not need explicit principles and divide datasets into training and testing sets for learning. It enables machines to "learn" and predict the future [7–9].

Several studies have lately been directed on modeling Bitcoin pricing as a growing market element with accurate, relevant rules. Machine learning techniques, for example, recurrent neural networks (RNNs) and long short-term memory (LSTM), are used to price Bitcoin, and the results are compared to those obtained with autoregressive integrated moving average (ARIMA) models, among other studies on data analysis or economic characteristics and representations of Bitcoin prices relate directly to its features and functionality as a financial asset. They exhibit low predictive performance accuracy through binomial logistic regression, support vector machine, and random forest [9, 10]. With the advancement of machine intelligence, forecasting Bitcoin has gained a lot of attention, and a lot of studies have evolved yet more improvement needs to be carried out using machine learning algorithms for the prediction of cryptocurrency system, in terms of the technical input parameters, existing continuous dataset, and machine learning algorithms for forecasting the performance. To provide improved projections, many studies have been used in modeling and forecasting stock prices. This work is, in fact, a crucial phase in financial decision-making involving portfolio optimization, risk assessment, and trading. The use of machine learning to analyze and predict cryptocurrency prices is one of the hottest topics in financial markets.

This study uses an innovative strategy based on machine learning to predict the price of Bitcoin cryptocurrency. In order to analyze the processes, an ICA with the Firefly method is used to choose relevant information from the given sample dataset and classify it using SVM kernels; the evaluation of the performance is investigated in terms of accuracy, sensitivity, and specificity, among others.

The paper is structured in five sections. The next section discusses about the related works. Section 3 discusses on the materials and methods used. In Sect. 4, the results are discussed. Section 5 summarizes and discusses the conclusion of the paper.

2 Review of Related Works

Machine learning algorithms are being used for predicting Bitcoin price fluctuations [4]; artificial neural networks (ANN), support vector machines (SVM), naïve Bayes (BM), and random forest (RF), in addition to the logistic regression model, were used to evaluate algorithm performance utilizing continuous and discrete datasets. The results of the ANN, SVM, NB, and RF were compared to the results of the LR using the t-test. According to empirical studies, the RF has the best prediction accuracy in the continuous dataset, whereas the NB has the worst. However, the ANN has the best performance in the discrete dataset, and the NB has the worst. Also, the algorithms (models) evaluated the discrete dataset and enhanced the overall predicting performance.

A microcredit scoring method using machine learning was suggested [11], evaluating different machine learning techniques on actual fintech information to see how effective they are in categorizing debtors into different credit classifications. They showed that using readily available information about customers and off-the-shelf multi-class classifier like decision tree methods may do excellently (such as age, occupation, and location). This provides microlending organizations in developing countries with a low-cost and effective means to evaluate creditworthiness in the lack of credit record or centralized credit repositories.

The use of Google trends and keywords in a deep learning model for global financial prognostication was suggested [12]; using Google Trends for Internet searches, they looked at the relationship involving Google movements and the Taiwan Weighted Stock Index. In the correlation test and the regression analysis, Google Trends search term volume was employed. The terms were then investigated in two studies: the first was machine learning, and the other was searching trends. Following empirical analysis, it was discovered that the neural network outperformed the support vector machine and decision trees in the experimental one. As a result, in the conducted test, the neural network was chosen in comparison with the search trend.

An empirical strategy to analyzing machine learning classification for stock markets was proposed [13]; through a series of exchange imitations in the FOREX marketplace, we examine the significance of simple machine learning models in achieving profitable trading. It assesses the models' effectiveness and the extent to which specific model parameters produce methodical trade projections. The use of feature extraction, frequently occurring retraining, and training set size is addressed in to get a combination of such parameters proficient with not only creating positive cumulative yields for machine learning models, and also demonstrating that simple algorithms initially prevented cash flow forecast. The study outlines how such an integration of features, in relation to technological markers which have been used as input data to machine learning-based predictive variables, such as cost-based information, seasonal fluctuations characteristics, and lagged values used in classical time series analysis, can be used to improve the classification capacities that directly affect the ultimate profitability.

Over the use of machine learning techniques in significant financial and banking industries, a study and classification of neural networks in financial services was presented [14], including an emphasis on modeling preparation, inputs, and parameter estimation. Furthermore, relevant components that would have an impact on the outcomes of economic deep neural networks are examined. This study informs scientists and professionals about the state of the art in the usage of machine learning techniques in investment banking.

Several cryptos prediction using machine learning and professional trading metrics [15], modeling cryptocurrency price increases with a neural network approach having the lowest method damage of over 100 epochs in training, and technical trade pointers charts that depict an actual BTC price of 5 to 10 times in 300 days of the recent economic year has bolstered trader confidence and shifted the universal cryptocurrency chart by predicted BTC values. In a similar vein, employing machine learning and sentiment analysis to analyze Bitcoin pricing. They looked at stock market movements to see whether they could forecast Bitcoin prices quantitatively. In addition, researchers investigated the impact of international currencies, such as the American dollars and currency exchange rates on Bitcoin rates, including whether Bitcoin really does have the consistency to depose the world's monetary systems and be the sole monetary system. The research seemed to be satisfactory to assist in the prediction models, with findings acquired that indicate an accuracy to predict price of Bitcoin and to use a machine learning-based neural network obtaining an accuracy of 94.89% in all constraints of conceptual market sign during estimation, significantly reducing its current value over 13.7% 2020 solely.

Machine learning is used to predict and trade cryptocurrencies amid the shifting economic conditions [16], The predictability of significant cryptocurrencies, such as Bitcoin, Ethereum, and Litecoin, and also the financial performance of investment strategies based on machine learning are explored in this research (e.g., linear models, random forests, and support vector machines). The results are developed in a time of profound upheaval and evaluated in a time frame of market corrections, enabling for the evaluation of whether the estimations are accurate and when business trajectory shifts among analysis and testing. The classification and regression methodologies utilize buying and selling and network connection qualities from August 2015 to March 2019, with the sample test starting in April 2018. Five of the 18 specific models tested had less than 50% success rates during the testing period. The exchange strategies depend on the assembly of models. For Ethereum and Litecoin, the ensemble that assumes five models yields undistinguishable signs (ensemble 5) and attains the best results, with annualized Sharpe ratios of 80% and 91%, respectively, and annualized revenues (subsequently relative round-trip exchange costs of 0.5%) of 9.62% and 5.73%, respectively. These encouraging findings back up the claim that machine learning can be used to investigate the predictability of cryptocurrencies and devise profitable trading strategies in these markets, even in the face of adversity.

Multiple machine learning technologies for predicting cryptocurrency price complexities were suggested [17]. The case diagram of cryptocurrencies is incredibly complicated, predicting the future hard. They used two popular machine learning algorithms, the adequately artificial neural network and the long short-term memory (LSTM), to model the value of numerous prevalent cryptocurrencies and Bitcoin, Ethereum, Ripple, Stellar Lumens, Litecoin, and Monero, in this study. They analyzed the model's performance and proceeded to run an analysis to learn more about our model's performance. They discovered that, while LSTM appears to be more relevant for time sequence prediction problems, ANN outperforms LSTM in general in our experiments. The use of price data from other cryptocurrencies for simultaneous training and prediction could make BTC prediction much easier. Furthermore, the predictive inaccuracy of the model is greatly dependent on the timescale of interest.

Data analysis in crypto exchanges was proposed [18], proposing a retraining machine learning-based actively traded model and applying it to five main cryptocurrencies in flux; they showed how, compared to a purchase strategy, this model generates greater risk returns and lowers downside risk. When actual transaction costs are taken into account, these conclusions hold true. They concluded that the model could be used in real-world asset management, but efficiency will vary depending on how it will be validated in test samples.

Financial markets and machine learning for cryptocurrency were suggested [19], determining how machine learning is used in the financial markets. Despite the fact that much artificial intelligence research has been published, they are concentrating on recent financial market studies with possible applicability to blockchain-based cryptocurrencies. While different machine learning techniques have varied goals, they all aim to make money and anticipate future pricing. Their conclusion demonstrates the advantages of employing such models to account for a variety of financial market occurrences. They come to the conclusion that artificial intelligence techniques for financial markets, namely, machine learnings, cannot acquire all features of the given data in the same way that general artificial intelligence can. As a result, depending on the dataset's structure and hierarchy, it's only natural to choose a desirable strategy. In addition, the researcher's perspective on the market influences the approach he or she takes.

Machine learning-based models for Bitcoin value predictions was proposed [20], by evaluating the efficiency of one-stage methodologies based on a single machine learning approach, such as the Bayesian neural network, feed forward, and long short-term memory neural networks, and two-stage bases constructed by the neural networks listed above in succession to support vector regression. The results show that the two-stage frameworks outperform the corresponding one-stage frameworks, except for the Bayesian neural network. The performance of the one-stage Bayesian neural network system is the best, and the demand of magnitude of the mean absolute percentage error calculated on the anticipated value by this framework agrees with those provided.

Using machine learning to analyze cryptocurrency values was proposed [21]; this method was presented to test the notion that the Bitcoin market's inefficiencies might be used to produce extraordinary gains. Between November 2015 and April 2018, they examined daily data for 1681 coins. They showed that established benchmarks are outperformed by simple trading techniques aided by cutting-edge machine learning methods. Their findings suggest that nontrivial but ultimately straightforward algorithmic processes can aid in forecasting the Bitcoin market's short-term evolution.

Predicting with deep learning in cryptocurrency elevated trades was suggested [22]. With 65,535 samples, they used a deep forward neural network (DFFNN) to analyze and forecast Bitcoin high-frequency price data. To see how standard numerical training procedures, such as the conjugate gradient with Powell-Beale resumes, the robust method, and the Levenberg-Marquardt method, affect the accuracy of DFFNN. The simulation findings demonstrate that the DFFNN trained with the Levenberg-Marquardt significantly outperforms the DFFNN trained with Powell-Beale restarts and the DFFNN trained with robust methods when using root mean squared errors (RMSEs). Furthermore, the robust method is quick, implying that it could be useful in online training and trading.

A review of huge developments and research gaps in blockchain-enabled IIoT networks supervised learning in blockchain-enabled IIoT networks was conducted [23]; though blockchain-enabled IIoT networks have the ability to meet the services and expectations of next-generation networks, the gap study offered in this paper identifies key areas that require development. The essay then proposes the use of reinforcement learning (RL) approaches to resolve some of the major difficulties of blockchain-enabled IIoT networks, for example, block time minimization and transaction throughput enhancement, based on these discoveries. A complete case study follows, in which a Q-learning methodology is utilized to reduce the transmission delays for a miner, hence limiting the occurrence of forking occurrences. Extensive simulations were carried out, and the results for the average transmission delay relating to forking events were obtained. The collected findings show that the Q-learning strategy beats the greedy policy while remaining relatively simple. Some future research directions are also documented in order to further create blockchain-enabled IIoT networks. While this research focuses on the use of RL approaches in blockchain-enabled IIoT networks, the insights and results presented could help accelerate the adoption of blockchain technology.

A targeted anomaly detection method based on machine learning for efficient blockchain transaction authentication was proposed [1]. Digitally signed transactions are used to store blockchain-related data in distributed ledgers. Its ledger data is saved once users with digital identities conduct a digital signing process. This procedure takes a long time and is not user-friendly. This paper presents a machine learning-based solution for automating the signing of blockchain transactions, as well as tailored anomaly detection. For testing the performance, an experiment and analysis were conducted using data from the Ethereum public main network, with positive findings paving the way for future integration of such a mechanism in specialized digital signing software for blockchain transactions.

The limitations of machine learning integration in blockchain-based smart technologies, and a way forward [24]. Blockchain technology (BT) has emerged as a one-of-a-kind, disruptive, and trending technology. Data security and privacy are prioritized in BT's decentralized database. It also has a consensus system that ensures data security and legitimacy. Nonetheless, it introduces new security concerns, such as majority attack and double-spending. Data analytics on blockchain-based secure data is required to address the aforementioned concerns. The value of emerging technology machine learning is highlighted through analytics on these data (ML). To create precise choices, machine learning uses a reasonable amount of data. In order to improve the accuracy of outcomes, data reliability and exchange are critical. When these two technologies (ML and BT) are combined, they can produce extremely exact results. We give a detailed analysis on machine learning adoption for making BT-based smart applications more resilient to attacks in this paper. Support vector machines (SVM), clustering, bagging, and deep learning (DL) algorithms, like convolutional neural network (CNN) and long short-term memory (LSTM), can all be used to analyze assaults on a blockchain-based network. They also discussed how knowledge could be utilized in various smart applications, including unmanned aerial vehicles (UAVs), smart grids, healthcare, and smart cities.

Security and privacy for blockchain technology in cyber-physical systems utilizing artificial intelligence – technologies, approaches, and issues – were suggested [25], Blockchain (BC) and cyber-physical system (CPS) applications are growing at a rapid pace. However, because of the intricacy involved, defining durable and correct smart contracts (SCs) for these smart applications is a difficult undertaking. Traditional industrial, technological, and business processes are being modernized by SC. It is self-executing, self verifiable, and fixed into the BC, obviating the necessity for trustworthy third-party systems and, as a result, saving both administration and service expenses. It also increases system efficiency while lowering security risks. SCs, on the other hand, are enthusiastic about Industry 4.0's new technology innovations, but there are still a number of security and privacy concerns to be solved. A study of SC security flaws in software encryption that might be readily hacked by a hostile operator or franchise the whole BC network is described in this study. According to the literature, the difficulties of SC safety and discretion are not well examined by investigators. According to the prevailing recommendations, constructing a complicated SC will not be able to address the privacy and security concerns. As a result, this study looks into several artificial intelligence (AI) methodologies and methods for protecting SC privacy. Then, for AI-based SC, open topics and problems are examined. Finally, a retail marketing case study was proposed, which employs AI and SC to maintain security and privacy.

Networking systems for blockchain and machine learning were suggested [26]. Infrastructural facilities, services, edge devices, and programs in communication and networking systems have become increasingly complex and varied as information and communication technologies advance at a breakneck pace. Furthermore, the vast amount of data and end devices could pose major privacy, security, service delivery, and network security issues. The combined evaluation of blockchain and machine learning with considerable profits has piqued the interest of both academics and industry to achieve distributed, encrypted, smart, and modern information system management and operation. On the one hand, blockchain can greatly facilitate the sharing of training data and machine learning models and decentralized intelligence, security, privacy, and trustworthy computational decision-making. On the other hand, machine learning will have a big impact on the advancement of blockchain in infrastructures and networking systems, especially in terms of energy and resource competence, scalability, safety, confidentiality, and intelligent smart agreements. However, numerous critical outstanding issues and problems, such as resource management, data processing, scalable operation, and security concerns, must be resolved before the widespread use of blockchain and machine learning. They offered a survey of existing works on the blockchain and machine learning technologies in this paper. We identify numerous key characteristics of combining blockchain and machine learning, including an overview, benefits, and applications. Then, we go over some of the open topics, obstacles, and broad views that must be solved in order to explore blockchain and machine learning for communications and networking systems together.

Integration of blockchain, IoT, and machine learning in industrial automation for multistage quality safety and access was carried out [27]. The demand for predictive equipment reliability and quality is driving the growth of smart manufacturing systems. To that goal, a variety of machine learning techniques are being investigated. Data security and management is another issue that is becoming increasingly significant in the industry. To address the issues raised above, we used a combination of blockchain and machine learning techniques to protect system transactions and handle a dataset to combat the false dataset. Big data approaches were employed to organize and evaluate the obtained dataset. The private Hyperledger Fabric platform was used to create the blockchain system. Consequently, the hybrid prediction methodology was used to assess the fault diagnosis prediction component. The system's worth control was assessed using nonlinear machine learning techniques, which represented the complex environment and determined the genuine positive rate of the quality control methodology used by the system.

Machine learning-based prediction of the Bitcoin prices in a short period was proposed [28]. They looked at the Bitcoin market's predictability over time horizons ranging from 1 to 60 min. They investigated a variety of machine learning models and discovered that, while all models outdo a random classifier, recurrent neural networks and gradient boosting classifiers are particularly well-suited for the prediction tasks under consideration. They employed a diverse feature set that included technical, blockchain-based, sentiment/interest-based, and asset-based elements. Their findings suggest that for most techniques, technical factors are the most important, followed by selected blockchain-based and sentiment/interest-based elements. They also discovered that predictability improves with longer prediction horizons. Although a quantile-based long-short trading strategy generates monthly profits of up to 39% before transaction costs, after transaction costs are factored in, the approach provides negative returns due to the extremely short holding periods.

Predictions of the Ethereum blockchain cryptocurrency value in a system of economic financing were suggested [29]. In the last decade, cryptocurrency has grown in popularity. Cryptocurrency's untraceable and uncontrollable nature appeals to millions of people all over the world. The goal of cryptocurrency research is to locate the ether and anticipate its price based on the coin's previous price inflations. Using a time series of daily ether cryptocurrency closing prices, price prediction is accomplished using two machine learning algorithms, namely, linear regression (LR) and support vector machine (SVM). By applying filters with varying weight coefficients, multiple window lengths are used in ether cryptocurrency price prediction. A cross-validation method is used in the training phase to build a high-performance model that is independent of the dataset. Two machine learning approaches are used to implement the suggested model. The SVM approach has greater accuracy (96.06%) than the LR approach when utilizing the proposed model (85.46%). Furthermore, by adding features to the SVM approach, the proposed model's accuracy score can be enhanced to 99%.

3 Materials and Methods

3.1 Blockchain, Cryptocurrency, and Bitcoin

Blockchain has sparked an influx of intrigue in both academia and industry. As a decentralized, immutable, shareable, and time-ordered ledger, it is a new technology. Transactions are saved in blocks with timestamps and references (i.e., the hash of the previous block) that are kept in a chain. Transactions, which indicate money transfers, are generated by anonymous individuals and competitively gathered by a worker to construct a new block in Bitcoin. The worker who creates a fresh and valid block can earn a reward, so the chain is continually lengthened by workers [30].

The blockchain systems, including Bitcoin and Ethereum, vary in various ways, including transaction mechanism and latency. The advantage of using this approach is that you can rest assured that the information you store is safe. Blockchain technology is the primary means of achieving this goal, as it distributes digital information lacking granting copy permission and administers the timestamp dataset in the network that connects the services and the system. The permissioned blockchain will be used to achieve this strategy [27]. In present networks, blockchain opens up new possibilities for coordinating several untrustworthy parties and enabling decentralized governance. The following are the main characteristics of blockchain: decentralization, transparency, immutability, security, suitability, anonymity, and autonomy [26]. Since growing blockchain size corresponds with more investors conducting Bitcoin transactions and Bitcoin market price is highly driven by consumer demand and interest, we selected blockchain size as a useful representative feature for our model.

Cryptocurrency is digital money that uses encryption to ensure that network transfers and exchanges are secure. As one of the most well-known cryptocurrencies, Bitcoin creates a distributed ledger that is preserved through a network. Decentralization is one of the most fundamental characteristics of Bitcoin, which distinguishes it from centralized, traditional currencies in that Bitcoin transactions and management are not governed by any central body. As a result, people solely trade between customers and vendors, rather than relying on a third-party trusted platform or other financial institutions. Bitcoin is also secure, thanks to cryptographic encryption technology. Even if some information is unnecessary, all transaction information is logged and cannot be changed [17].

Bitcoin (□) is a sort of electronic money that is based on digital currency. It is a distributed advanced exchange without a central bank or a solitary chairman that can be directed from one client to another across the Bitcoin network without the necessity for a middleman. System hubs use cryptography to verify exchanges, which are then stored in an openly disseminated record known as a blockchain. Bitcoin was created by an unidentified individual or cluster of persons under the name Satoshi Nakamoto and unconfined as open-source programming in 2009. Bitcoins are created as a reward for performing a task known as mining. They can be applied to a variety of financial structures, items, and organizations. According to research by the University of Cambridge, there were 2.9 to 5.8 million notable clients using a digital currency wallet in 2017, with the most of them using Bitcoin [31].

3.2 *Dataset*

In this investigation, one of the most popular blockchain currencies in existence, Bitcoin (BTC), was used with a sampling period from June 2015 to June 2021 (1447 data points) and was obtained from the Yahoo Finance current data for Bitcoin from the Coin Market Cap website [32]. The dataset includes the day-to-day price in US dollars, market capitalization, and the trading volume for 2194 cryptocurrencies, where market capitalization is the merchandise of price and circulating stock, with the volume number of coins traded every day. Figure 1 shows the Bitcoin current price data.

Date	Open	High	Low	Close*	Adj. close**	Volume
02-Jun-2021	36,628.57	37,906.16	36,036.41	37,906.16	37,906.16	32,08,28,98,944
01-Jun-2021	37,293.79	37,896.73	35,787.09	36,684.93	36,684.93	34,63,94,23,297
31-May-2021	35,658.59	37,468.25	34,241.95	37,332.86	37,332.86	39,00,98,47,639
30-May-2021	34,607.41	36,400.67	33,520.74	35,678.13	35,678.13	31,64,60,80,921
29-May-2021	35,684.16	37,234.50	33,693.93	34,616.07	34,616.07	45,23,10,13,335
28-May-2021	38,507.08	38,856.97	34,779.04	35,697.61	35,697.61	55,20,01,91,952

Fig. 1 Bitcoin current price data

3.3 Machine Learning, Independent Component Analysis (ICA), Firefly Algorithm, and Support Vector Machine Classifier

Machine Learning

In the last 15 years, machine learning has become one of the most rapidly increasing technologies. It has a wide variety of applications in industries, for example, computer vision, bioinformatics, business analytics, healthcare, finance, fraud detection, and trend prediction, to name a few. Machine learning is a technique that permits a computer to learn from large data samples and predict patterns in the data. In various study fields, machine learning algorithms are used to anticipate and classify test data in order to produce reliable results. Machine learning classifier models are becoming increasingly popular, and they have proven to be extremely useful in a variety of financial datasets [8, 33].

To aid with executive and business decisions, machine learning algorithms are employed to extract important dataset patterns. Because it can infer data associations that are frequently not readily visible by humans, machine learning is an effective technique for designing Bitcoin and other cryptocurrency trading techniques [34]. Machine learning, at its most basic level, is based on the definition of two fundamental components: input features and objective function. Dimensionality reduction techniques like independent component analysis and Firefly can drastically reduce the time complexity of the training phase of machine learning algorithms, lowering the strain on the algorithms [35]. The primary goal of this research is to see how dimensionality reduction techniques affect the performance of the commonly used support vector machine classifier with the Bitcoin blockchain price prediction from Yahoo Finance.

Independent Component Analysis (ICA) Algorithm

ICA is considered an extension of the principal component analysis (PCA) approach. On the other hand, ICA optimizes higher-order statistics like kurtosis, while PCA optimizes the data's covariance matrix, which represents second-order statistics. As a result, PCA identifies uncorrelated components, whereas ICA identifies independent components. As a result, when the higher-order correlations of mixture data are minimal or insignificant, PCA can recover separate sources. Fast-ICA, projection pursuit, and Infomax are only a few of the ICA algorithms. These algorithms' main purpose is to extract independent components by maximizing non-Gaussian, reducing mutual information, or employing the maximum likelihood (ML) estimation approach. However, ICA has a number of flaws, including over-complete and under-complete ICA [36].

Independent component analysis is a technique for recovering the original independent variables from observations that have been transformed linearly. The majority of ICA methods are formulated as an optimization of a contrast function that minimizes component cross-dependency [37]. The observable stochastic signal x is assumed to obey the model $x = As$, where s is the unidentified source signal, its components are independent of one another, and A is an unknown mixing matrix. The ICA's objective is to estimate the mixing matrix A and the source signal s solely by looking at x.

Firefly Algorithm

The Firefly algorithm uses flashing lights to simulate the mating system and data. The behavior of fireflies, binary fireflies, the artificial Firefly algorithm, and the Firefly algorithm are explored. There are about 2000 kinds of fireflies on the planet, and the great majority of them produce brief, pattern flames. The interest of accomplices in mating, such as correspondence, potential, and a component of mechanism, is the major cause of this debt.

Most fireflies can only be seen at a given distance, away from the bat, due to two factors: the brightness of a source at a specific distance conforms to the law of the opposing square, implying that the power of light with expansion decays somewhere $I \propto 1/r^2$. The assimilation of light, which is recognized surrounding it, is the next factor, which reduces the force as the separation rises.

The Firefly algorithm is a "nature-inspired" algorithm that is based on flies' behavior. Nature-inspired algorithms are widely employed in the machine learning process at various stages. The natural lights that fireflies emanate from their bodies aid them in attracting or finding other flying mates. It also aids them in catching prey and defending themselves against predators. Three main assumptions underpin the algorithm [38, 39]:

(i) The Firefly algorithm is a versatile capacity that promotes the population growth's advantages.
(ii) The Firefly algorithm can effectively deal with multi-model problems in a few steps by dividing the population and gradually limiting the vision of each leaflet to provide them subdivisions in the study region.

(iii) Setting random and enchantment limits for the FA throughout the stress cycle can boost the calculation's collection rate.

This Algorithm 1's core architecture is as follows:

Algorithm 1: Firefly Pseudocode

```
Input: X = Sample populations of Fireflies
        X Iterations = Samples of Optimizing the Iterations
        β0 = attraction criterion
        ∝ = criterion selection
Output: Best Fireflies Algorithm fit
        create a colony of x fireflies using the arbitrary firefly algorithm
        identify the optimal fitness-related option
        While stop parameters not met do
          For individual Firefly a do
          For individual Firefly b do
            If Firefly b is better than Firefly a then
              Move Firefly a to Firefly b
                Update Intensity of Firefly
              End
                Evaluate each Firefly best stand
              End
          End
        Loop
```

Support Vector Machine (SVM)

The SVM was created to solve complicated pattern recognition problems. For two-dimensional classification, SVMs were provided, and for high-dimensional difficulties, they were improved [40, 41]. SVM (support vector machine) is a discriminative classifier that creates a separating hyperplane. In the case of inseparable class data, an error tolerance budget is introduced to make the separating hyperplane robust. Kernel implementation transforms linear decision boundaries into more complicated boundary shapes (e.g., polynomial, Gaussian, and radial kernel). SVMs have gotten a lot of attention because of their remarkable classification performance.

3.4 *Evaluation Metrics for the Model and the System Configuration*

The performance of the suggested model using ICA-Firefly with SVM classification for blockchain Bitcoin price prediction was assessed using some assessment indicators. The most basic evaluation metric is accuracy, which is a percentage of

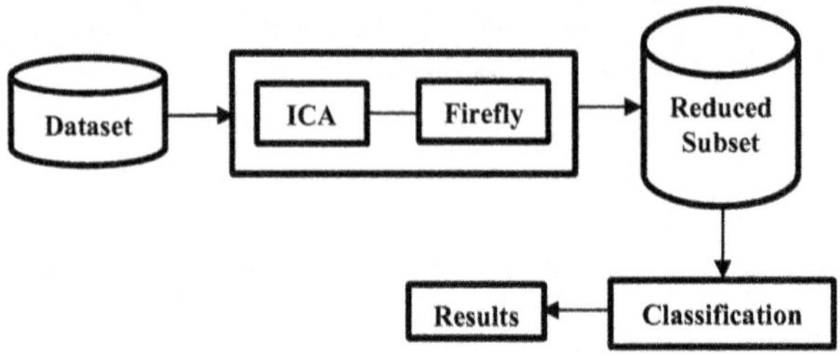

Fig. 2 Model workflow of the study

correctly identified messages, among additional metrics like sensitivity, specificity, recall, precision, and f-score [7, 42].

The Yahoo Finance repository provided access to the blockchain BTC dataset. As a result, the suggested research is carried out on the MATLAB environment. The MATLAB platform's significantly increased computational power and data handling capabilities aided in processing dataset inputs for the application of machine learning algorithms. The workflow model of the study is shown in Fig. 2.

4 Results and Discussions

An ICA-Firefly algorithm was used to generate a prediction for blockchain Bitcoin price prediction in this study. The datasets were divided into two portions in the proposed models: training and testing. The proposed models with six attributes were built using a total of 2193 sample data; Fig. 3 shows the dataset features. By comparing the corresponding values of the evaluation performance, the models' performances were determined. An adaptive search technique employing the ICA-Firefly algorithm is applied to overcome missing data in a dataset and select the most appropriate imputation value. The general result demonstrated that the model algorithm is an efficient methodology based on the simulation results from the datasets employed.

Figures 4 and 5 show confusion matrix of how the linear SVM and Sigmoid SVM classifiers performed on the blockchain Bitcoin price dataset when combined with the ICA-Firefly algorithms. Performance evaluations in terms of accuracy, sensitivity, and specificity, among other metrics, are the most commonly used metrics to analyze the performance of machine learning algorithms. The evaluation metrics of the proposed ICA-Firefly and SVM kernel classification models are compared in Table 1.

Table 1 summarizes the findings of this study's experimentation. SVM kernels were used further to investigate the performance of ICA-Firefly on the datasets. The Sigmoid SVM outperformed the L-SVM by around 97% accuracy. Several studies

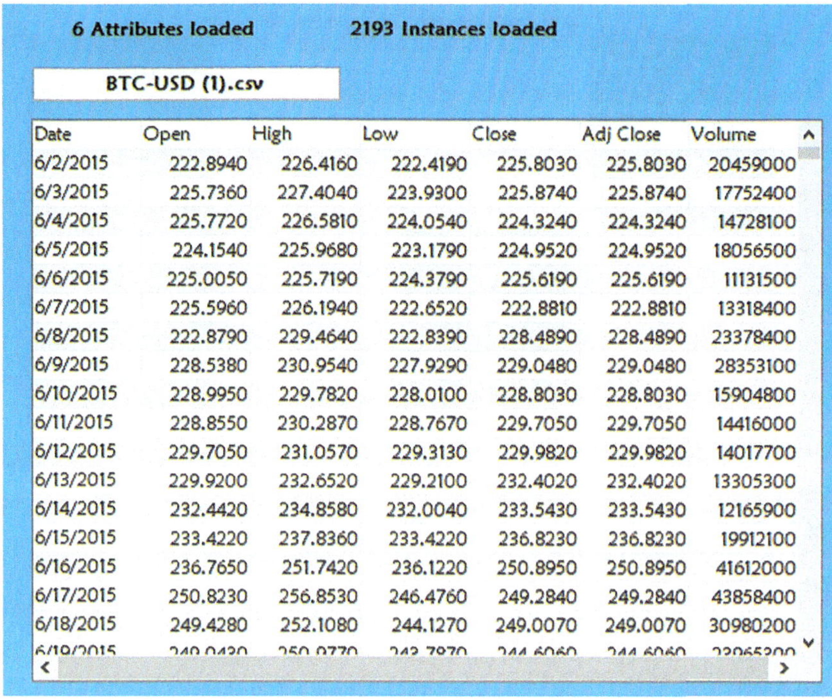

Date	Open	High	Low	Close	Adj Close	Volume
6/2/2015	222.8940	226.4160	222.4190	225.8030	225.8030	20459000
6/3/2015	225.7360	227.4040	223.9300	225.8740	225.8740	17752400
6/4/2015	225.7720	226.5810	224.0540	224.3240	224.3240	14728100
6/5/2015	224.1540	225.9680	223.1790	224.9520	224.9520	18056500
6/6/2015	225.0050	225.7190	224.3790	225.6190	225.6190	11131500
6/7/2015	225.5960	226.1940	222.6520	222.8810	222.8810	13318400
6/8/2015	222.8790	229.4640	222.8390	228.4890	228.4890	23378400
6/9/2015	228.5380	230.9540	227.9290	229.0480	229.0480	28353100
6/10/2015	228.9950	229.7820	228.0100	228.8030	228.8030	15904800
6/11/2015	228.8550	230.2870	228.7670	229.7050	229.7050	14416000
6/12/2015	229.7050	231.0570	229.3130	229.9820	229.9820	14017700
6/13/2015	229.9200	232.6520	229.2100	232.4020	232.4020	13305300
6/14/2015	232.4420	234.8580	232.0040	233.5430	233.5430	12165900
6/15/2015	233.4220	237.8360	233.4220	236.8230	236.8230	19912100
6/16/2015	236.7650	251.7420	236.1220	250.8950	250.8950	41612000
6/17/2015	250.8230	256.8530	246.4760	249.2840	249.2840	43858400
6/18/2015	249.4280	252.1080	244.1270	249.0070	249.0070	30980200

Fig. 3 The dataset features

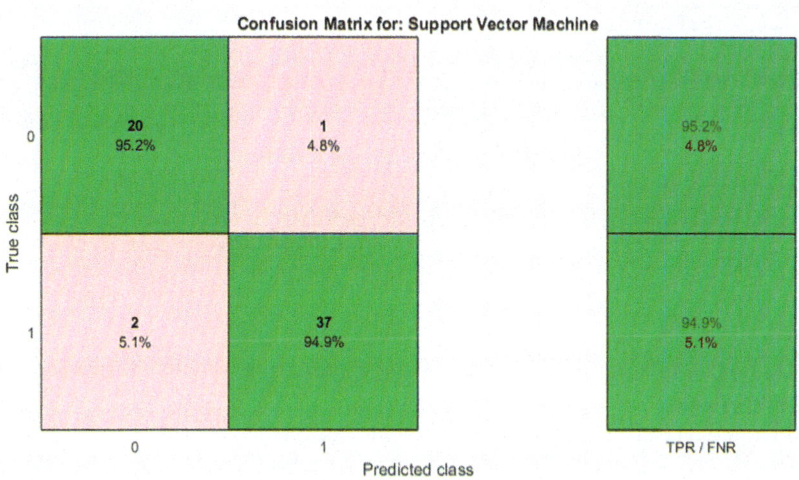

Fig. 4 ICA-Firefly with L-SVM classification TP = 37 TN = 20 FP = 2 FN = 1

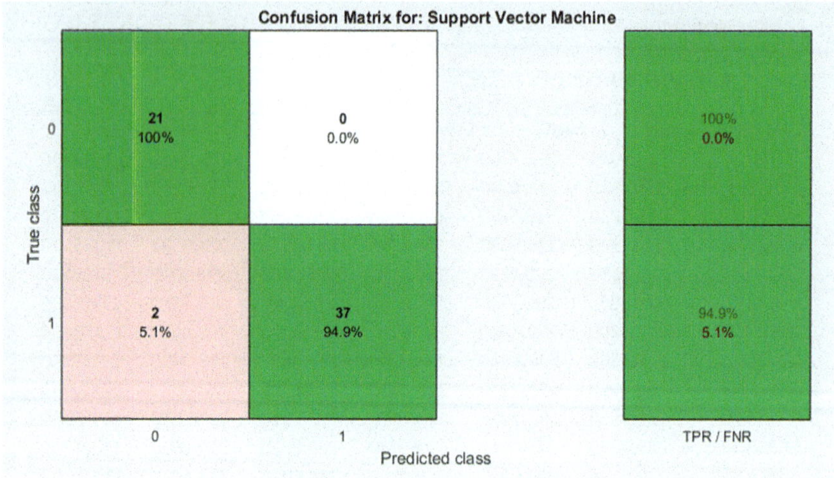

Fig. 5 ICA-Firefly with L-SVM classification TP = 37 TN = 21 FP = 2 FN = 0

Table 1 Performance evaluation

Performance metrics (%)	ICA-Firefly-L-SVM (%)	ICA-Firefly-Sigmiod-SVM (%)
Accuracy	95.00	96.67
Sensitivity	97.37	100
Specificity	90.91	91.30
Precision	94.87	94.87
F1-score	96.1	97.37
Matthews correlation coefficient	89.19	93.07

Table 2 Comparison with existing models

Authors	Work done	Accuracy results (%)
Mohammed et.al [43].	ANN, SANN, SVM, and LSTM	65
Yan and Dai [44].	CNN-LSTM	64
Sebastiao and Godinho [16]	Linear models, RF, and SVMs	80.17
Zhengyang et al. [17]	Artificial neural network (ANN) and the long short-term memory (LSTM)	90

have been conducted in literature; Table 2 compares the results of this study with existing models.

The effects of using ICA-Firefly for dimensionality reduction were measured in this study. The components of ICA-Firefly were utilized to forecast BTC price increases and decreases by capturing variances in the data. The categorization models, on the other hand, outperformed various other models shown in Table 2. For example, the accuracy of Sigmoid SVM was 97%, and L-SVM was 95%. The analysis revealed that Sigmoid-SVM is quite similar to the models in Table 2.

The increase and decrease of the price, as well as the factual price, are two components of the BTC pricing model. Internal and external criteria were used to classify the increase/decrease in BTC price in this study. BTC prices are stochastic, and no single set of characteristics can be used to predict the future. Nonetheless, researchers have had varying degrees of success estimating BTC values based on various feature sets. We have included characteristics that are directly related to the blockchain in this article. If numerous individuals are using it for transactions, then the relevant features such as dynamic addresses and number of transactions would be high.

5 Conclusion

BTC price estimates utilizing machine models were carried out in this study using BTC price estimates from 2015 to 2021. The best overall performance was achieved with ICA-Firefly with SVM kernels and ICA-Firefly algorithm with Sigmoid SVM. All of the models developed are satisfactory and perform well, with classification models recording up to 97% accuracy for next-day forecasting. The findings show that the provided models are unproductive and undependable cryptocurrency price forecasters, owing to the complexity of the problem, which even advanced deep learning techniques like LSTM and CNNs are unable to address efficiently. According to our findings and investigations into the cryptocurrency pricing problem, cryptocurrency prices follow a nearly random walk process, with a few underlying patterns that an intelligence framework must uncover in order for a prediction model to create accurate and dependable projections. As a result, novel algorithmic methodologies, alternative methodologies, and new validation measures should be investigated.

The findings reveal that while it is easy to estimate the real BTC price with very low error rates, forecasting its rise and decrease is far more difficult. The performance scores for the classification models offered are the top in the works. Having said that, more research on Bitcoin classification models is required. Hourly BTC prices and technical indicators, as well as decision tree models that mix several types of models for forecasting, could be used in future work. Artificial intelligence can be used to model the price of cryptocurrencies as a basis for calculating the risk issue associated with the financial use of blockchain technology.

References

1. Podgorelec, B., Turkanović, M., & Karakatič, S. (2019). A machine learning-based method for automated Blockchain transaction signing including personalized anomaly detection. *Sensors, 20*, 147. https://doi.org/10.3390/s20010147
2. Żbikowski, K. (2016). *Application of machine learning algorithms for bitcoin automated trading*. Presented at the (2016). https://doi.org/10.1007/978-3-319-30315-4_14.

3. Gulihar, P., & Gupta, B. B. (2019). A taxonomy of bitcoin security issues and defense mechanisms. In *Machine learning for computer and cyber security* (pp. 209–232). CRC Press. https://doi.org/10.1201/9780429504044-9. Taylor & Francis Group, [2019] | "A science publishers book".
4. Pabuçcu, H., Ongan, S., & Ongan, A. (2020). Forecasting the movements of bitcoin prices: An application of machine learning algorithms. *Quantitative Finance and Economics, 4*, 679–692. https://doi.org/10.3934/QFE.2020031
5. Li, X., & Wang, C. A. (2017). The technology and economic determinants of cryptocurrency exchange rates: The case of bitcoin. *Decision Support Systems, 95*, 49–60. https://doi.org/10.1016/j.dss.2016.12.001
6. Casino, F., Dasaklis, T. K., & Patsakis, C. (2019). A systematic literature review of blockchain-based applications: Current status, classification and open issues. *Telematics and Informatics, 36*, 55–81. https://doi.org/10.1016/j.tele.2018.11.006
7. Arowolo, M. O., Adebiyi, M. O., Ariyo, A. A., & Okesola, O. J. (2021). A genetic algorithm approach for predicting ribonucleic acid sequencing data classification using KNN and decision tree. *TELKOMNIKA (Telecommunication Computing Electronics and Control), 19*, 310. https://doi.org/10.12928/telkomnika.v19i1.16381
8. Alabi, K.O., Abdulsalam, S.O., Ogundokun, R.O., Arowolo, M.O. (2021). Credit risk prediction in commercial Bank using Chi-Square with SVM-RBF. Presented at the (2021). https://doi.org/10.1007/978-3-030-69143-1_13.
9. Jang, H., & Lee, J. (2018). An empirical study on modeling and prediction of bitcoin prices with Bayesian neural networks based on Blockchain information. *IEEE Access, 6*, 5427–5437. https://doi.org/10.1109/ACCESS.2017.2779181
10. Gupta, A., Nain, H. (2021). *Bitcoin price prediction using time series analysis and machine learning techniques*. Presented at the (2021). https://doi.org/10.1007/978-981-15-7106-0_54.
11. Ampountolas, A., Nyarko Nde, T., Date, P., & Constantinescu, C. (2021). A machine learning approach for micro-credit scoring. *Risks, 9*, 50. https://doi.org/10.3390/risks9030050
12. Fan, M.-H., Chen, M.-Y., & Liao, E.-C. (2021). A deep learning approach for financial market prediction: Utilization of Google trends and keywords. *Granular Computing, 6*, 207–216. https://doi.org/10.1007/s41066-019-00181-7
13. Gerlein, E. A., McGinnity, M., Belatreche, A., & Coleman, S. (2016). Evaluating machine learning classification for financial trading: An empirical approach. *Expert Systems with Applications, 54*, 193–207. https://doi.org/10.1016/j.eswa.2016.01.018
14. Huang, J., Chai, J., & Cho, S. (2020). Deep learning in finance and banking: A literature review and classification. *Frontiers of Business Research in China, 14*, 13. https://doi.org/10.1186/s11782-020-00082-6
15. Khalid Salman, M., & Abdu Ibrahim, A. (2020). Price prediction of different cryptocurrencies using technical trade indicators and machine learning. *IOP Conference Series: Materials Science and Engineering, 928*, 032007. https://doi.org/10.1088/1757-899X/928/3/032007
16. Sebastião, H., & Godinho, P. (2021). Forecasting and trading cryptocurrencies with machine learning under changing market conditions. *Financial Innovation, 7*, 3. https://doi.org/10.1186/s40854-020-00217-x
17. Zhengyang, W., Xingzhou, L., Jinjin, R., & Jiaqing, K. (2019). Prediction of cryptocurrency price dynamics with multiple machine learning techniques. In *Proceedings of the 2019 4th International Conference on Machine Learning Technologies – ICMLT 2019* (pp. 15–19). ACM Press. https://doi.org/10.1145/3340997.3341008
18. Koker, T. E., & Koutmos, D. (2020). Cryptocurrency trading using machine learning. *Journal of Risk and Financial Management, 13*, 178. https://doi.org/10.3390/jrfm13080178
19. Cho, H., Lee, K.-H., Kim, C. (2021). *Machine learning and cryptocurrency in the financial markets*. Presented at the (2021). https://doi.org/10.1007/978-981-33-6137-9_13.
20. Cocco, L., Tonelli, R., & Marchesi, M. (2021). Predictions of bitcoin prices through machine learning based frameworks. *Computer Science – PeerJ, 7*, e413. https://doi.org/10.7717/peerj-cs.413

21. Alessandretti, L., ElBahrawy, A., Aiello, L. M., & Baronchelli, A. (2018). Anticipating cryptocurrency prices using machine learning. *Complexity, 2018*, 1–16. https://doi.org/10.1155/2018/8983590
22. Lahmiri, S., & Bekiros, S. (2021). Deep learning forecasting in cryptocurrency high-frequency trading. *Cognitive Computation, 13*, 485–487. https://doi.org/10.1007/s12559-021-09841-w
23. Jameel, F., Javaid, U., Khan, W. U., Aman, M. N., Pervaiz, H., & Jäntti, R. (2020). Reinforcement learning in Blockchain-enabled IIoT networks: A survey of recent advances and open challenges. *Sustainability, 12*, 5161. https://doi.org/10.3390/su12125161
24. Tanwar, S., Bhatia, Q., Patel, P., Kumari, A., Singh, P. K., & Hong, W.-C. (2020). Machine learning adoption in Blockchain-based smart applications: The challenges, and a way forward. *IEEE Access, 8*, 474–488. https://doi.org/10.1109/ACCESS.2019.2961372
25. Gupta, R., Tanwar, S., Al-Turjman, F., Italiya, P., Nauman, A., & Kim, S. W. (2020). Smart contract privacy protection using AI in cyber-physical systems: Tools, techniques and challenges. *IEEE Access., 8*, 24746–24772. https://doi.org/10.1109/ACCESS.2020.2970576
26. Liu, Y., Yu, F. R., Li, X., Ji, H., & Leung, V. C. M. (2020). Blockchain and machine learning for communications and networking systems. *IEEE Communication Surveys and Tutorials, 22*, 1392–1431. https://doi.org/10.1109/COMST.2020.2975911
27. Shahbazi, Z., & Byun, Y.-C. (2021). Integration of Blockchain, IoT and machine learning for multistage quality control and enhancing security in smart manufacturing. *Sensors, 21*, 1467. https://doi.org/10.3390/s21041467
28. Jaquart, P., Dann, D., & Weinhardt, C. (2021). Short-term bitcoin market prediction via machine learning. *The Journal of Financial Data Science, 7*, 45–66. https://doi.org/10.1016/j.jfds.2021.03.001
29. Sharma, M. P., Bhardwaj, A. V. V., Sharma, V., Iqbal, A. P., & Kumar, R. (2020). Prediction of the price of Ethereum blockchain cryptocurrency in an industrial finance system. *Computers and Electrical Engineering, 81*, 106527. https://doi.org/10.1016/j.compeleceng.2019.106527
30. Weng, J., Weng, J., Zhang, J., Li, M., Zhang, Y., & Luo, W. (2019). DeepChain: Auditable and privacy-preserving deep learning with Blockchain-based incentive. *IEEE Transactions on Dependable and Secure Computing, 1–1*. https://doi.org/10.1109/TDSC.2019.2952332
31. Negar, M., Alireza, N., Masoud, R., & Yasser, Z. (2020). Bitcoin price prediction based on other cryptocurrencies using machine learning and time series analysis. *Iranian Journal of Science and Technology*. https://doi.org/10.24200/SCI.2020.55034.4040
32. Yahoo, F. *Bitcoin price prediction*, https://in.finance.yahoo.com/quote/BTC-USD/history?p=BTC-USD&guce_referrer=aHR0cHM6Ly9naXRodWIuY29tL3N1cmFqYmFoYWR1cjU5MS9CaXRjb2luLVByaWNlLVByZWRpY3Rpb24tVXNpbmctUk5OLS0tTFNUTQ&guce_referrer_sig=AQAAAHL6O0g7rnrvKBzzqIlA6B4fdHCk0Vp_mdcZPKO9V7W9d0d6cMkBLA1DxSSlcvb49Oq_m9nOh9y-Ay6s73nEk31hxma615IMw9okiq3sqeK9L9rUxlqvPMrhvOjARP7X96ckdYbjO0HzbsPe9VXZXn4YqkOZe-dkS-dia7qDMB-3&guccounter=2.
33. Reddy, G. T., Reddy, M. P. K., Lakshmanna, K., Kaluri, R., Rajput, D. S., Srivastava, G., & Baker, T. (2020). Analysis of dimensionality reduction techniques on big data. *IEEE Access, 8*, 54776–54788. https://doi.org/10.1109/ACCESS.2020.2980942
34. Awotunde, J.B., Ogundokun, R.O., Jimoh, R.G., Misra, S., Aro, T.O. (2021). *Machine learning algorithm for cryptocurrencies price prediction*. Presented at the (2021). https://doi.org/10.1007/978-3-030-72236-4_17.
35. Vaddi, L. (2020). Predicting crypto currency prices using machine learning and deep learning techniques. *International Journal of Advanced Trends in Computer Science and Engineering, 9*, 6603–6608. https://doi.org/10.30534/ijatcse/2020/351942020
36. Tharwat, A. (2021). Independent component analysis: An introduction. *Applied Computing and Informatics, 17*, 222–249. https://doi.org/10.1016/j.aci.2018.08.006
37. Maghrebi, H., Prouff, E. (2018). *On the use of independent component analysis to denoise side-channel measurements*. Presented at the (2018). https://doi.org/10.1007/978-3-319-89641-0_4.

38. Gadekallu, T. R., Khare, N., Bhattacharya, S., Singh, S., Maddikunta, P. K. R., Ra, I.-H., & Alazab, M. (2020). Early detection of diabetic retinopathy using PCA-firefly based deep learning model. *Electronics, 9*, 274. https://doi.org/10.3390/electronics9020274
39. Veysel, A., Ahmet, N., & T., Farah, Hatem, K., Bashar, Ahmed, K. (2020). Wrapper feature selection approach based on binary firefly algorithm for spam E-mail filtering. *Journal of Soft Computing and Data Mining, 1*, 44–52. https://doi.org/10.30880/jscdm.2020.01.02.005
40. Ferdiansyah, F., Negara, E. S., & Widyanti, Y. (2019). Bitcoin-USD trading using SVM to detect the current DAY'S trend in the market. *Journal of Information Systems and Informatics, 1*, 70–77. https://doi.org/10.33557/journal-isi.v1i1.7
41. Ali Alahmari, S. (2020). PREDICTING THE PRICE OF CRYPTOCURRENCY USING SUPPORT VECTOR REGRESSION METHODS. J. Mech. Contin. *The Mathematical Scientist, 15*. https://doi.org/10.26782/jmcms.2020.04.00023
42. Majeed, Y., Zhang, S., & Ren. (2021). A big data-driven framework for sustainable and smart additive manufacturing. *Robotics and Computer-Integrated Manufacturing, 67*, 102026. https://doi.org/10.1016/j.rcim.2020.102026
43. Mudassir, M., Bennbaia, S., Unal, D., & Hammoudeh, M. (2020). Time-series forecasting of bitcoin prices using high-dimensional features: A machine learning approach. *Neural Computing and Applications*. https://doi.org/10.1007/s00521-020-05129-6
44. Li, Y., & Dai, W. (2020). Bitcoin price forecasting method based on CNN-LSTM hybrid neural network model. *Journal of Engineering, 2020*, 344–347. https://doi.org/10.1049/joe.2019.1203

Blockchain-Based Framework for Secure Medical Information in Internet of Things System

Joseph Bamidele Awotunde, Sanjay Misra, Oluwafisayo Babatope Ayoade, Roseline Oluwaseun Ogundokun, and Moses Kazeem Abiodun

1 Introduction

Like other fields, healthcare system has benefited from the blockchain technology due to its built-in features like authentication, security, distributed ledger, and immutability. The blockchain have moved beyond cryptocurrency to practical application in other fields especially in healthcare system [1–2]. Due to severe regulatory restrictions, such as the Health Insurance Portability and Accountability Act of 1996 (HIPAA), in healthcare sector, the application of blockchain needs severe record sharing authentication and interoperability requirements. Researchers in academia and industry have begun to investigate solutions aimed toward healthcare use, based on existing blockchain technologies. Smart contracts, fraud detection, and identity verification are examples of these applications.

Blockchain stores information in decentralized recording ledgers that are distributed across all computer devices that are part of the blockchain architecture [3]. The blockchain works having both network users who take part in transactions and facilitate the transactions in a distributed ledger; thus, the infrastructure is peer-to-peer networks. Cryptographic techniques are employed by all miner, and the

J. B. Awotunde (✉) · O. B. Ayoade
Department of Computer Science, University of Ilorin, Ilorin, Nigeria
e-mail: awotunde.jb@unilorin.edu.ng; 15-68hg004.pg@students.unilorin.edu.ng

S. Misra
Department of Computer Science and Communication, Østfold University College (HIOF), Halden, Norway
e-mail: sanjay.misra@covenantuniversity.edu.ng

R. O. Ogundokun · M. K. Abiodun
Department of Computer Science, Landmark University, Omu Aran, Nigeria
e-mail: ogundodun.roseline@lmu.edu.ng; moses.abiodun@lmu.edu.ng

S. Misra, A. Kumar Tyagi (eds.), *Blockchain Applications in the Smart Era*, EAI/Springer Innovations in Communication and Computing,
https://doi.org/10.1007/978-3-030-89546-4_8

transaction is maintained in a decentralized set of nodes built by all these miners [4]. Furthermore, because it is built utilizing consensus methods, digital signatures, and hash chains, the blockchain ledger provides extremely reliable storage capabilities [5]. The services deliver by blockchain are in various ways like security, integrity, traceability, and non-repudiation using privacy-preserving manner while keeping all information in a public and in a decentralized manner [6].

In recent years, blockchain technology has demonstrated its tremendous adaptability as a range of healthcare sectors have found ways to incorporate its capabilities into their operations. Although much of the emphasis has been on the financial services sector so far, many projects are starting to shift in other service-related fields to included blockchain technologies [4]. Blockchain is a rapidly evolving technological innovation that has piqued the imagination of people all around the world. This technology enables computerized medical information transfer easier and safer than the traditional technique. It's a common knowledge that blockchain can make healthcare data more secure and accessible. For healthcare system, blockchain help in building a secure application because of the increased security and privacy provided in healthcare platforms. A decentralized database that is continuously kept up-to-date transactions information provides the healthcare sector with many benefits. When various parties require access to the same information, these benefits become particularly interesting. For instance, in the area of healthcare monitoring, access to healthcare records, transfers of vital documents like x-ray are vital areas in healthcare system where Blockchain technology can create additional level of security [4].

There has been a huge research breakthrough in finance and banking sectors unlike healthcare that has lately begun to gain significant interest in terms of blockchain-based applications [7–10]. Various researchers and medical scholars have highlighted the potential of blockchain application in the healthcare sector, and there is evidence that it can help to solve current security problems [7] [11–12]. The healthcare special security and privacy issues were able to be resolved using additional legal responsibilities in securing patients' medical information using blockchain technology. The risk of malicious attacks keeps on increasing in this era of Internet connectivities most especially as cloud storage and the proliferation of mobile health devices increase in sharing medical records and data. This has exposed the chance of private information during sharing to be compromised [13].

The sharing and privacy become an issue as smart devices are used to access health information, but there is no doubt these have help in reducing the number of patients that visit doctors. The healthcare sector is facing various challenges like data sharing, authentication, interoperability, and the transfer of medical information using mobile health applications [14]. Medical data from body sensors and other applications include patient physiological signs and symptoms and patient files and medical data. There is a need for proper security as medical records transition from paper to digital formats; role-based privileges must be implemented to preserve data and the security of healthcare records. There is a need to make sure that only authorized users are allowed to access medical data and records on the cloud databases, for example, and such access should be enforced and monitored. The query methods must be rigorous and must be audited regularly to reduce the danger of tampering or copying healthcare records, and rigorous access controls must be implemented [15–16].

Therefore, this chapter presented a blockchain-based framework for a secure healthcare system. When collaborating with smart healthcare systems, privacy and authenticity are crucial. The concept employed the blockchain distributed ledger to provide authenticity and endorsement while maintaining anonymity through approved management of consortia and anonymized accounts. The chapter is prearranged as follows: The IoT-based applications in healthcare system are discussed in Sect. 2. In Sect. 3, the chapter look at how blockchain can be used in the healthcare sector. Section 4 discusses various challenges of implementing blockchain technology in IoT-based system to secure the medical information. The architecture for a secure smart healthcare monitoring system using blockchain is presented in Sect. 5. Finally, Section 6 brings the chapter to a close by discussing the future work on implementing the framework.

2 Application of the Internet of Things in Healthcare System

Digital wellness advances offer significant incentives for reshaping existing healthcare programs. From the advent of automated therapeutic annals to portable medical equipment to other new technology, digital health advances have enhanced the quality of care at a lower cost. Politicians are constantly exploring, embracing, and adopting information and communication technologies as part of healthcare policies (ICT) [13]. It influences how individuals and patients see and communicate with eHealth system. The path to digital medical care (eHealth) is a systemic evolution of the traditional medical care system that includes a variety of features, such as universal access to automated medical records, online tracking systems, inmate services, wearable devices, portable medical apps, data analytics, and other transformative innovations [13–14].

Due to the global spread of the pandemic, it is critical to make an effective use of contemporary technologies. The Internet of Things (IoT) is widely recognized as one of the most revolutionary breakthroughs, with enormous potential for combating disease outbreaks [17]. The IoT consists of a sparse network, where the IoT systems feel the world and transmit valuable data across the network. The IoT-based system generates a massive amount of data known as big data, which influences the development and expansion of more personalized healthcare systems. Active surveillance capabilities in wearable medical devices can collect a large quantity of medical data, resulting in big data, from which clinicians can predict the patient's future state [18]. These observational study and information extraction are a dynamic process that necessitates improved security approaches [19]. The use of AI on generated big data from IoT-based systems opens up a number of possibilities for healthcare systems ([19]. The use of AI in the big data generation process has the potential to greatly improve global healthcare systems [15].

The Internet of Things-based system has been utilized to lower the worldwide cost of disease prevention. The IoT-based technology can help patients with self-administration therapies by capturing data in real time. In IoT-based sensor data collecting for telemedicine and mHealth systems, mobile app integration is a commonplace [20]. One of the important tasks in creating health fairness is to use IoT-based expertise to swap different sections of present medical services. Cloud and IoT-based systems meet consumer demand in a timely manner, take into account the patient's current state of health, improve contact between physicians and sick people, and reduce the time spent waiting for therapeutic care, all of which will increase client loyalty while also maximizing hospital performance. With the right telemedicine, a standardized standard might be achieved.

Wearable technology for the IoT-based system has opened up a new potential in the medical area, thanks to the new emerging technologies such as medical sensors for remotely monitoring patients. WBANs (wireless body area networks) are a type of IoT healthcare pattern. Various embedded and implanted technologies have lately been utilized to monitor the essential physiological parts of the human body, such as detecting heart rates and glucose levels in real time. Other devices and sensors, similar to an actuator's measurement, can provide automated care and therapy. The data report sent to a mobile phone functions as a storage device and sends the information to healthcare staff in real time, allowing them to respond quickly to users' demands. This remote monitoring eliminates the need for doctors' visits and allows patients to move around more freely in their daily life [21].

Remote patient monitoring is becoming more common; in 2016, 7.1 million out-patients in the United States outsourced their health care plan to remote monitoring, with that number predicted to rise to 50.2 million by 2030 [14, 22]. Furthermore, the US Centers for Medicare and Medicaid Programs (CMSP) announced the new payment incentives on January 1, 2018, to encourage the use of "active feedback loop" devices that provide real-time observation [23]. As the field of remote patient monitoring expands, there are worries about the accuracy and security of medical data transfer. To enable integrated health monitoring, measurements from numerous sensors must be aggregated, structured, and analyzed together. Because health data is a primary target for hackers, there is a need for government regulation to protect the transmission of personal health information (PHI). As a result, patient privacy must be protected, and electronic health records (EHRs) must be easily controllable and portable.

With pinpoint accuracy and eluate in the data collected, an IoT has the capacity to monitor specimens, equipment, people, supplies, and even service animals. To measure various vital signs, sensors can be fitted on the patient's body with various biometric data. This allows physician to provide better care to the patients, allowing problems to be diagnosed more quickly and resources to be used more efficiently. To detect body temperature and blood pressure from any patient, sensors can be placed in the patient's room in a hospital or home care setting. These sensors can also be used to detect the odor of vomit within an area in hospital or home care premises. The use of sensors and IoT devices can detect fast walking activity against the normal walk habit and the excessive cardiac training. This information could be useful in the diagnosis and treatment of the condition.

In today's healthcare, safety and violence are the major concerns. There have been numerous reports of horizontal violence, including nurse against nurse, as well as violence directed at healthcare providers or patients by visitors or family members. An IoT can be used to enforce a zero-tolerance policy as video surveillance systems are installed in healthcare facilities to perform these vital functions. Tracking the movements of employees, patients, and visitors, for example, could provide early alerts of unusual or threatening conduct. People visiting or residing in these situations could be monitored using biometric sensors to detect indicators of aggressiveness or stress. To charge patient account becomes easier with the use of barcode tags or low-cost RFID tags by a pharmacy. This helps to tag for scanning for an acute or long-term care setting for their various supplies. The IoT-based system can also be used to track and check such supplies from a repository or administered to a patient. An item could be located more rapidly in some circumstances where an RFID tag is utilized. For instance, goods like bandages, catheters of various types, and personal care items, are likely to be trackable. Medical products could be labeled with RFID tags in a home environment to track usage and warn the home care team when an item is being overused or the supply is running short. Many more IoT healthcare applications, according to researchers and practitioners, might significantly improve patient care, maximize resource usage, and save large sums of money if only the systems could be developed.

Recent technological advancements have drastically altered people's perceptions of how they should go about their daily lives. In the real world, the IoT has to be a growing trend in various industries, including healthcare. This rapid IoT revolution, however, has raised several questions and worries regarding the security of data held in various linked devices. It gets more difficult to ensure comprehensive data protection and privacy when the number of items, such as sensors and computers, grows. These security and privacy issues are the result of a decrease in the efficacy of IoT-based healthcare systems, which has a negative impact on individual's sensitive health information. Because healthcare data is so valuable and sensitive, the IoT healthcare paradigm's security and privacy protections exacerbate the situation. While growing IoT paradigms in the medical system help to develop the present healthcare systems, end users must face a number of privacy and security concerns. End users may be vulnerable to malicious threats if they grant authorization for potentially insecure or leaky third-party applications. Because the data is sent to the cloud, it travels across insecure communication networks, many of which are vulnerable to attack [19]. Furthermore, when data is uploaded to the owner's cloud storage facility, there are additional data security concerns.

However, the sheer number of connected devices (Fig. 1) and the massive amounts of sensory statistics generated by those devices have created new issues in terms of information security and confidentiality. Cyberattacks have evolved in tandem with the rapid development of IoT, resulting in a new channel of intrusion and risk for the whole medical business. Many research investigated IoT's multiple privacy and security vulnerabilities, as well as device flaws in

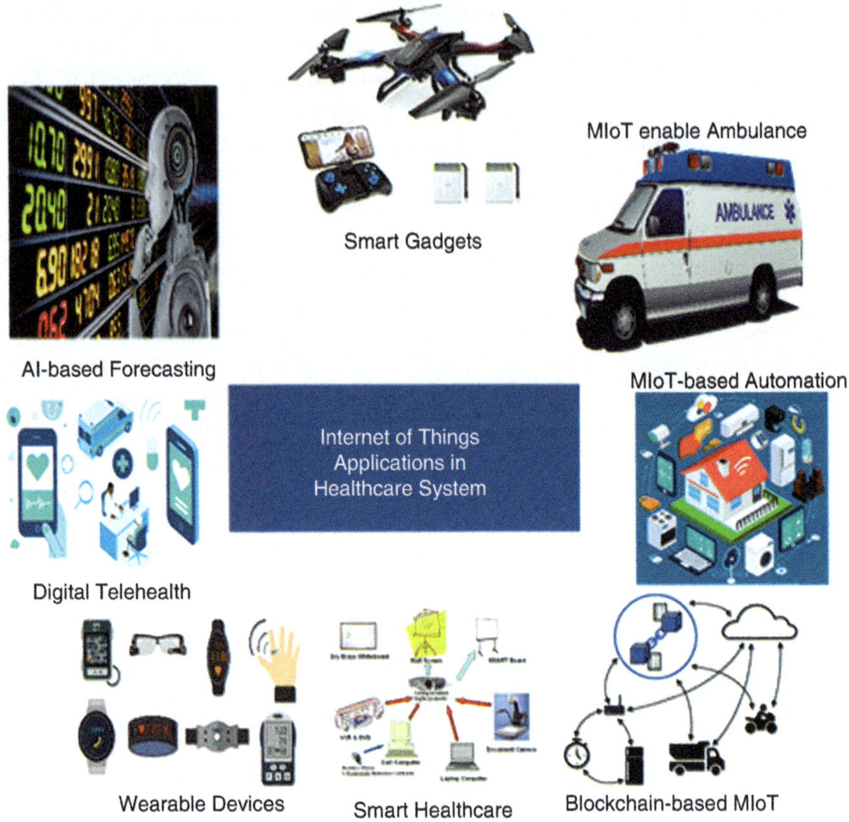

Fig. 1 Applications of the Internet of Things in healthcare system

cloud and fog computing settings pertinent to the IoT-based medical management gadgets [24–25]. The security and confidentiality of patient records are two critical considerations. When we talk about record safety, we mean that records are stored and communicated in a secure manner to preserve their absoluteness, genuineness, and legitimacy. The term "record confidentiality" refers to the fact that records can only be accessed and used by those who are allowed to see and use them [25–27]. An aggregate signature-based trust routing for data gathering in sensor networks can be used to create security and communication networks. Security and Communication Networks. With distinct objectives and specifications in mind, more reasonable security methods may be devised. The widespread use of IoT devices provides better guarantee of an individual's health [28], but it also creates a high demand on record security and concealment.

3 Applications of Blockchain in Internet of Things

The various aspects of healthcare systems can be improved using blockchain technology, and the security and privacy of patients with their well-being can be expanded and has the potential to improve various aspects of healthcare and well-being. Device tracking, clinical trials, pharmaceutical tracing, and health insurance are just a few examples. Hospitals can track their assets on a blockchain infrastructure, including throughout the device's whole life cycle. The information gathered can then be used to improve patient safety and do post-market analysis to save money. Current research has focused on pharmaceutical traceability, data interchange, clinical trials, and device tracking. With its immutability, fraud prevention, and ability to transport data between firms without requiring trust, healthcare is primed for revolution.

Bell et al. (2018) [29] improved the device identification and tracking that is a critical problem in healthcare sectors tagging medical equipment with a usable ID. For instance, device tracking can be used to disclose the cause of problem when a device malfunctions, thus saving cost and unnecessary repurchasing of lost items. A strong trust infrastructure based on the identification of medical equipment is expected to decrease these risks. With hospitals due to security and privacy concerns, just 20% to 30% of medical equipment are connected, according to the survey. Blockchain can assist the pharmaceutical business in overcoming the rising risks of counterfeit and unapproved pharmaceuticals. Smart contacts for pharmaceuticals can be formed with integrated GPS and chain-of-custody logging and then identified, just like a device tracking.

Blockchain can be used within clinical trials to address issues such as falsified results and data removal by researchers that contradict the funding source's objective. Clinical studies will be more reliable as a result of this. It also enables for the creation of an irreversible log of trial subject consent. Almost $200 billion was saved using a chain of custody in the supply chain in the pharmaceutical sector [29]. A trustworthy record of events around the patient journey would be beneficial to several sectors of health insurance like improved incident reporting and automated underwriting operations. Contracts could also be carefully written and then executed, such as automated payments for stages of the patient journey.

In healthcare services, information security, privacy, completeness, and access must be considered very seriously. Another area that the huge healthcare business might look into is the increased necessity of healthcare cost control. With the promise of blockchain mixed with IoT-based application layers built atop, healthcare services have enjoyed ultimate security and privacy, ensuring that applicable users may access a continuous record of information. By eliminating the third-party brokers' involvement in any financial transactions, blockchain has enhanced stakeholders' access to medical information and lowered costs, potentially lowering healthcare expenses and providing better results [30]. The researchers are interested in using blockchain technology to solve real-world issues, such as healthcare diagnostic and monitoring systems, centralizing research data, lowering healthcare overhead costs, and organizing patient data from massive input big data. The abovementioned examples of blockchain technology deployment in healthcare systems touch on near-term potential and obstacles [31].

Blockchain technology has been utilized for money exchange transactions to eliminate the requirement for a trustworthy third party to validate and notarize transactions, as well as to protect data confidentiality and privacy throughout those transactions. The innovation has been restructured to meet the needs of various industries, such as healthcare, education, transportation, electricity, and tourism. Over the next decade, healthcare systems based on the IoT are expected to generate trillions of dollars [32]. More importantly, smart healthcare has resulted in a significant decrease in death rates and healthcare costs, as well as enhancing the quality of the healthcare system and reducing emergency room visits and hospital stays [33].

Medical records are saved in a cloud database that is weighty to allow knowledge sharing and quick access among many healthcare stakeholders [34]. Cloud storage also offers security and privacy features, which are bolstered by data longevity. There is no interoperability between different healthcare providers and treatments in cloud storage. Furthermore, there is no way to confirm the data's quality or veracity. Blockchains play an important role in improving the trustworthiness, accuracy, and validity of medical data that is stored and exchanged. By monitoring and ensuring allowed access to personal medical information, blockchains ensure the security of sensitive data [35]. Blockchains operate as a distributed database to protect medical data from modification [36–37]. To addresses the safety concerns in IoT-based systems, blockchains used a distributed trust mechanism to distribute patient records on the cloud storage database that could be handled by various users and advisors like caregivers, physicians, clinic experts, pharmacies, patients themselves, and insurance providers.

Blockchains rely on hashing and public cryptography techniques to preserve confidentiality, integrity, and accessibility of past transactions relating to the records of scattered patients. This prevents unauthorized users from destroying, falsifying, or accessing the papers. Patient records in blockchains can only be appended to the database, not deleted. Cryptographic hashing allows new data to be securely linked to a previous record. The majority of miners in the network must agree before records may be added to the blockchain. Miners are a group of special nodes that work together to verify new transactions added to a blockchain. Miners compete to solve a difficult mathematical task known as proof of work (POW), which takes an average of 10 minutes to add a record to a blockchain. This will help ensure that no single entity is able to alter or tamper with checked records. Furthermore, caregivers will be able to supply patients with encrypted alias focused on personalized health advice without having to reveal their names, thanks to blockchains.

Blockchain technology is still in its infancy and, particularly in the healthcare industry, must be linked with existing policies and processes. The National Research Council of Canada's Industrial Research Assistance Program (NRC-IRAP) has used the blockchain and its associated immutability, clarity, and distribution to coordinate and disseminate public knowledge about its operations and companies, recognizing that operating within government restrictions is a significant challenge in and of itself [38]. The success of the effort demonstrates that public blockchain may be used to protect government data, resolve administrative issues, and pave the way for more complicated data integration, particularly in smart healthcare [39]. The projects' enormous success creates a productive approach to record important data, exchange valuable data, and serve as a crucial building stone for future, more sensitive initiatives.

The distributed database management system (DDBMS) is technically centralized (i.e., users believe a centralized database is running, but the underlying machines can be physically distributed), whereas blockchain is a peer-to-peer, decentralized database management system (i.e., each node runs independently while adhering to the protocols) [40]. As a result, blockchain is excellent for applications in which biomedical/healthcare stakeholders (e.g., hospitals, suppliers, patients, and payers) may communicate with one another without relying on a central management middleman [41–42].

IoT-based devices must be securely logged using orders issued to actuator nodes, in addition to maintaining the integrity of patients and maintaining an accurate timeline of occurrences, as both records and treatment for a patient must be approved by medical specialists [43–44]. When it comes to wearing medical devices, this solution would provide patients with piece of mind by offering an immutable ledger and automatic health incident updates in a secure manner. Medical experts receive real-time information on their patients, furthering the practice of precision medicine. Smart contacts aid in the automation of health alarms from multiple devices into a centralized cloud storage location, resulting in a game-changing solution that allows healthcare practitioners to easily implement new medical technologies.

By placing data in the hands of individuals, blockchain has the potential to change healthcare. Patients and physicians can access an immutable log of medical records using MedRec is one particularly interesting step in this direction [45–46]. In exchange for maintaining the network, miners are compensated with anonymized healthcare data, which is a novel technique of incentivizing miners. MedRec maps patient-provider relationships (PPRs) using smart contracts, in which the contract displays a list of references indicating the relationships between nodes on the blockchain [29]. It also gives patients control over PPRs, allowing them to accept, reject, or change partnerships with healthcare providers like hospitals, insurers, and clinics.

By generating a decentralized ledger of acknowledged fact in medical records that is available to all healthcare practitioners, blockchain enables interoperability in healthcare systems [42, 47]. This means that, while user interfaces may vary, all providers' basic ledgers will remain the same. The current state of health records across providers, which contain large amounts of the same data under different IDs that may or may not be linked, is a roadblock. As the blockchain expands in size, this produces duplication, and performance degrades as a result. Deduplication would be required to maintain a reasonably performant system with unique, anonymized identities to identify patients across all services [48]. Implementing a distributed ledger medical record is a practical difficulty in and of itself, but it's important to note that health data would not be created from the ground up as they'd have to replace the old infrastructure, which raises challenges [49–51].

Another option is drug monitoring on the blockchain, which takes advantage of the data integrity of nodes that are connected for tracking and chain of custody from the maker to patient. Discover, a chain of custody model that shows where a medicine was created, is being developed by Chronicled, a technological business. It has been proved that leveraging on blockchain's error-handling capabilities can prevent pharmaceutical fraud during distribution of drugs to various clients and patients [52]. This allows hospitals to meet current medical criteria in terms of pharmaceutical sustainable development, with a focus on provider connectivity. The Counterfeit Medicines Project was recently formed by Hyperledger [53], to combat the problem of illegal drugs; the Open-Source Blockchain Working Group was also formed. Blockchain can be used to track down the origins of counterfeit pharmaceuticals and eliminate them from the supply chain. The inherent democratization of faith and legitimacy in the technology's principles gives blockchain an advantage over traditional techniques in drug monitoring. While central authority can be influenced or faked, influencing a distributed ledger unanimity is significantly more challenging.

4 The Challenges of Using Blockchain in the Internet of Things in Healthcare Systems

Medical information, as well as medical information such as clinical information, can be obtained using body sensors and other applications. Additional security and participation credentials must be established as health information shifts from traditional to digital versions in order to maintain data and the confidentiality of health information. Only authorized individuals should be allowed to access healthcare records housed in databases, for example, and such access should be enforced and monitored. To decrease the risk of interfering with or duplicating hospital documents, as well as requests to get those records, the query must be audited and rigorous access controls must be implemented [16].

Confidentiality of patient history (e.g., electronic patient records (EHR, EMR) and personal health record (PHR)) can also be difficult if conventional cryptographic standards are utilized in multiple platforms [54–56]. Current methods for protecting and securing records have proven ineffective, and the public disclosure of a patient's medical information might have real-world ramifications (for instance, challenges to clients' anonymity in the form of hostile assaults, which can impact the status and financially linked with those records) [57–59].

Some of the privacy concerns that connected health solutions confront include identification confidentiality, identity management, enquiry privacy, trace privacy, and proprietor privacy [60–65]. Third-party cloud providers face a variety of

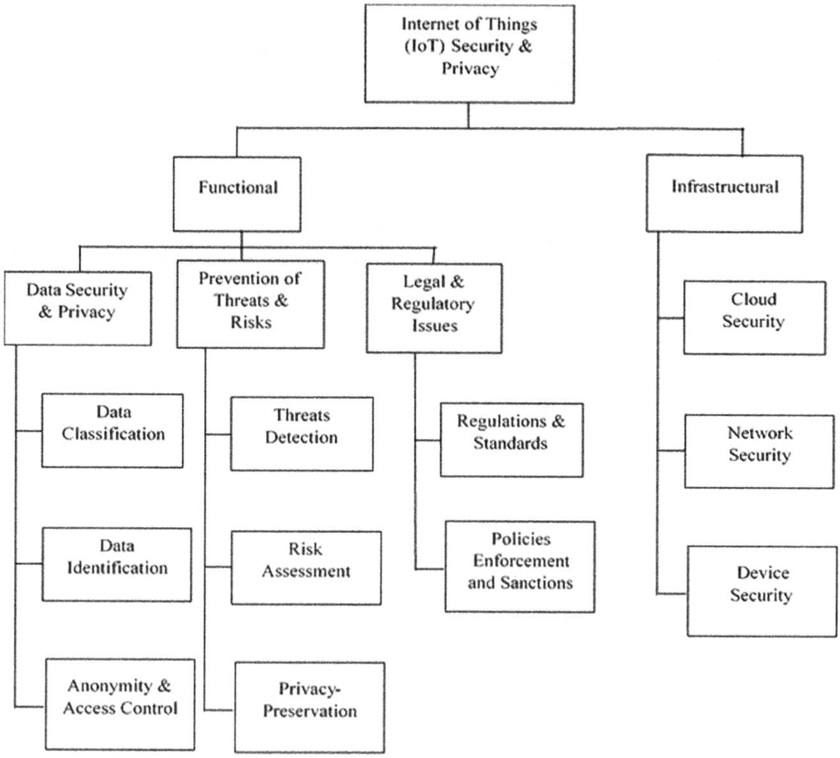

Fig. 2 The security and privacy in smart healthcare system

privacy issues when it comes to sharing medical information between various medical institutions [61]. One of these privacy issues is unauthorized access of medical information and patient data that is used and handled by third-party service providers [54, 61]. In addition to the privacy considerations of access control, IoT technologies also pose a risk of inference assaults [66]. Malicious actors utilize a mixture of wireless interception techniques and data mining to infer the value of a particular communication or signal, which is known as an inference attack [66]. The inferred information can then be utilized to further breach the account by using a phishing attack to get beyond authentication obstacles [66]. Cyber criminals can use data tampering, deception, spying, and material replay to target wireless devices in both active and passive ways [67]. A graphical demonstration of IoT security and privacy issues in smart healthcare system is shown in Fig. 2.

The use of blockchain technologies in healthcare is only getting started, and there are a lot of roadblocks to overcome and huge decisions to make in the future. In light of the issues that we faced a decade ago, our social understanding of privacy has evolved, and blockchain technology has the potential to uphold these boundaries if accepted. If deciding whether or not to use blockchain-based solutions, the trade-off between the risk of data loss and the ability to control one's data (assuming no big data leaks) should be examined. The new repositories created by cloud computing have given birth to big data, which can then be analyzed by AI to create a personalized healthcare plan that doctors and policymakers can use.

If low-quality and wrong data is published on the blockchain, the blockchain will remain fair to its users; the chain will remain with low-quality and inaccurate information because immutability and decentralization can be trusted [68–69]. There are a number of new options for blockchain and supporting technologies, but attention must be paid to the implementation process as well as what information has been gained [70–71]. Vigilance about the information being processed is one of the interoperability options, as is providing responses to solve inconsistencies and distribute confidence to diverse forms of information. With the arrival of quantum computing and its predicted ability to overcome current encryption mechanisms, there is a new restricted possibility in blockchain technology [72]. Although it is unclear when this will occur, it appears to be within the next decade. If quantum computing's resistance to encryption is not resolved by then, we will face a number of problems, because storing all health data on publicly accessible servers on blockchain puts the data at danger.

In blockchain, a key with a certain sequence of characters is the ability to access data. However, if a key is lost, the information accessed by it becomes irretrievable. Then, it becomes unfair since consumers lose access to a lifetime's worth of health records simply because one of these keys is missing. Then, in order to reconnect users with their data, new approaches or techniques must be established. With existing solutions creating back doors to accessing the blockchain's private data, these methods will now be substituting one question with another. Another issue with blockchain technology is that if the decentralization of a blockchain is disturbed, one agent will become the only consensus agent and would be able to change the blockchain keys, which is in violation of the virtue of immutability. To guard against this possibility, new consensus mechanisms and government oversight of blockchain monopolization may be necessary [73].

The goal of blockchain technology is to allow for efficient information sharing with stakeholders while guaranteeing data confidentiality and patient privacy. This will motivate and empower individuals all over the world to make healthy choices in order to improve their health. With the blockchain model, the world's data is being protected more than ever before. Beyond the hoopla, skeptics are concerned about the complexities, and many established and invested parties are likely to oppose the shift, not to mention legal, regulatory, and technological aspects that have yet to be determined.

If the problems of standardization are continually overcome, reliable privacy established anonymization mechanisms constructed, and consensus gained on the kind of contracts required to manage information, a new era of healthcare may be on the horizon. These are significant obstacles, but as previously said, corporations have already made significant progress toward overcoming them. The use of artificial intelligence to learn from data has already demonstrated that the technology is prepared to provide revolutionary new insights with the massive data created by the healthcare system, with privacy and patient control as a fundamental premise. The sectors are moving toward a disruptive event known as the health singularity, in which personalized healthcare is provided based on a comprehensive understanding of each individual's biology.

Other significant hurdles of implementing blockchain in the healthcare field are transparency and confidentiality. Increased openness and reduced secrecy, such as open data transparency during the transition, are commonly considered blockchain limitations because "everyone can see everything on a blockchain network" [74–75]. Also, even though a user is "anonymized" by using hash values as addresses, the user can still be identified by reviewing and analyzing publicly accessible transaction information on the blockchain. Because patient-related information (protected health information) is crucial, this issue is significant for healthcare applications.

One of the most distinguishing characteristics of blockchain systems is their immutability, and decentralized storage, which allows users to transfer data across several applications without relying on a centralized service provider [76]. A fundamental disadvantage of hierarchical structures is the prospect for privacy leakage from the public ledger that is propagated across the blockchain system. When a user receives his or her data, he or she is required to submit a private key in order to verify and decode the information from cypher text to plain text, potentially exposing personal information. Because the data is not stored locally, as it would be in a centralized database, the public key must be present on the network when the verification and decryption process begin. Due to the stringent requirements of the healthcare industry, this is a concern.

Because blockchain technology is still undeveloped and restively a new technology, there is no standardization, which impedes adoption and slows development [77]. Blockchain technology is being considered by many countries for use in government contexts, such as voting [78–79]. Countries like Estonia are seeking to achieve e-residency by combining residence rules with blockchain technology. This is the process of setting up an online account to verify a citizen's citizenship in a certain state and enable them to vote using that account [7]. To support all of these varied datacenters, there must be a high level of standardization across the numerous parties involved. The issue of standardization and regulations will become increasingly more crucial as more governments use blockchain as a solution [80, 82].

5 Blockchain-Based Framework for Secure Medical Information in Internet of Things System

The crust of the entire framework is the combination of the detection approach in the behavior of the patient's health data using the IoT, blockchain, and machine learning (ML). The shown methodology is essentially a system that requires the

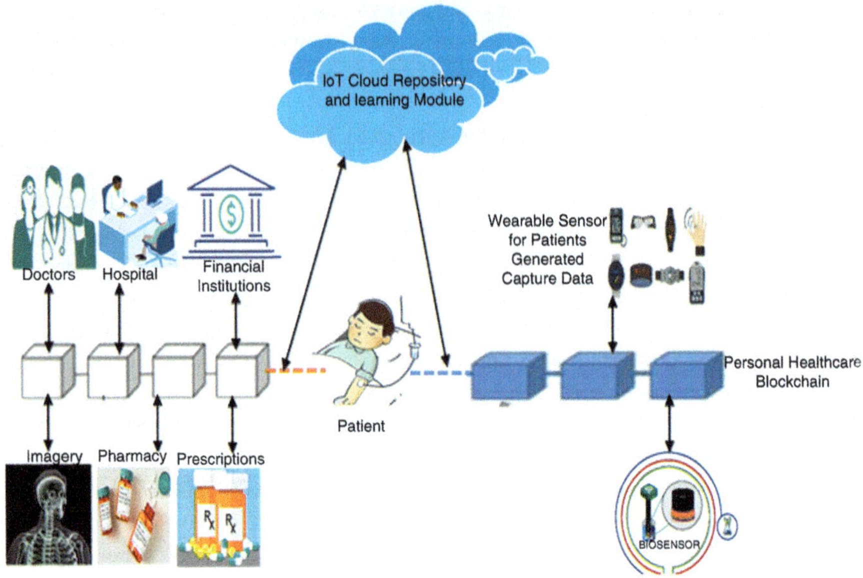

Fig. 3 The proposed blockchain-based Internet of Things for healthcare system

usage of an IoT module to intercept and retrieve data generated by the patient's wearable gadgets. The blockchain system presented is ideally suited for storing and keeping patient data in the form of multiple transactions, as well as providing access control to various stakeholders [81]. Furthermore, the blockchain framework is utilized to support medical research by maintaining the pseudo-anonymity of the patient's identity while yet providing permitted and reliable data for more accurate research. The ML model is mostly utilized for the detection of anomalies and the forecasting of future scenarios by evaluating data based on parameters provided by doctors for the basic diagnosis of the diseases that patients encounter (Fig. 3).

As a result, an effective IoT development must place a high priority on security and secrecy. Despite the fact that most healthcare organizations do not allocate sufficient funding to protect safety and secrecy, there is no doubt that safety and confidentiality play an important role in the IoT. IoT devices generate an increasing number of increasingly complicated real-time records, which is exceedingly delicate. On the one hand, the collapse of health organizations or system security could be disastrous. On the other hand, all levels of record processing, record transfer, cloud storage, and record republication have access to the patient's personal information. The framework was made up of three components, each of which has a distinct role to play:

IoT-Based Wearable Devices: These tools are used in real time to capture the symptoms of patients and monitor their status. These devices are made up of a number of sensors that detect the patient's vitality and atmosphere (like temperature, blood pressure, pulse rate, heart rate, humidity, ECG, etc.). When these criteria are violated, physicians and clinicians are notified in real time (from the permitted limits). Short message services are utilized to report any cases to the appropriate physicians, and the messages are delivered via smart devices. In a more obvious sense, if a patient is getting some moderate therapy and is being followed up with some medical tests, a wearable sensor is an excellent approach to track the data generated by the patient at every second. Heart rate, calorie release, breath strengthening, and sleep stage monitoring are examples of data that might be considered based on the wearable worn by the patient. If blood pressure sensors are employed, or if pacemakers are installed in the patient, such data can also be accessed remotely via the IoT application module. Now, if the patient is bedridden or confined to the hospital, there is a huge demand for IoT sensors or biosensors that can recognize environmental conditions and take appropriate actions.

Blockchain Transaction and Access Management: The storage of the massive amount of data created by the patient must be managed and processed while adhering to a secure methodology. Furthermore, when there are several stakeholders involved with the data being generated, a vital module called an access management system must be created, which the blockchain network addresses. We've outlined the use of two critical blockchain networks in the suggested architecture: The personal healthcare (PHC) blockchain and the external record management (ERM) blockchain. The patient typically maintains the personal healthcare blockchain since it perceives and gathers data via personal wearable devices. The doctor will be given access to the data, which will be used for proper medicine and comprehension of the disease that the patient is suffering from. The data created by the wearable devices is then kept in a third-party cloud database that is governed by the blockchain network. Immutable storage blocks are used to hold transactional data. Only authorized users have access to the information. We can use blockchain technology to create privacy-preserving and fundamentally secure data exchange networks that allow participating agencies to readily access archived and real-time patient data using smart contracts that eliminate the need for data reconciliation completely. In a typical blockchain, there is no single administrator; therefore, it is a distributed system of control and access with some level of interest in each member, and everyone has equal rights and power.

Machine Learning (ML) Layer: The ML layer examines the data generated by the patient to look for anomalies. Anomaly detection may be greatly improved by using the model to extract abnormalities from the data being generated. When an abnormality is discovered, a notification is sent to the doctor, who can then take appropriate action based on the situation. The suggested system employs two-level blockchain technology. Internal healthcare agencies, such as service providers, physicians, inventory, and other internal stakeholders, employ a private blockchain. A public blockchain is utilized to communicate with other entities, such as patients, pharmacies, insurance providers, and so on. The usage of a two-level blockchain implementation allows for separation of distinct entities, resulting in a safe, privacy-preserving, consistent, and transparent workflow.

To discover responses to security breaches or system coercions, the blockchain layer processed data collected from overall terminal status data as well as network traffic. This was done to discover various attack circumstances to device trends in real time and set up safeguards against them. This can be done by combining the incursion activity pattern with an access control strategy based on the IoT-based environment's acquired protection status data. The analysis tool searches for events or trends that may indicate that a device is vulnerable to security attacks. At this stage, malicious conduct analysis and rule-based analysis are carried out.

6 Results and Discussion

The implementation of the proposed framework was executed using Core i5 processor system with 8 GB RAM, running on Windows 8 with a 64-bit operating system. NetBeans 8.2, JDK 1.8, Tomcat 8.0.15, Jelastic cloud platform, and MySql 5.7 were used for the development of the framework. The proposed technique is compared to the closest traditional approaches using the Yahoo! Cloud Serving Benchmark (YCSB) and small bank datasets.

The proposed system was evaluated and expresses using average delay, success rate, and system execution time for the proposed mechanism. In comparison to conventional methods, it was discovered that using the proposed strategies reduces average latency and SET (system execution time) and enhances SR (success rate). How can we attain privacy efficiency while maintaining system compatibility, with the lowest possible error rate, the shortest possible execution time, and the highest possible success rate?

Table 1 displays the result of the proposed framework with the traditional methods. As shown in Table 1, the traditional approaches used are Ethereum [83], Hyperledger [83–84], and Parity [83] with the used metrics success rate (SR), system execution time (SET), and average delay (AD) all in (%) values from various research studies. From the obtained results, the proposed framework performed better than the existing methods based on the several metrics used for the observation like average delay, system execution time, and success rate for small bank dataset and the YCSB.

From Fig. 4, the results show that the proposed system performed best when compared with the conventional methods used on the YCSB and small bank dataset based on success rate (%).

Table 1 The performance of the proposed system against the traditional methods

	YCSB			Small bank		
Methods	SR (%)	AD (%)	SET (%)	SR (%)	AD (%)	SET (%)
Ethereum	27.4	10.5	12.52	29.5	13.4	X
Hyperledger	46	4.9	4.01	49	6.9	4.32
Parity	65	4.8	2.9	69	6.1	3.05
Proposed method	89	1.6	1.03	93	2.95	1.04

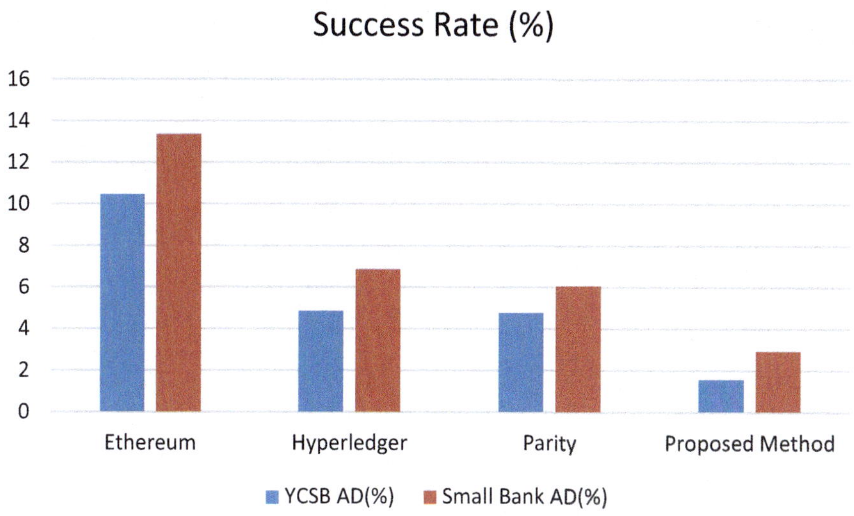

Fig. 4 The results of YCSB and small bank dataset by success rate (%)

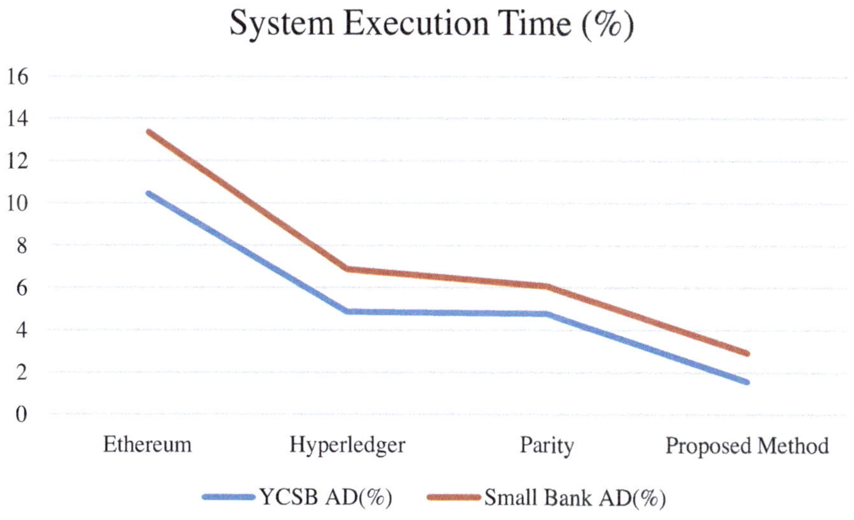

Fig. 5 The results of the YCSB and small bank dataset by system execution time (%)

From Fig. 5, the results show that the proposed system performed best when compared with the conventional methods used on the YCSB and small bank dataset based on system execution time (%).

From Fig. 6, the result show that the proposed system performed best when compare with the conventional methods used on the YCSB and small bank dataset based on Average Delay (%).

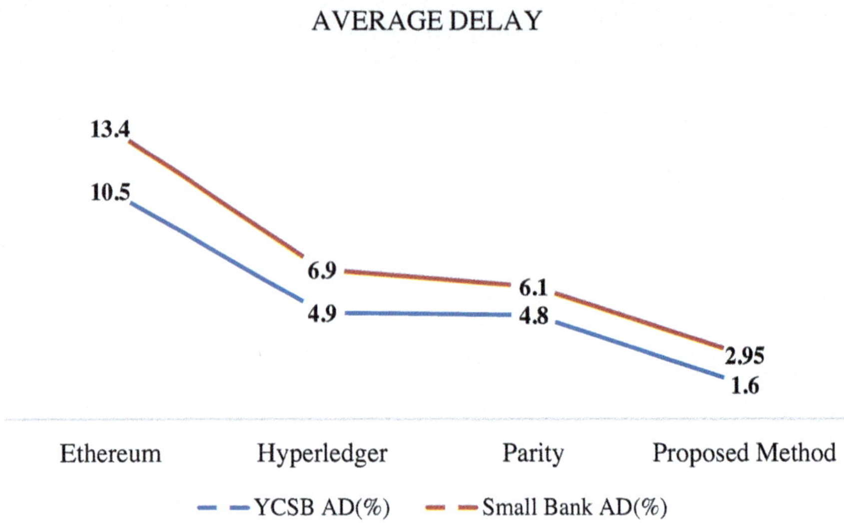

Fig. 6 The results of YCSB and Small Bank dataset by Average Delay (%)

7 Conclusion and Future Directions

In today's healthcare sector, the application of blockchain in healthcare systems is crucial. It can lead to automated data collecting and verification processes, as well as correct and aggregated data from diverse sources that is immutable, tamper-resistant, and safe, with a lower risk of cybercrime. It also allows distributed data, as well as system redundancy and failure tolerance. As a result, using blockchain technology, this chapter presented a secure smart healthcare system. The proposed approach was used to transform a concentrated and vulnerable smart system into a distributed, transparent, and safe system, thereby raising the standard of medical-related services on the smart healthcare system. There are various theories on why blockchain could be used to improve the healthcare system. First, it provides clear data to all stakeholders while safeguarding the privacy of patients. It also safeguards sensitive medical records from theft and eavesdropping by malicious attackers. In the proposed system, mathematical derivation is used to evaluate the efficiency, security, and cost-effectiveness of sharing healthcare data. The suggested framework is compatible with a cloud platform and completely independent for secure data transmission and recovery. The proposed system has reduced 1.6 AD in seconds and 1.03 SET in seconds and improves 25% SR. Finally, when compared to the traditional methods, the suggested methodology outperforms them on each parameter and dataset. The proposed framework's complete implementation will be carried out in the future. The lack of blockchain awareness among healthcare stakeholders is a key roadblock to its implementation, which will be addressed in the future to ensure that blockchain is properly implemented in the healthcare system.

References

1. McGhin, T., Choo, K. K. R., Liu, C. Z., & He, D. (2019). Blockchain in healthcare applications: Research challenges and opportunities. *Journal of Network and Computer Applications, 135*, 62–75.
2. Awotunde, J. B., Ogundokun, R. O., Jimoh, R. G., Misra, S., & Aro, T. O. (2021). Machine Learning Algorithm for Cryptocurrencies Price Prediction. In Artificial Intelligence for Cyber Security: Methods, Issues and Possible Horizons or Opportunities (pp. 421–447). Springer, .
3. Zheng, Z., Xie, S., Dai, H., Chen, X., & Wang, H. (2017). An overview of blockchain technology: Architecture, consensus, and future trends. In *2017 IEEE international congress on big data (BigData congress)* (pp. 557–564). IEEE.
4. Awotunde, J. B., Ogundokun, R. O., Misra, S., Adeniyi, E. A., & Sharma, M. M. (2020). Blockchain-Based Framework for Secure Transaction in Mobile Banking Platform. Advances in Intelligent Systems and Computing, 2021, 1375 AIST, pp. 525–534.
5. Nawari, N. O., & Ravindran, S. (2019). Blockchain and the built environment: Potentials and limitations. *Journal of Building Engineering, 25*, 100832.
6. Tian, Z., Li, M., Qiu, M., Sun, Y., & Su, S. (2019). Block-DEF: A secure digital evidence framework using blockchain. *Information Sciences, 491*, 151–165.
7. Sullivan, C., & Burger, E. (2017). E-residency and blockchain. *Computer Law & Security Review, 33*(4), 470–481.
8. Awotunde, J. B., Bhoi, A. K., & Barsocchi, P. (2021). Hybrid cloud/Fog environment for healthcare: an exploratory study, opportunities, challenges, and future prospects. Intelligent Systems Reference Library, 2021, 209, pp. 1–20.
9. Kshetri, N. (2017). Blockchain's roles in strengthening cybersecurity and protecting privacy. *Telecommunications Policy, 41*(10), 1027–1038.
10. Yüksel, B., Küpçü, A., & Özkasap, Ö. (2017). Research issues for privacy and security of electronic health services. *Future Generation Computer Systems, 68*, 1–13.
11. Shae, Z., & Tsai, J. J. (2017). On the design of a blockchain platform for clinical trial and precision medicine. In *2017 IEEE 37th international conference on distributed computing systems (ICDCS)* (pp. 1972–1980). IEEE.
12. Dubovitskaya, A., Xu, Z., Ryu, S., Schumacher, M., & Wang, F. (2017). How blockchain could empower ehealth: An application for radiation oncology. In *VLDB workshop on data management and analytics for medicine and healthcare* (pp. 3–6). Springer.
13. Sanjukta, B., Sourav, B., & Chinmay, C. (2019). IoT-based smart transportation system under real-time environment. *IET: big data-enabled internet of things: Challenges and opportunities* (Ch. 16) (pp. 353–373). ISBN 978-1-78561-637-2.
14. Chakraborty, C., Banerjee, A., Kolekar, M. H., Garg, L., & Chakraborty, B. (Eds.). (2020). *Internet of things for healthcare technologies*. Springer.
15. Ho, C. W., Ali, J., & Caals, K. (2020). Ensuring trustworthy use of artificial intelligence and big data analytics in health insurance. *Bulletin of the World Health Organization, 98*(4), 263.
16. Suzuki, S., & Murai, J. (2017). Blockchain as an audit-able communication channel. In *2017 IEEE 41st annual computer software and applications conference (COMPSAC)* (Vol. 2, pp. 516–522). IEEE.
17. Farahani, B., Firouzi, F., Chang, V., Badaroglu, M., Constant, N., & Mankodiya, K. (2018). Towards fog-driven IoT eHealth: Promises and challenges of IoT in medicine and healthcare. *Future Generation Computer Systems, 78*, 659–676.
18. Chen, P. T., Lin, C. L., & Wu, W. N. (2020). Big data management in healthcare: Adoption challenges and implications. *International Journal of Information Management, 53*, 102078.
19. Abdulraheem, M., Awotunde, J. B., Jimoh, R. G., & Oladipo, I. D. (2021). An efficient lightweight cryptographic algorithm for IoT security. *Communications in Computer and Information Science, 2021*(1350), 41–53.
20. Albahri, A. S., Alwan, J. K., Taha, Z. K., Ismail, S. F., Hamid, R. A., Zaidan, A. A., ... Alsalem, M. A. (2021). IoT-based telemedicine for disease prevention and health promotion: State-of-the-art. *Journal of Network and Computer Applications, 173*, 102873.

21. Akkaş, M. A., Sokullu, R., & Çetin, H. E. (2020). Healthcare and patient monitoring using IoT. *Internet of Things, 11*, 100173.
22. Lee, S. M., & Lee, D. (2021). Opportunities and challenges for contactless healthcare services in the post-COVID-19 era. *Technological Forecasting and Social Change, 167*, 120712.
23. Daniel, J. G., & Uppaluru, M. (2017). New reimbursement for remote patient monitoring and telemedicine.
24. Alsubaei, F., Abuhussein, A., & Shiva, S. (2017). Security and privacy on the internet of medical things: Taxonomy and risk assessment. In *2017 IEEE 42nd conference on local computer networks workshops (LCN workshops)* (pp. 112–120). IEEE.
25. Mutlag, A. A., Ghani, M. K. A., Arunkumar, N. A., Mohammed, M. A., & Mohd, O. (2019). Enabling technologies for fog computing in healthcare IoT systems. *Future Generation Computer Systems, 90*, 62–78.
26. Sun, W., Cai, Z., Li, Y., Liu, F., Fang, S., & Wang, G. (2018). Security and privacy in the medical internet of things: A review. *Security and Communication Networks, 2018.*
27. Tang, J., Liu, A., Zhao, M., & Wang, T. (2018). An aggregate signature-based trust routing for data gathering in sensor networks. *Security and Communication Networks, 2018.*
28. Sun, W., Cai, Z., Liu, F., Fang, S., & Wang, G. (2017). A survey of data mining technology on electronic medical records. In *2017 IEEE 19th international conference on e-health networking, applications and services (Healthcom)* (pp. 1–6). IEEE.
29. Bell, L., Buchanan, W. J., Cameron, J., & Lo, O. (2018). Applications of blockchain within healthcare. *Blockchain in healthcare today, 1*(8).
30. Jamil, F., Ahmad, S., Iqbal, N., & Kim, D. H. (2020). Towards a remote monitoring of patient vital signs based on IoT-based blockchain integrity management platforms in smart hospitals. *Sensors, 20*(8), 2195.
31. Basha, S. M., Janet, J., & Balakrishnan, S. (2020). A study on privacy-preserving models using blockchain technology for IoT. In *Blockchain, big data and machine learning* (pp. 265–290). CRC Press.
32. Cancarevic, I., Plichtová, L., & Malik, B. H. (2021). Healthcare systems around the world. In *International medical graduates in the United States* (pp. 45–79). Springer.
33. Islam, M., Usman, M., Mahmood, A., Abbasi, A. A., & Song, O. Y. (2020). Predictive analytics framework for accurate estimation of child mortality rates for internet of things enabled smart healthcare systems. *International Journal of Distributed Sensor Networks, 16*(5), 1550147720928897.
34. Trivedi, S. A., Patel, M., & Patel, S. (2021). Health care cube integrator for health care databases. In *Web semantics* (pp. 129–151). Academic Press.
35. Shi, S., He, D., Li, L., Kumar, N., Khan, M. K., & Choo, K. K. R. (2020). Applications of blockchain in ensuring the security and privacy of electronic health record systems: A survey. *Computers & Security, 101966.*
36. Jia, Q. (2021). Research on medical system based on blockchain technology. *Medicine, 100*(16).
37. Rajput, A. R., Li, Q., & Ahvanooey, M. T. (2021). A Blockchain-based secret-data sharing framework for personal health records in emergency condition. In *Healthcare* (Vol. 9, p. 206). Multidisciplinary Digital Publishing Institute.
38. Mettler, M. (2016). Blockchain technology in healthcare: The revolution starts here. In *2016 IEEE 18th international conference on e-health networking, applications, and services (Healthcom)* (pp. 1–3). IEEE.
39. Khurshid, A. (2020). Applying blockchain technology to address the crisis of trust during the COVID-19 pandemic. *JMIR Medical Informatics, 8*(9), e20477.
40. Esmat, A., de Vos, M., Ghiassi-Farrokhfal, Y., Palensky, P., & Epema, D. (2021). A novel decentralized platform for peer-to-peer energy trading market with blockchain technology. *Applied Energy, 282*, 116123.
41. Arani, S. A., Nawab, M. R. I., Rahman, M. T., & Zaman, M. (2020). A blockchain-based approach to prevent hidden contagion of COVID-19. *Compiler, 9*(2), 71–84.

42. Dagher, G. G., Mohler, J., Milojkovic, M., & Marella, P. B. (2018). Ancile: Privacy-preserving framework for access control and interoperability of electronic health records using blockchain technology. *Sustainable Cities and Society, 39*, 283–297.
43. Hewa, T., Ylianttila, M., & Liyanage, M. (2020). Survey on blockchain based smart contracts: Applications, opportunities and challenges. *Journal of Network and Computer Applications, 102857.*
44. Habibzadeh, H., Dinesh, K., Shishvan, O. R., Boggio-Dandry, A., Sharma, G., & Soyata, T. (2019). A survey of healthcare internet of things (HIoT): A clinical perspective. *IEEE Internet of Things Journal, 7*(1), 53–71.
45. Stafford, T. F., & Treiblmaier, H. (2020). Characteristics of a blockchain ecosystem for secure and sharable electronic medical records. *IEEE Transactions on Engineering Management, 67*(4), 1340–1362.
46. Chanchaichujit, J., Tan, A., Meng, F., & Eaimkhong, S. (2019). Blockchain technology in healthcare. In *Healthcare 4.0* (pp. 37–62). Palgrave Pivot.
47. McGhin, T., Choo, K. K. R., Liu, C. Z., & He, D. (2019). Blockchain in healthcare applications: Research challenges and opportunities. *Journal of Network and Computer Applications, 135*, 62–75.
48. Sun, Z., Sun, R., Lu, L., & Mislove, A. (2021). Mind your weight (s): A large-scale study on insufficient machine learning model protection in mobile apps. In *30th {USENIX} Security Symposium ({USENIX} Security 21).*
49. Dubovitskaya, A. (2021). Blockchain applications in healthcare. In *The emerald handbook of Blockchain for business*. Emerald Publishing Limited.
50. Duy, P. T., Hien, D. T. T., Hien, D. H., & Pham, V. H. (2018). A survey on opportunities and challenges of Blockchain technology adoption for revolutionary innovation. In *Proceedings of the Ninth International Symposium on Information and Communication Technology* (pp. 200–207).
51. Chang, V., Baudier, P., Zhang, H., Xu, Q., Zhang, J., & Arami, M. (2020). How Blockchain can impact financial services–the overview, challenges and recommendations from expert interviewees. *Technological Forecasting and Social Change, 158*, 120166.
52. Bocek, T., Rodrigues, B. B., Strasser, T., & Stiller, B. (2017). Blockchains everywhere-a use-case of blockchains in the pharma supply-chain. In *2017 IFIP/IEEE symposium on integrated network and service management (IM)* (pp. 772–777). IEEE.
53. Sukhwani, H., Martínez, J. M., Chang, X., Trivedi, K. S., & Rindos, A. (2017). Performance modeling of PBFT consensus process for permissioned blockchain network (hyperledger fabric). In *2017 IEEE 36th Symposium on Reliable Distributed Systems (SRDS)* (pp. 253–255). IEEE.
54. Xu, J. J. (2016). Are blockchains immune to all malicious attacks? *Financial Innovation, 2*(1), 1–9.
55. Abouelmehdi, K., Beni-Hssane, A., Khaloufi, H., & Saadi, M. (2017). Big data security and privacy in healthcare: A review. *Procedia Computer Science, 113*, 73–80.
56. Khan, S. I., & Hoque, A. S. L. (2016). Privacy and security problems of national health data warehouse: a convenient solution for developing countries. In *2016 International Conference on Networking Systems and Security (NSysS)* (pp. 1–6). IEEE.
57. Vithanwattana, N., Mapp, G., & George, C. (2016). mHealth-Investigating an information security framework for mHealth data: Challenges and possible solutions. In *2016 12th International Conference on Intelligent Environments (IE)* (pp. 258–261). IEEE.
58. Xia, Q. I., Sifah, E. B., Asamoah, K. O., Gao, J., Du, X., & Guizani, M. (2017). MeDShare: Trust-less medical data sharing among cloud service providers via blockchain. *IEEE Access, 5*, 14757–14767.
59. Xu, J. J. (2016). Are blockchains immune to all malicious attacks? *Financial Innovation, 2*(1), 1–9.
60. Ding, D., Conti, M., & Solanas, A. (2016). A smart health application and its related privacy issues. In *2016 Smart City Security and Privacy Workshop (SCSP-W)* (pp. 1–5). IEEE.

61. Fernando, R., Ranchal, R., An, B., Othman, L. B., & Bhargava, B. (2016). Consumer oriented privacy preserving access control for electronic health records in the cloud. In *2016 IEEE 9th International Conference on Cloud Computing (CLOUD)* (pp. 608–615). IEEE.
62. Sajid, A., & Abbas, H. (2016). Data privacy in cloud-assisted healthcare systems: State of the art and future challenges. *Journal of Medical Systems, 40*(6), 155.
63. Dinev, T., Albano, V., Xu, H., D'Atri, A., & Hart, P. (2016). Individuals' attitudes towards electronic health records: A privacy calculus perspective. In *Advances in healthcare informatics and analytics* (pp. 19–50). Springer.
64. Panigrahi, R., Borah, S., Bhoi, A. K., & Mallick, P. K. (2020). Intrusion detection systems (IDS)—an overview with a generalized framework. Cognitive Informatics and Soft Computing, 107–117.
65. Srinivasu, P. N., Bhoi, A. K., Nayak, S. R., Bhutta, M. R., & Woźniak, M. (2021). Blockchain Technology for Secured Healthcare Data Communication among the non-terminal nodes in IoT architecture in 5G network. *Electronics, 10*(12), 1437.
66. Torre, I., Koceva, F., Sanchez, O. R., & Adorni, G. (2016). A framework for personal data protection in the IoT. In *2016 11th International Conference for Internet Technology and Secured Transactions (ICITST)* (pp. 384–391). IEEE.
67. Al-Muhtadi, J., Shahzad, B., Saleem, K., Jameel, W., & Orgun, M. A. (2019). Cybersecurity and privacy issues for socially integrated mobile healthcare applications operating in a multi-cloud environment. *Health Informatics Journal, 25*(2), 315–329.
68. Abeyratne, S. A., & Monfared, R. P. (2016). Blockchain-ready manufacturing supply chain using a distributed ledger. *International Journal of Research in Engineering and Technology, 5*(9), 1–10.
69. Kshetri, N. (2018). 1 Blockchain's roles in meeting key supply chain management objectives. *International Journal of Information Management, 39*, 80–89.
70. Reyna, A., Martín, C., Chen, J., Soler, E., & Díaz, M. (2018). On blockchain and its integration with IoT. *Challenges and opportunities. Future generation computer systems, 88*, 173–190.
71. Mougayar, W. (2016). *The business blockchain: Promise, practice, and application of the next internet technology*. Wiley.
72. Awotunde, J. B., Chakraborty, C., & Adeniyi, A. E. (2021). Intrusion Detection in Industrial Internet of Things Network-Based on Deep Learning Model with Rule-Based Feature Selection. Wireless Communications and Mobile Computing, 2021, 7154587
73. Song, H., Zhu, N., Xue, R., He, J., Zhang, K., & Wang, J. (2021). Proof-of-contribution consensus mechanism for blockchain and its application in intellectual property protection. *Information Processing & Management, 58*(3), 102507.
74. Greenspan, G. (2015). Multichain private blockchain-white chapter. URL: http://www.multichain.com/download/MultiChain-White-Chapter.pdf.
75. De Filippi, P. (2016). The interplay between decentralization and privacy: The case of blockchain technologies. *Journal of Peer Production, Issue, 7.*
76. Liang, X., Shetty, S., Tosh, D., Kamhoua, C., Kwiat, K., & Njilla, L. (2017). Provchain: A blockchain-based data provenance architecture in cloud environment with enhanced privacy and availability. In *2017 17th IEEE/ACM International Symposium on Cluster, Cloud and Grid Computing (CCGRID)* (pp. 468–477). IEEE.
77 Kouhizadeh, M., Saberi, S., & Sarkis, J. (2021). Blockchain technology and the sustainable supply chain: Theoretically exploring adoption barriers. *International Journal of Production Economics, 231*, 107831.
78. Kshetri, N., & Voas, J. (2018). Blockchain-enabled e-voting. *IEEE Software, 35*(4), 95–99.
79. Shahzad, B., & Crowcroft, J. (2019). Trustworthy electronic voting using adjusted blockchain technology. *IEEE Access, 7*, 24477–24488.
80. Savelyev, A. (2018). Copyright in the blockchain era: Promises and challenges. *Computer law & security review, 34*(3), 550–561.
81. Chakraborty, S., Aich, S., & Kim, H. C. (2019). A secure healthcare system design framework using blockchain technology. In *2019 21st International Conference on Advanced Communication Technology (ICACT)* (pp. 260–264). IEEE.

82. Awotunde, J. B., Jimoh, R. G., Folorunso, S. O., Adeniyi, E. A., Abiodun, K. M., & Banjo, O. O. (2021). Privacy and security concerns in IoT-based healthcare systems. Internet of Things, 2021, pp. 105–134.
83. Dinh, T. T. A., Liu, R., Zhang, M., Chen, G., Ooi, B. C., & Wang, J. (2018). Untangling blockchain: A data processing view of blockchain systems. *IEEE Transactions on Knowledge and Data Engineering, 30*(7), 1366–1385.
84. Mubarakali, A., Bose, S. C., Srinivasan, K., Elsir, A., & Elsier, O. (2019). Design a secure and efficient health record transaction utilizing block chain (SEHRTB) algorithm for health record transaction in block chain. *Journal of Ambient Intelligence and Humanized Computing*, 1–9.

Blockchain Technology and Organizational Practices: The Case of Nigerian Academic Libraries

Rebecca Chidimma Ojobor, Cletus Ifeanyichukwu Ojobor, and Jonathan Oluranti

1 Introduction

The emergence of modern technologies has modernized most organizational practices in the digital field and a new paradigm of the automated era with regards to service-oriented evolved. Considering these changes, organizations need to change significantly to accommodate the prevailing needs of their clientele. Similarly, the academic library is a vibrant organ known for the provision of intellectual resources and service-oriented to support and promote quality teaching, learning, and research product of its parent institution which is expected to follow suit; its personnel is equally expected to adapt to the new normal of working and providing services with the emerging technologies.

The academic library is an intellectual resource center established in tertiary institutions to play a supportive role in enhancing the knowledge frontier of students, teaching, and nonteaching staff of the institution [1]. Although the aims and objectives of the academic library remain unchanged, the development of information and communication technology (ICT) redefined its mode of services by ushering in new paradigms in its practices which requires new approaches in meeting the ever-increasing needs of information seekers. Authors in [2] note that academic libraries are affected by environmental and technological changes and will need to cope with the changing needs of their clientele due to the emergence of high-powered new technologies. The advent of ICTs induced the adoption and

R. C. Ojobor (✉) · C. I. Ojobor
University of Nigeria, Nsukka, Nigeria
e-mail: rebecca.ojobor@unn.edu.ng; cletus.ojobor@unn.edu.ng

J. Oluranti
Covenant University, Ota, Nigeria
e-mail: jonathan.oluranti@covenantuniversity.edu.ng

S. Misra, A. Kumar Tyagi (eds.), *Blockchain Applications in the Smart Era*, EAI/Springer Innovations in Communication and Computing,
https://doi.org/10.1007/978-3-030-89546-4_9

use of the Internet, computer, e-mail, projectors, video CDs, social media apps, and many other technology-related applications in library practice. The impact of these technologies in library practices cannot be overvalued. They have redefined the ideology of library as a pool of information resources to an interactive, socializing, learning, knowledge creation, service-oriented, and information provision center. However, there are some accrued challenges resulting from such development that need to be addressed. Some of such challenges include changing the legal paradoxes of copyright, doubts on online payment for digital resources and databases, and poor record management due to the increasing volume of information resources and many other upheavals resulting from ICT advancement in the library. As life increasingly moves online, the greatest challenge facing Internet users is conducting financial transactions in a setting where the parties involved are unknown and do not trust each other [3]. Similarly, lack of reliable data on authorship and copyrights, together with the inequitable contractual terms authors are subjected to, lead to disjointed, inaccurate, and incomplete information on the original authors of most of the online resources and payment of royalties Ito and O'Dair in [4]. These challenges have been relegating library practices until the advancement of the semantic Web technologies popularly known as Web 3.0. Web 3.0 is conceptualized as the third-generation technology upgrade through 2010–the 2020s [5]. The technologies are designed to improve the services of the previous Web technologies and overcome various challenges characterizing the previous technologies. They include advanced technologies like wireless networks, the Internet of Things (IoT), and blockchain technology, which this paper intends to publicize its potentials to librarians.

Blockchain is a distributed ledger technology (DLT). It was first introduced in 2008 by Satoshi Nakamoto – a pseudonymous mastermind behind the theory of Bitcoin cryptocurrency. It is a new technical infrastructure for cash payments and document storage. It is more secure, traceable, and transparent as all transactions stored in it are equally saved on various computers and are verifiable by others reliably and securely [6]. Furthermore, blockchain technology can also be used to correct the poor voting system in Nigeria. As observed [7], the m-voting system was not reliable in solving the problems of voting system due to its inability of securing and storing the casted votes. But the blockchain, according to them, was proposed to eliminate the problems characterizing the m-voting system in Nigeria. Nevertheless, blockchain has other concealed potentials, which could significantly undermine the most challenging factors that subverted library practices. Malyarov [8] reveals that blockchain has the potentials to help libraries and academics mitigate Internet-induced risks. But ignorance of its potentials in most organizational practices precludes its adoption and use especially in Nigeria.

According to [9], Nigeria is characterized by inadequate ICT infrastructure. Although the United Nations (UN) ranks her high in the Online Services Index (OSI) and e-Government Development Index (EGDI), yet she does not feature among the top ten in Africa [10]. In 2019, Nigeria was ranked 75th in the Global Connectivity Index (GCI), which places her below the global average. This implies that Nigeria has fallen behind others in terms of broadband penetration [11]. Infrastructural deficiencies and lack of technological competencies in Nigeria affect most organizations in the country, academic libraries in particular. In this regard, the focus of this article is on academic libraries, specifically those in public institutions of higher learning in Southeast Nigeria. This is because public institutions/organizations have more advantage and opportunity to obtain government subventions and subsequently have greater chances in adoption and use of emerging technologies regardless of their cost.

It is evidence-based that both the local and global competitiveness of any organization depend on how the organization deploys technology to transform its various sectors. Since technology plays a key role in enhancing organizational goals, there is a need for Nigerian organizations to reorient their practices and deploy ICT and digital technology to ensure the realization of global best practices. The library is one of the most prominent organizations that render services to many other organizations that need to be on the lead toward the development of ICT capacity and infrastructure to best serve the nation. However, the infrastructural deficiencies and lack of technological competencies in Nigeria affect academic libraries as most of their staff are not encouraged or sponsored to upgrade their ICT skills. Consequently, greater percentages of librarians are technologically illiterates and could hardly influence library practices with the innovative technologies of the time. The assertion advocates [12] that it is very important for librarians/information professionals to change with the system by ensuring knowledge and skills update as well as the utilization of emerging Web technologies in library service delivery. For this reason, coping with the best global library practices becomes a greater challenge to Nigerian libraries. The scenario calls for the need to create awareness on emerging technology such as blockchain, which has the potential to subvert the most challenging issues confronting libraries, and enhance and promote its activities to achieve global best practices.

Although scholars such as [13–15] have carried out studies on blockchain and its implications for libraries yet, literature in the area is still very slim. Besides, none of the previous studies investigated blockchain technology in organizational practices. It is on this platform that the researcher deems it necessary to embark on this study with regard to the following specifications:

To identify library practices that require the application of blockchain technology
To determine the extent to which blockchain technology enhances library practices
To determine the ways through which Nigerian libraries could support and encourage librarians to embrace emerging technologies

Identify the challenges associated with the adoption and use of blockchain technologies in Nigerian libraries.

The expected outcomes of the study are anticipated to be useful to libraries, librarians, authors and publishers, library administration, Nigerian youths, and future researchers.

It is assumed that the findings of the study will pave the way for greater improvement in library practices. This is because the technology is adopted and used in libraries which will get rid of the current deadlock among libraries concerning resource sharing and service delivery. Librarians are also to benefit from the findings of the study, because the adoption and use of blockchain in the library will lead to effectiveness and efficiency in the system. This will remove too much stress on the side of the librarians and gives them a sense of belonging. It will also encourage knowledge and skill update among librarians. Authors and publishers will also appreciate the outcome of this study. This is because their intellectual property right (IPR) will fully be protected if libraries adopt and use blockchain technology. The result of the study will equally be beneficial to Nigerian citizens, especially the youths. If blockchain is used to store staff personal data, it will give no room for falsification of employment history, thereby ensuring that staff retires at the appropriate time. Consequently, there will be the need for employment, which will subsequently reduce the massive unemployment rate in the country. Library administrators will also find this work useful. The findings of the study will expose to the administrators the need to adopt this technology; this will encourage the management to provide adequate funds and other necessary support for adoption and use of the technology to enhance library activities toward achieving global best practices. Finally, the expected outcome of the study will add to the existing literature on blockchain technology in libraries and hence serve as a reference material to future researchers who will embark on a similar study.

In as much as no previous study was found to have examined blockchain technology in organizational practices in Nigerian academic libraries, the study is worth conducting. In so doing, the paper is sectioned into four. The succeeding section provides the review of related literature while the next to it (sect. 3) presents the research methodology. The last section that is the fourth section presents and analyzes the data collected for the study. The section has three subsections; while the first subsection discusses the major findings, the second subsection highlights the implications of the study, and the last subsection concludes the paper.

2 Review of Related Literature

Blockchain technology is an innovative technology with the capacity to transform organizational practices using new approaches. Blockchain technology as opines [16] is a database containing all the transactions ever executed in a peer-to-peer network system. It is a time-stamped series of immutable records of data that is managed by a cluster of computers not owned by any single entity [13]. It consists of blocks of data that are secured and bound to each other using cryptographic principles; and once these blocks are collected in a chain, they cannot be changed or deleted by a single actor; instead, they are verified and managed using automation and shared governance protocols [17]. Furthermore, [18] describes blockchain technology as a shared electronic database that contains immutable and encrypted data records, which could be shared within a group of people, organizations, or a community; however, the authenticity of the data could be verified using a unique key associated with the data. This implies that blockchain database is not stored in any single location but rather on millions of computers indicating that the records are transparent, easily verifiable, and accessible to anyone on the Internet. This feature makes the information stored in the blockchain difficult for hackers to corrupt as no centralized version of the information exists.

The technology according to [19] bears three distinct properties, which have necessitated its widespread consideration in organizational practices. These properties include the following: decentralization, transparency, and immutability. Because most developing nations are used to the centralized system of service, all their records are stored in an entity, and such data could be accessible only when there is close contact with the host. The banking system is a good example of a centralized system. Access to one's money stored in the bank is possible by going through the bank or bank-related apps. Besides, the traditional client-server model according to [19] is another example of a centralized system. Mougayar explains that in a Google search, the searcher sends a query to the Google server, which later gets back to him/her with the relevant information. Although a centralized system has been in use for decades of years yet, several vulnerabilities have been recorded. First and foremost, the system is an easy target spot for potential hackers. Secondly, any software upgrading will halt the entire system. Finally, if the centralized entity shuts down for whatever reason or gets corrupted and malicious, the system losses its entire records or data forever. However, an organization can get rid of these anomalies through the adoption and use of blockchain technology. Supportively, [20] submit that blockchain technology eliminates centralized risks, low efficiency, and high transaction cost. With the decentralized feature in blockchain, information is not stored by one single entity; therefore, everyone in the network owns the information, and interaction could be made without going through a third party.

- Transparency: With this feature, information stored in blockchain has wider visibility and traceability except for personal identity (user privacy), which can be protected by adding anonymity protections in the blockchain using the CoinJoin method [21]. According to Bünz et al., CoinJoin method is an attractive means of anonymizing blockchain transactions which require no modification to the Bitcoin protocol.
- Immutability: The cryptographic hash function in blockchain makes the data stored in it immutable. This implies that once data or transactions are appended, accepted, and confirmed by the nodes on the blockchain, it is not easily changed [22]. Barnes and Xiao's [23] also affirmed that because the blockchain ledgers cannot be modified or deleted the data is immutable. The feature safeguards all documents stored in the blockchain and subsequently makes the technology valuable to financial institutes and other institutions or organizations that deal with financial issues, since it has the potential to checks financial frauds.

With these inherent features, blockchain can potentially promote and enhance processes and service provision within various organizations [24]. However, this is not to claim that blockchain is a panacea to all organizational shortcomings, but it has the capacity for greater achievement for the organization of all kinds. The technology according to [25] is valuable in the financial organization as its contributions to financial organizations include but are not limited to eliminating the need for intermediation and enhancing direct transactions between trading parties. Blockchain's usefulness is indispensable in the music industry [26]. In the same vein, an empirical study conducted by [24] on "the impact of blockchain technology on business models – a taxonomy and archetypal patterns" identify five archetypal patterns, which enhance the understanding of how blockchain technology affects existing and creates new business models.

2.1 *Library Practices that Require Application of Blockchain Technology*

Record Management
It is not an overstatement stating that record management is an integral function of any organization that deals with humans. Library as an organization cannot exempt from this role. The library has long been known for keeping document and are expected to provide high-quality information service to their patron at any point in time. Library records according to [22] are an integral form of evidence concerning accountability to the citizens, including such diverse categories of records as administrative records, registration files, financial records, and historical records. For these records to be used in their evidentiary capacity, they must be created, managed, and preserved, respectively, applicable policies, legislation, regulations, standards, codes of practice, procedures, and community expectations [27]. The tremendous increase in information resources and the explosive growth of digital devices and related applications as noted by [28] have collectively altered the traditional resource management practice in the library beyond recognition. Ensuring effective financial records of e-resources and databases, which are usually transact with unknown and untrusted party, is a challenging issue to most libraries. Besides, managing large volume of scholarly production, disseminating them to the wider user community, and preserving them for future use are other areas of concern to the library.

Above all is the ineffectiveness in securing the available records. Rein and Peterson [29] allege that despite libraries' best efforts to protect their systems, criminals might gain access to their databases and steal or manipulate records. Blockchain technology can help libraries in digitizing existing records and manage them within a secure infrastructure; once digital information is committed onto a blockchain, it is permanently stored and impossible to manipulate or hack [22].

Protection of Copyright
The potentiality of blockchain technology to serve as an effective digital right management tool qualify it to protect the copyright of materials stored in it. Because digital resources are inherently reproducible, they are indiscriminately reproduced by most users without the consent of the author or publisher. This has often led to the problem of piracy in the economy. This attitude has prompted publishers to

impose a digital management tool known as "draconian" on libraries and consumers to prevent copying their materials, but unfortunately, it was unworkable [30]. This heinous act debasing academic exercise can be overcome through blockchain technology. Because the blockchain creates a unique, verifiable record that can be accessed by anyone, it could be tied to digital materials and used as a method to show "provable scarcity" of that resource; this would allow digital materials to be uniquely identified, controlled, and transferred [30].

Resource Sharing Among Libraries

Resource sharing is the process by which resources facilities and services of the member institutions are shared effectively [31]. The practice of sharing resources among libraries has been an old and important activity because no library can independently satisfy the information needs of its users with its collection. Resource sharing activities usually include but are not limited to interlibrary lending, sharing of expertise, union list of periodicals, union catalog, directory of resource person, directory of research, training programs, and providing online open-access catalog (OPAC). This activity has some challenging factors, like poor funding, inadequate communication system, an uncooperative attitude of librarians, lack of trained staff, inadequate available resources, lack of mailing or transportation system, and inadequate security of materials, which usually undercut sharing practice among libraries, thereby hindering effective dissemination of information resources among information seekers. Although some libraries at their various capacities have deployed modern technologies to strengthen the weaknesses and improve on resource sharing practice, [32] argued that blockchain technology could serve best for this purpose because it possesses the following advantages – anonymity, efficiency, cost savings, security, flexibility, and many others. With the outlined advantages, blockchain technology can enable libraries to facilitate peer-to-peer sharing beyond just books, which could help members of the community authenticate the availability of different tools or services for a more efficient sharing economy. Though blockchain makes it harder to change these shared documents, it can help in making them more secure. This is in advocacy with [33] that user-centric blockchain applications could enable end users to control, trace, and claim ownership of every piece of content they share.

Partnership with Other Organizations

Successful collaboration among organizations is not easily achievable [34]. This is because most partners do lack commitment, lie, or cheat. Moreover, at times poor communication gap and transportation system together with other related issues may lead to uncompromising the terms of the agreement among the collaborative parties. However, [34] aver that blockchain can highly support collaborations, because it is a digital ledger where several people have joint control over shared information. The authors further explain that such a feature makes the blockchain technology ideal for situations where trust and information sharing are important. The outcome of an investigation on the applications of blockchain in libraries revealed that the technology among other things greatly facilitates partnership across centers/organizations [14]. The investigators explain that libraries can partner with museums, universities, and government agencies to share machine-readable catalog (MARC) records, authority control, and user-generated content through a blockchain framework.

Endorsement of Personal Data
This is also and another crucial activity of the library. As an entity, the library keeps the personal data of its employees. These data include the credentials and employment records of every employee in the establishment. It is these records that determine the retirement age or period of disengagement of an employee. With the high rate of corruption in most organizations, which the library is not an exception, some staff do falsify their record to enable them to stay longer than the required time in the system. Such attitude consequently retards the progress of the organization, because such employees may no longer be strong enough to contribute effectively to the progress of the organization. Consequentially, an attitude of such kind, in the long run, leads to ineffectiveness and the inefficiency of the organization. Moreover, it contributes to the high rate of unemployment on the part of the youths, as the able men and women who have the energy and zeal to contribute positively toward increased productivity of the organization find it difficult to secure employment opportunities. However, such a devastating scenario could be ameliorated through the introduction or adoption of blockchain technology in the organization to checkmate such practices.

2.2 *Challenges Associated with Adoption and Use of Blockchain in Libraries*

Although scholars from different perspectives have revealed the abilities of blockchain technology in various organizations, most organizations in Nigeria are yet to benefit from it due to the following constraints.

Poor Funding: As noted [35], inadequate government funding inhibits IT application by African libraries. Poor funding is a major factor confronting Nigerian organizations in deploying modern technologies, because most of the organizations cannot afford the huge amount of money for the provision of the technologies. In libraries, for instance, the 10% budgetary allocation allotted to them by their parent institutions cannot sufficiently run the affairs of the library and afford to purchase modern technology such as blockchain. The situation in the library is even more alarming as [36] reveals that most libraries get lesser than 10% of their institution's budgetary allocation due to slim internal generated revenue and high rate of expenditure.

Lack of In-Service Training: Although the core knowledge and skills of traditional librarianship are still useful in this digital era, they need to be augmented by new technological knowledge [37]. Upgrading of skills and knowledge could be possible through in-service training. In-service training is a means of addressing weaknesses in staff performance, but, more importantly, it enables the staff to upgrade their knowledge and skills to contribute effectively in an environment of ongoing change. Mthembu and Ocholla [38] are of the view that poor or nonexistent training affects the use of digital technologies and the difficulties in retaining qualified staff in most libraries deter adoption of modern technologies for managing the affairs of the library.

Lack of Management Support: Management support and commitment are proportional to the effective adoption and implementation of new technologies in any organization. In this regard, [39] maintain that management has the power to adopt or reject new technologies from being used by the organization. In the case of blockchain technology, most administrators are scared of adopting it, because it has high values of visibility and transparency, which makes it to lack privacy. In advocacy, [40] state that not every administrator supports these values, as it is capable of exposing their illegal practices and decisions.

Poor Broadband Connectivity: The issue of broadband connectivity is worsening daily in the country. Zubairu et al. [41] **reported that** Nigeria has fallen behind others in terms of broadband penetration. Actually, with the increasing rate of Internet connectivity and frequent vandalization of broadband infrastructure, the broadband connectivity in most organizations in the country is usually low.

Irregular Power Supply: Irregular power supply is another issue of great concern concerning the effective functioning of technological appliances in Nigeria. The intermittent power supply does not only cause malfunctioning of information and communication technologies but also leads to poor broadband connectivity, which is very irritating and disruptive.

Data Storage Capacity Limitation: Data storage capacity limitation in blockchain factor limiting its implementation in most organizations. Considering its cost efficiency, performance, and flexibility, the real design challenge is to decide what data and computation should be placed on-chain and what data should be stored off-chain. A common practice for storing data in the blockchain ledger is to store raw data off-chain and to store meta-data, small critical data, and hashes of the raw data only on-chain [42].

In summary, this section has extensively reviewed various literature concerning blockchain technology for a better understanding of the technology. The section highlights various areas of library practices that require the application of blockchain technology and however identify the most challenging factors to the implementation of the technology in the library. Of all the literature reviewed, none is conducted on blockchain technology in library practices. This, therefore, creates the gap which this study intends to fill.

3 Research Method

The study adopts a descriptive survey design. The design is deemed appropriate, because the study intends to describe the existing state of blockchain technology in libraries in the study area without manipulating the variables. The study is concluded

Table 1 Distribution of population size according to gender

S/N	Institutions	Academic librarians		
		Male	Female	Total
1	Libraries	31	37	68
2	Nnamdi Azikiwe library, UNN	17	13	30
3	Festus Aghagbo Nwako library, NAU	5	8	13
4	MOUAU library	6	10	16
5	FUTO library	4	2	6
Total		63	70	133

in Southeast Nigeria, which consists of five states – Enugu, Imo, Abia, Anambra, and Ebonyi. Each of these states has a federal university and their libraries are used for this study (Table 1). The population of the study is 133. It comprises all the academic librarians of the libraries under study. The study is a census study; thus, it requires no sample. A questionnaire and focus group discussion were used for data collection. A total of 133 questionnaires was administered to the respondents, and a returned rate of 86.5% was obtained. Data collected was analyzed using mean.

An item with a mean score below 2.5 is rejected, while items with a mean score of 2.50 and above are accepted.

3.1 *Presentation and Analysis of Data*

Table 2 displays respondents' mean responses on the library practices that require the application of blockchain technology. The data display in the table shows that resource sharing, protection of copyright, and record management with mean scores of 3.06, 3.05 and 2.92, respectively, require application of blockchain technology, whereas library tour ($\overline{X}$ = 2.23) and user education ($\overline{X}$ = 2.03) barely require application of blockchain technology.

Table 3 shows the extent to which blockchain technology could enhance library practices. The data displayed on the table indicates that blockchain technology can enhance the protection of copyright ($\overline{X}$ = 3.69), record management ($\overline{X}$ = 3.59), and resource sharing among libraries ($\overline{X}$ = 3.56) to a very high extent. The table equally reveals that the technology has the potential to enhance endorsement of personal data and encourage partnership between libraries and other organizations to a high extent. The information on the table also indicates that blockchain technology is not an enhancing mechanism for library tours ($\overline{X}$ = 2.37) and user education ($\overline{X}$ = 2.41) as their mean score is below the criterion mean of 2.50.

Table 4 displays the respondents' mean responses on the various ways Nigerian libraries could support and encourage librarians to embrace emerging technologies. As indicated in the table, organizing seminars and workshops ($\overline{X}$ = 3.3), in-service training ($\overline{X}$ = 2.97), sponsoring conferences and workshops ($\overline{X}$ = 2.92), and research grants ($\overline{X}$ = 2.75) are the most supportive measures for encouraging

Table 2 Library practices that require the application of blockchain technology

No	Item statement	Strongly agree	Agree	Disagree	Strongly disagree	Mean	Decision
1	Endorsement of personal data	27	48	27	13	2.77	Agree
2	Record management	41	38	22	14	2.92	Agree
3	Protection of copyright	53	30	17	15	3.05	Agree
4	Resource sharing	41	53	8	13	3.06	Agree
5	Encouraging partnership with other organization	46	29	23	17	2.9	Agree
6	Library tour	20	18	45	32	2.23	Disagree
7.	User education	20	12	34	49	2.03	Disagree

Table 3 Extent blockchain technology could enhance library practices

S/N	Item statement	Very high extent	High extent	Low extent	Very low extent	Mean	Decision
1	Endorsement of personal data	29	37	19	30	2.57	High extent
2	Record management	78	27	10	0	3.59	Very high extent
3	Protection of copyright	85	27	0	3	3.69	Very high extent
4	Resource sharing among libraries	67	45	3	0	3.56	High extent
5	Encouraging partnership with other organization	40	36	27	12	2.9	High extent
6	Library tour	8	38	58	11	2.37	Low extent
7.	User education	36	15	24	40	2.41	Low extent

librarians to embrace emerging technology. However, the respondents disagree on the employment of skilled workers and staff promotion as supportive measures for librarians to embrace emerging technologies. The low mean score of 2.32 and 2.03, respectively, against the items indicated this.

Table 5 above discloses the mean response of the librarian on the challenges associated with the adoption and use of blockchain technology in Nigerian libraries. As indicated in the table, the respondents agree on all the items as factors confronting the adoption and use of blockchain technology in Nigerian libraries. The respondents' high mean scores above the criterion mean of 2.50 on all the items on

Table 4 Ways through which Nigerian libraries could support and encourage librarians to embrace emerging technologies

No	Item statement	Strongly agree	Agree	Disagree	Strongly disagree	Mean	Decision
1	In-service training	30	55	27	3	2.97	Agree
2	Research grant	28	45	27	15	2.75	Agree
3	Organizing seminars and workshop	64	30	13	8	3.3	Agree
4	Promotion	20	12	34	49	2.03	Agree
5	Employing skilled workers	45	36	14	20	2.32	Agree
6	Sponsorship for conferences and workshop	36	10	24	45	2.92	Agree

Table 5 What are the challenges associated with the adoption and use of blockchain technology in Nigerian libraries?

No	Item statement	Strongly agree	Agree	Disagree	Strongly disagree	Mean	Decision
1	Poor funding	68	6	12	29	2.98	Agree
2	Lack of in-service training	24	44	34	13	2.69	Agree
3	Lack of management support	36	16	45	18	2.61	Agree
4	Poor broadband connectivity	34	48	14	19	2.84	Agree
5	Irregular power supply	50	20	10	35	2.74	Agree
6	Lack of skilled technicians	37	53	12	13	2.99	Agree
7	Poor attitude of librarians toward knowledge update	10	12	44	49	1.85	Disagree

the table are proof. The low mean of 1.85 on item no.7 indicates a negative response on the item.

4 Discussion of Major Findings

From the data collected and analyzed in Table 2, it was discovered that various library practices, such as resource sharing among libraries, endorsement of personal data, record management, partnership with other organization, and protection of copyright, require application of blockchain technology. These findings are corresponding with [30], who reports that the problem of piracy is a serious issue between the publishers and libraries. However, he reveals that such a problem can be controlled through the use of blockchain technology. According to [30], blockchain creates a unique, verifiable record that can be accessed by anyone; it could be tied to digital materials and used as a method to show "provable scarcity" of that resource; this would allow digital materials to be uniquely identified, controlled, and transferred [30]. The finding is also in line with [32], who argued the numerous challenges (lack of trained staff, inadequate available resources, lack of mailing or transportation system, inadequate security) characterizing resource sharing can be eliminated through the use of blockchain technology as the technology possesses distinct features, such as anonymity, efficiency, cost savings, security, flexibility, and many others. The finding to the study also relates with [22], who found out that blockchain technology can help libraries in digitizing existing records and manage them within a secure infrastructure; once digital information is committed onto a blockchain, it is permanently stored and impossible to manipulate or hack. It was also discovered that user education and library tours do not require blockchain technology. Response from the focus group discussion in this regard reveals that user education and library tour are programs through which librarians educate users on how to make efficient use of library resources and for the fact that blockchain is usually for safeguarding documents and enhancing online financial payment it does not have much influence on instruction programs.

Data presented in Table 3 reveals that blockchain technology could enhance most library practices to a high extent. This finding corresponds with [34] that blockchain can highly support collaborations, because it is a digital ledger where several people have joint control over shared information. It is also in line with the findings of [14], who after conducting a study on the applications of blockchain in libraries reveals that the technology among other things greatly facilitates partnership across centers/organizations.

The study also discovered various supportive measures that could encourage librarians to embrace emerging technology. These measures among other things are in-service training, organizing seminars and workshops, and providing research grants. This finding agrees with [37]; although the core knowledge and skills of traditional librarianship are still useful in this digital era, they need to be augmented by new technological knowledge. It also relates with the finding of [38] that poor or nonexistent training affects the use of digital technologies and the difficulties in retaining qualified staff in most libraries deter adoption of modern technologies for managing the affairs of the library. The respondents, through a focus group discussion, revealed that research grants, sponsoring of conferences, and workshops are very good measures for skill update, but unfortunately, they are not awardable by most libraries due to insufficient budget. The respondents also report that at times sponsoring conferences and workshops is not certain, because the library administration may not be buoyant at the time of conference or workshop.

Finally, the study identified various factors delaying the adoption and use of blockchain technology in Nigerian libraries. These factors among others include poor funding, inadequately skilled technicians, poor broadband connectivity, lack of in-service training, irregular power supply, and lack of management support. The finding coincides with [35] that inadequate government funding inhibits IT application by African libraries. It also corresponds with [41] that the issue of broadband connectivity is worsening daily in the country. The respondents in a focus group discussion also report on the poor state of Internet connectivity. They explain that poor Internet connectivity and intermittent power supply hinder most of the online conferences and seminars, which would have been a great help to them to upgrade their knowledge and skill.

Implication of the Study
The findings of the study show that various practices of the library need to be enhanced to achieve greater productivity, standardization, and reliability. Most irregularities, like delay in service delivery, resource sharing, poor record management, and inadequate protection of authors' copyright, are issues of great concern. The implication is that the library may lose its worth and users will develop negative thinking and attitudes toward its usage. This will consequently lead to the underutilization of the library and its resources.

5 Conclusion

Based on the research findings, the paper concludes that the poor state of most library practices in the study area is not encouraging. But related literature reviewed shows that blockchain could serve best in enhancing the various practice of the library. However, some factors were identified as constraints hindering the adoption and use of blockchain in libraries. These factors among others are poor funding, lack of management support, irregular power supply, and poor broadband connectivity. There is, therefore, a need to take necessary action to encourage the adoption and use of blockchain technology in the library, since it can improve most library activities and achieve global best practices.

Acknowledgments It is a pleasure to acknowledge the support provided by Nnamdi Azikiwe Library, University of Nigeria, through its computer center and cataloging section. My thanks also go to my mentors, who always encouraged me to add to existing knowledge. To the authors, especially [43] whose article titled "A Step-by-Step Guide for Choosing Project Topics and Writing Research Papers in ICT Related Disciplines" serves as the pillar for building this paper, and librarians of various universities that were used for this research study, I am very grateful to you all.

References

1. Anyim, W. O. (2017). Improving reference services in Federal University Libraries in Southeast Nigeria using interpersonal communication mechanism. *Review of Information Science and Technology Journal*, 2(1), 27–38.
2. Noh, Y., & Chang, R. (2020). A study on the factors of public library use by residents. *Journal of Librarianship and Information Science 2020*, *52*(4), 1110–1125.

3. Bohannon, J. (2016). The bitcoin busts. *Science, 351*(6278), 1144–1146.
4. Treiblmaier, H., & Beck, R. (2019). *Business transformation through blockchain* (Vol. 2, pp. 3–8). Springer.
5. Balaji, P. B., Vinay, M. S., Shalini, B. G., & Mohan, J. R. (2018). An integrative review of web 3.0 in academic libraries. *Library Hi Technical News, 35*(4), 13–17.
6. Lnes, S., Ubacht, J., & Janssen, M. (2017). Blockchain in government: Benefits and implications of distributed ledger technology for information sharing. *Government Information Quarterly, 34*, 355–364.
7. Abayomi-Zannu, T. P., Odun-Ayo, I., Tatama, B. F., & Misra, S. (2020). Implementing a mobile voting system utilizing blockchain technology and two-factor authentication in Nigeria. In *Proceedings of first international conference on computing, communications, and cyber-security (IC4S 2019)* (pp. 857–872). Springer.
8. Malyarov, N. (2019). The impacts of advanced technology on libraries and education. *The Insider: The Future of Libraries Magazine.*
9. Gillwald A., Odufuwa, F., & Mothobi, O. (2018). *The state of ICT in Nigeria.*
10. United Nations. (2016). *E-government survey*. UNO.
11. Zubairu, H. A., Oyefolahan, I. O., Babakano, F. J., Etuk, S. O., & Mohammed. (2016). I.: Assessing the e-readiness of Nigeria for digital economy. *American Journal of Computer Science and Information Technology, 8*(2), 50.
12. Ayo-Olafare, F. R. (2020). Global trends and emerging technologies in libraries and information science. *Library Philosophy and Practice (e-journal), 3835.*
13. Hoy, M. B. (2017). An introduction to the blockchain and its implications for libraries and medicine. *Medical Reference Services Quarterly, 36*(3), 273–279.
14. Brown, J. L. (2018). Blockchain in the library? Researchers explore potential applications. *Digital Learning in Higher Education.*
15. Coghill, J. G. (2018). Blockchain and its implications for libraries. *Journal of Electronic Resources in Medical Libraries, 15*(2), 1–5.
16. Atozori, M. (2017). Blockchain technology and decentralized governance: Is the state still necessary? *Journal of Governance and Regulation, 6*(1), 45–62.
17. Blockgeeks, S. (2016). *What is blockchain technology?* A step-by-step guide for beginners http://blockgeeks.com/guides/what-is-blockchaintechnology.
18. Shaw (2016). *Smart surveillance: An enterprise-grade video surveillance solution for small and medium-sized businesses.* United State of America: California.
19. Mougayar, W. (2016). *The business blockchain: Promise, practice, and application of the next internet technology*. Wiley.
20. Ying, W., Jia, S., & Du, W. (2018). Digital enablement of blockchain: Evidence from HNA group. *International Journal of Information Management, 39*, 1–4.
21. Bünz, B., Bootle, J., Boneh, D., Poelstra, A., Wuille, P., & Maxwell, G. (2017). Bulletproofs: Short proofs for confidential transactions and more. In *Institute of Electrical and Electronics Engineers (IEEE) symposium on security and privacy (SP)* (pp. 315–334).
22. Masenya, T. M. (2020). Application of modern technologies in the management of records in public libraries. *Journal of the South African Society of Archivists, 53*, 65–79.
23. Barnes III, B. W., Xiao, B. (2019). *Organizational adoption of blockchain technology: An ecosystem perspective proceeding of the twenty-fourth digital workshop.* Mumch, Germany, December.
24. Weking, J., Mandalenaki, M., Hein, A., Hermes, S., Böhm, M., & Krcmar, H. (2019). The impact of blockchain technology on business models – A taxonomy and archetypal patterns. *Electronic Markets, 2020*(30), 285–305.
25. Short, C. (2018). Blockchain applications. *Modern Trader, 1*(539), 21.
26. Silvia, A. C (2019). *Blockchain challenges to copyright revamping the online music industry.* Master of Laws in European Intellectual Property Law Stockholm University.
27. Shepherd, E., & Yeo, G. (2003). *Managing records; a handbook of principles and practice.* Facet publishing.

28. Raju, J. (2014). Knowledge and skills for the digital era academic library. *The Journal of Academic Librarianship, 40*, 163–170.
29. Rein, L., & Peterson, A. (2015). *What you need to know about the hack of government background investigations*. The Washington Post.
30. Griffey, J. (2016). Blockchain and intellectual property. *Internet Librarian*.
31. Rifaudeen, M. M. (2008). The problems of resource sharing in Sri-Lanka: the case among scientific and technical libraries. A conference paper presented at the *International conference on information and knowledge management at Healthnet Nepal*, T.U. Central Library, Nepal March.
32. Bauerle, N. (2017). *How does blockchain technology work?* https://www.coindesk.com/information/how-does-blockchain-technology-work.
33. Chakravorty, A., Liang, C.R. (2017). Ushare: Use controlled social media based on blockchain. *Proceedings of the 11th international conference on ubiquitous information management and communication*. IMCOM.
34. Lumineau, F., Wang, W., Schilke, O., & Huang, L. (2021). How blockchain simplify partnerships. *Harvard Business Review*.
35. Irenoa, K. O., Emilian, B., & Eru, J. (2019). Funding academic libraries in Nigeria for effective services: Alternative to resource development and library management. *Journal of Library and Information Science, 21*(1), 104–111.
36. Alabi, A. T., Mojisola, O. O., & Abdulkareem, A. Y. (2013). Budgeting systems in universities in Southwest Nigeria. *Makere Journal of Higher Education, 4*(2), 203–219.
37. Dalkir, K. (2017). *Knowledge management in theory and practice* (3rd ed.). Cambridge MIT Press.
38. Mthembu, M. S., & Ocholla, D. (2019). Perceptions on job requirements of LIS graduates in public libraries: A reflection on public libraries in KwaZulu-Natal. *South Africa Mousaion, 36*(4), 1–5.
39. Amron, M. T., Ibrahima, R., Bakara, N. A., & Chuprata, S. (2019). Determining factors influencing the acceptance of cloud computing implementation. The fifth information systems international conference. *Procedia Computer Science, 161*, 1055–1063.
40. Leible, S., Schlager, S., Schubotz, M., & Gipp, B. (2019). A review on blockchain technology and blockchain projects fostering open science. *Frontiers Blockchain, 2*, 16.
41. Zubairu, H. A., Oyefolahan, I. O., Babakano, F. J., Etuk, S. O., & Mohammed, I. (2020). Assessing the e-readiness of Nigeria for digital economy. *American Journal of Computer Science and Information Technology, 8*(2), 50.
42. Xu, Q., Aung, K. M., Zhu, Y., & Yong, K. L. (2028). A Blockchain-based storage system for data analytics in the internet of things. *Studies in Computational Intelligence, 715*, 119–138.
43. Misra, S., Muhammad-Bello, B. (Eds.) (2021) *ICTA 2020, CCIS 1350* (pp. 727–744). Springer Nature Switzerland AG.

A Comparative Study of Regression Analysis for Modelling and Prediction of Bitcoin Price

Yakub Kayode Saheed, Raji Mustafa Ayobami, and Terdoo Orje-Ishegh

1 Introduction

Bitcoin is an electronic peer-to-peer cash system [1] conceived in 2008 by Satoshi Nakamoto, as the world's first cryptocurrency based on blockchain technology to address the inherent shortcoming of the trust transaction-based model [2]. In 2008, an unknown scientist known as Satoshi Nakamoto published an article entitled "Bitcoin: A Peer-to-Peer (P2P) Electronic Currency System" introducing the notion of peer-to-peer (P2P) cash transfers for Internet payments that do not require the involvement of any financial intermediaries [3]. He showed the concept of a decentralized chain of legitimate transactions, referred to as a chain of blocks, that is spread across all network peers. It can be achieved using consensus procedures based on time stamps and hashes that are based on proof of work (PoW). Because the chain is disseminated to all nodes or peers, it is transparent in nature. This introduces the concept of cryptocurrency, a new type of digital currency [3].

According to the CoinMarketCap, Bitcoin is an asset traded in the world with more than 16,000 markets [2]. It is known to be the most controversial digital currency [4]. There are other virtual currencies which have been introduced in the past such as Liberty Reserve (between 2006 and 2013) and eGold (between 1996

Y. K. Saheed (✉)
School of IT & Computing, American University of Nigeria, Yola, Nigeria

Al-Hikmah University, Ilorin, Nigeria
e-mail: yakubu.saheed@aun.edu.ng

R. M. Ayobami
Al-Hikmah University, Ilorin, Nigeria

T. Orje-Ishegh
School of IT & Computing, American University of Nigeria, Yola, Nigeria

S. Misra, A. Kumar Tyagi (eds.), *Blockchain Applications in the Smart Era*, EAI/Springer Innovations in Communication and Computing,
https://doi.org/10.1007/978-3-030-89546-4_10

and 2014), but they were not successful in developing as a results of laundering issues [5]. Traditional economies rely entirely on third-party financial organizations (that is banks) to process all forms of payment, both cash and electronic. These institutions function as intermediaries between parties trading funds and maintain total control over transactions. While it is effective for financial transactions, it allows for just a limited amount of money to be transferred and lacks transparency, trust, flexibility, and security. To overcome the difficulties, we require a system that can eradicate financial transaction intermediaries, allowing for direct cash transfer between parties, so altering the way the economy operates [6]. Bitcoin is a well-known crypto currency that has the top-notch market investment among all other available crypto currencies [7]. Consequently, many investors are intense to capitalize in Bitcoin than any other crypto-currencies, which has attracted more research efforts on the price of Bitcoin predictions [8]. There are heated deliberations in reply to two important questions that why did Bitcoin have value and what determines Bitcoin value? In financial innovation field, the value of Bitcoin indicates the confidence level of stockholders in cryptocurrency [9]. The advantages of Bitcoin are that in the traditional currency transactions, there is a third party which is the bank and has the fundamental problem of central authority taking massive commission and long duration of time for transaction. The Bitcoin blockchain assists in removing the central authority by sharing the control of the Bitcoin system among all the users [5]. Nowadays, cryptocurrencies are used as a medium of exchange for associated expenses, which is the primary purpose Bitcoin was created, as well as for speculation [10–12]. Additionally, payment rails are used for low-cost cross-border money transfers and a variety of non-monetary purposes, like time stamping [13].

Therefore, several past studies focus on the determinants of the price of Bitcoin. The price fluctuation patterns as a result of the inherent volatility of Bitcoin have overwhelmed stockholders since the time Bitcoin has been traded [14, 8]. McNally et al. [15] proposed various model using the recurrent neural network (RNN), the long short-term memory (LSTM), and Arima models. The models are gotten from the features high, open, close, and low data features of the coin desk and data of hash rate from blockchain. The obtained findings indicated a high accuracy of 52.78% with RMSE of 5.45%. Greaves et al. [16] presented transaction data graph for prediction of Bitcoin prices. They used logistic regression, linear regression, neural network, and SVM models. The accuracy of their results achieved 55%. The authors [17] used OkCoin dataset for Bitcoin price prediction. They divided the data into three series: 30 minutes, 60 minutes, and 120 minutes. SVM, random forest, and binomial logistic regression were utilized to forecast the prices of Bitcoin with an obtained accuracy of 97 percent and 55 percent. However, their proposed models are prone to overfitting. The study [18] utilized numerous regression models, such as random forests [19], neural networks, linear regression, and gradient boosting [20]. They used the attributes that are interrelated to the price of Bitcoin to construct the regression models. Li et al. [21] and Kim et al. [22] used social media data for the prediction of Bitcoin price fluctuations. Greave et al. [16] used SVM and ANN to predict the price of Bitcoin. The findings of their work showed an accuracy of

55%. Phaladisailoed et al. [23] proposed a ML model for Bitcoin price prediction. They used the Huber regression, LSTM, and gated recurrent unit (GRU) models for the prediction of Bitcoin prices. The results obtained indicated that the GRU has the highest accuracy with the lowest MSE and R^2 values. An ensemble of neural network approach was presented in [24]. The authors presented MLP for the Bitcoin prices prediction with an accuracy of 58% to 63%. Geourgoula et al. [25] presented the price of Bitcoin determinants and also implemented a sentiment analysis using the SVM. Nevertheless, all the studies mentioned above did not consider various regression models for Bitcoin price prediction, and none has applied the Theil-Sen regressor for Bitcoin price prediction. In addition, majority of the works also did not consider the RMSE, RMSLE, and MAPE as the performance metrics for evaluating the performance of regression model for Bitcoin prices prediction. The changes in Bitcoin price are very sacrosanct to predict from time to time.

Several scholars gave an account of the demanding properties of Bitcoin and its interaction with other resources [26–30]. The valuation of the price-changing aspects of Bitcoin is very challenging as a result of the nonlinear, nonstationary, influence of several uncontainable factors and volatile manners [31, 32]. This gave opportunity for further research efforts in the Bitcoin price prediction [33]. Many of the reported works have studied the issues that influence the price of Bitcoin and the fluctuation forms based on experimental and analytical approaches [34]; an example are the works by the author [35] and researcher [36]. As a result of the progress in ML, several DL methods for Bitcoin prices prediction have also been reported in [1, 18, 37–41].

Additionally, most reported studies addressed only the classification issue s[42], where the model forecasts if the prices of the next day will go down or up in respect of the past prices. Further, for some of the reported regression problem, the issue of performance of the models is done in respects of the MAPE between the predicted real values and the predicted values or in root mean square error (RMSE). Nevertheless, a low MAPE or RMSE does not mean the models are very effective. This study fulfilled this gap by utilizing and introducing more performance metrics to evaluate the performance of Bitcoin price prediction based on six regression models. The central contribution of this research is that in the data preprocessing stage, we used z-score for normalization after which six models were utilized to predict the daily Bitcoin prices and the prices for the next 5 days, respectively. Additionally, the evaluation metrics for the performance of Bitcoin prices was extended in this paper.

This paper aims to propose six regression models for Bitcoin price prediction. This paper is structured as follows. Section two is the related work. We highlight the methodology employed in Sect. 3 and present in Sect. 4 the results with the discussion. The conclusion with the future work is presented in Sect. 5.

2 Literature Review

The authors [43] combined ANN with genetic algorithm optimization. To prevent achieving a local minimum outcome, the initial weights of ANN were optimized using a genetic approach. Ref. [24] used the ensemble of ANN methodology known as the GASEN ensemble to forecast Bitcoin values. The authors [44] employed an artificial neural network to forecast cryptocurrency closure prices, and they compared the price movements to those seen on traditional stock exchanges. Ref. [45] utilized tweets to do Bitcoin analysis of sentiment and make predictions.

The researchers in [46] forecasted Ethereum, Bitcoin Cash, and Bitcoin prices using tweets and historical data. To forecast Bitcoin values, Wu et al. [47] utilized an LSTM model. They created a new method for forecasting bitcoin prices as well as a solution to the LSTM input variable selection problem. To anticipate Bitcoin price changes, Mohanty et al. [48] combined daily generated data such as block size, price, number of the transactions per block, and other Twenty-six (26) attributes of the Bitcoin blockchain with twitter data. The ARIMA model was used by Roy et al. [49] and the study [50] to predict Bitcoin prices. The ARIMA model was used by the authors [51] to estimate the Bitcoin close price for 545 days. The percentage mean error is derived once the acquired results are compared to actual prices. For most of the data, the current mean error is less than 6%. The authors [52] examined the accuracy of decision trees and linear regression in predicting Bitcoin prices using historical time series data. The authors [53] presented three different ML model known as LSTM, ARIMA, and GRU for predicting the Bitcoin prices. The obtained results showed that the ARIMA model gave 302.53% for the RMSE and 2.76% for the MAPE. Hence, the ARIMA model gave outstanding results than the deep neural network-based models. The authors [54] used LSTM and ANN to predict the prices of Monero, Bitcoin, Stellar Lumens, Ripple, Litecoin, and Bitcoin. They evaluate the performance of the two models and performed sensitivity test to study the behavior of the model. The findings showed that ANN surpasses LSTM. Ref. [55] upgraded the SVM with PSO for the Ethereum, Bitcoin, Stellar, and Ripple prices prediction. Aggarwal et al. [56] attempted to quantify the effect of socioeconomic variables on Bitcoin prices. The authors [57] utilized the HMMs in combination with LSTM and N-LSTMs to produce better outcomes in comparison to conventional ARIMA and LSTM for the Bitcoin prices for 3 days.

A new cryptocurrency prediction method was developed in [6], and it utilizes a GRU-based LSTM neural network and a GRU hybrid network to forecast only two cryptocurrencies, namely, Monero and Litecoin. The findings show that the proposed strategy works as intended, as demonstrated by its ability to forecast multiple cryptocurrencies with impressive accuracy. Table 1 depicts the existing strategies and literatures for predicting cryptocurrency prices.

Table 1 Existing methods for cryptocurrency price prediction

Research paper	Year	Methodology	Cryptocurrency	Forecast duration	Results
[43]	2017	BP2NN, NEAT, GABPNN	Bitcoin	One (1) day	MAPE and BPNN = 1.998 ± 0.038, GABPNN = 1.883 ± 0.066, NEAT = 2.175 ± 0.096
[24]	2017	GASEN	Bitcoin	Fifty (50) days	Accuracy = 58% - 63%
[44]	2018	Rprop algorithm and ANN	Bitcoin, BitcoinCash, Dash	Thirty-one (31) days and 150 h	Accuracy = 75% to 97.3% for Bitcoin
[45]	2018	Multi-linear Regression	Bitcoin, and Litecoin	Three (3) days	R^2score = 44% for Litecoin and 59% for Bitcoin
[46]	2018	ANN	Bitcoin, Ethereum, BitcoinCash	Three (3) months	Accuracy: Bitcoin = 85%, Ethereum = 93.33%, BitcoinCash = 70%
[47]	2018	LSTM, AR	Bitcoin	Seventy-one (71) days	RMSE: 247.33
[48]	2018	Bidirectional LSTM, Word2vec	Bitcoin	Six months	AR = 50% Precision = 60.99%
[49]	2018	ARIMA	Bitcoin	Ten days	ARIMA model = 90.31%, AR = 89.25%, MA = 87.58%
[52]	2018	ARIMA	Bitcoin	Five hundred and forty-five (545) days	Accuracy = 60–70%
[53]	2019	Regression, decision tree	Bitcoin	Five (5) days	Accuracy; decision tree = 95.88013, regression = 97.59812
[54]	2019	ARIMA, LSTM, GRU	Bitcoin	Four hundred and nine-two (492) days	RMSE = ARIMA: 302.53; LSTM = 603.68. GRU = 381.34
[55]	2019	ANN, LSTM	Bitcoin, Ethereum, ripple, stellar	One (1), ten (10), twenty (20), thirty (30) days	RMSE = 53.30: One day, 67.99: Ten days, 91.41: Twenty days; 45.71: Thirty days for bitcoin

(continued)

Table 1 (continued)

Research paper	Year	Methodology	Cryptocurrency	Forecast duration	Results
[56]	2019	SVM with PSO	Bitcoin, Ethereum, Litecoin, Nem, ripple, stellar	One (1) year	Bitcoin = 90.4, Ethereum = 97, Litecoin = 92.1, Nem = 57.8, ripple = 82.8, stellar = 64.5
[57]	2019	CNN, LSTM, GRU	Bitcoin	One (1) and Three (3) months	RMSE: Gold prices CNN = 201.34, LSTM = 151.67, GRU = 179.23 twitter sentiments LSTM = 32.98
[50]	2019	HMM, LSTM, GA	Bitcoin	Three (3) days	RMSE:LSTM = 7.006; HMM:LSTM = 5.821
[51]	2019	ARIMA	Bitcoin	Seven (7) days	Least MAPE: 0.87 (1 day), 5.98 (7 days)
[6]	2020	LSTM, GRU	Litecoin and Monero	One (1), three (3), seven (7) days	RMSE: 1 day: L = 2.2986, M = 3.2715, 3 days: L = 2.0327, M = 5.5005, 7 days: L = 4 0.5521, M = 20.2437

Table 2 The extracted features

Features name	Feature description
Date	Price of bitcoin for a particular date
Open	The price opening
High	The price high
Low	The price low
Close	The closed bitcoin price at particular day
Adjacent close	The adjacent closed bitcoin price at particular day
Volume	The top exchange volume

3 Materials and Methodology

We describe the methodology used in this section by adopting the method of research paper methodology writing reported in [58]. The dataset utilized in this research was obtained from the Kaggle dataset for a period of 2014–2020 [59]. The dataset consists of seven (7) features and two thousand one hundred and twenty (2123) instances. In this study, we first performed data preprocessing using the concept of z-score method of normalization, where the mean value is 0 and the standard deviation is 1. The data preprocessing is very vital as it helps in eliminating outliers and removing redundant attributes. The output of the z-score normalization is fed into the regression models. The preprocessed dataset is then train by the CatBoost regressor, GBR, ETR, AdaBoost regressor, K-neighbors regressor, and the Theil-Sen regressor. The information of the dataset utilized in this research is given in Table 2.

3.1 Data Preprocessing

This is the proposed method for the initial step. The aim of preprocessing data is to convert raw data into a format that is simpler and more convenient to use for subsequent processing stages [60]. In this initial step, we normalize the data using the z-score normalization technique.

3.2 Z-Score Normalization

Z-score normalization is also referred to as the zero-mean normalization [61]; it is a technique which gives the range of data from the unstructured data utilizing the concept of standard deviation and mean [62]. Z-score normalization allowed attributes to be reordered with their standard deviation equal 1 and mean equal 0. The data can be normalized utilizing z-score as:

$$z' = \frac{z - mean\ (z)}{\rho} \tag{1}$$

3.3 CatBoost Regressor

CatBoost is referred to as the boosting classification algorithm that is ordered [63] and based on the gradient boosting classification algorithm [64] that utilizes the decision tree (DT), which is oblivious as the predictor base [65]. The DT are utilized for the regression where each tree means a partition of the attribute space and the output value [66]. Moreover, in CatBoost model, the defined parameters controlled the number of trees. To avoid overfitting, the CatBoost model utilizes the detector for overfitting that exist in the algorithm [67].

3.4 Gradient Boosting Regressor

GBR is a very influential approach for uniting numerous individual classifiers to yield a committee in which the performance is far better than any of the individual classifier [68]. The chief idea of the boosting algorithm is to complement novel models to the ensemble model sequentially. GBR allows optimization of loss function that are arbitrary differentiable [69]. GBR is also a generalization boosting algorithm that consists of three (3) elements [70], a loss function, an additive model, and a weak learner [71].

3.5 Extra Trees Regressor

Extra tree is made up of a difference of RFS in which the whole dataset is utilized at each of the instance [72], and the tree splits are selected at random completely [73].

3.6 AdaBoost Regressor

The AdaBoost algorithm was announced by Freund and Schapire in 1995 [74]. AdaBoost.R2 [75, 76] boosting is an improvised alteration of the AdaBoost introduced by Freund and Schapire that is an extension of AdaBoost.m2 for regression problems. AdaBoost is a successful procedure in the area of ML that add weak learners to construct a robust classification algorithm [77]. AdaBoost was introduced to address regression and feature classification. But as a result of its outstanding classification performance, it is been used for image processing [78].

3.7 K-Neighbor Regressor

The KNN is a method of learning that is instant based [79], where the function is estimated and all the computation is delayed till the classification phase [80]. The same approach can be utilized for regression problem by conveying the value of the thing to the regular of values of KNN. This can be used to weigh contributions of neighbors in which the closer neighbors commit to average, the more the distant ones. For regression problem, the prediction is the average of the closer neighbor outcomes [81].

3.8 Theil-Sen Regressor

The Theil-Sen regressor is an estimator that is nonlinear and less sensitive to the outliers. The Theil-Sen regressor can achieve precise and robust results [82]. It is also known to be median estimator based that is robust contrary to multivariate outliers. The Theil-Sen robustness estimator reduces with respect to the dimensionality of the problem [83].

3.9 Performance Metrics Adopted for All the Models

Mean Absolute Error

The MAE agrees to the value of the absolute error (L_1) and can be represented as follows.

$$\text{MAE}\left(\text{k},\check{\text{k}}\right)=\frac{1}{v}\sum_{i=1}^{j}\mid ki-\check{\text{k}}\mid \tag{2}$$

The v refers to the samples number in k, and the ki to ith sample of k.

3.10 Mean Squared Error

The MSE correlate with the value expected in the error of the quadratic. Neglecting the term of $\frac{1}{n}$, the MSE would become L2 function loss, and it is utilized as a function cost for the purpose of optimization in which both yield the comparable results. The MSE can expressed as follows.

$$\text{MAE}\left(\text{k}, \breve{\text{k}}\right) = \frac{1}{v}\sum_{i=1}^{j}\left(ki - \breve{k}\right)^2 \tag{3}$$

3.11 Root Mean Squared Error

The RMSE is the root square of the square of the mean of the difference between the predicted and those of the actual observation. RMSE is a statistical metric that is standard to assess the performance of a model [84].

$$\text{RMSE} = \sqrt{\frac{1}{mx\Pi}\sum_{1=1}^{N}\sum_{j=1}^{M}\left(y_{\ddot{y}} - yiJ\right)^2} \tag{4}$$

3.12 Coefficient of Determination (R^2)

The R^2 is the amount of variance in the attribute dependent that is predictable from the feature independent in the sample which is known as R^2. The R^2 provides a means of how good the future samples would likely be predicted by the model [85]. The R^2 can be given as follows.

$$\text{R2} = \frac{SSbetween}{SStotal} \tag{5}$$

3.13 Root Mean Squared Logarithmic Error

The RMSLE finds out difference of values forecast by the model and the real values. The RMSLE can be expressed as follow [86].

$$\sqrt{\frac{1}{N}\sum_{i=1}^{N}(\log(yi + 1) - \log(yi + 1))^2} \tag{6}$$

3.14 Mean Absolute Percentage Error

The MAPE can be referred to as the metric to evaluate how accurate the forecast model performed. This accuracy measures as a percentage.

$$M = \frac{1}{v} \sum_{i=1} v \mid \frac{Ai - Fi}{Ai} \mid \quad (7)$$

4 Results and Discussion

4.1 *Correlation Analysis of the Bitcoin Dataset Features*

We report the experimental analysis in this section. First, we perform correlation on the dataset in other to examine the features that are highly correlated between the independent features and the dependent features. The finding of our experiment is depicted in Table 3.

As can be seen in Table 3, the assessment of the correlation was carried out corresponding to all features. All the features are really correlated; therefore, we used the volume and close as the features to fit in the model.

4.2 *Performance of the Six Models Using the Bitcoin Daily Price*

We used the six models to perform analysis on the Bitcoin daily price in this section. As seen in Table 4, the CatBoost regressor gave MAE of 2.592, MSE of 3.585, RMSE of 5.854, R^2 of 0.7116, RMSLE of 1.264, and MAPE of 3.27. The gradient boosting regressor gave a MAE of 3.049, MSE of 3.981, RMSE of 6.196, R^2 of 0.6779, RMSLE of 1.361, and MAPE of 3.681. The extra tree gave a MAE of 2.468, MSE of 4.153, RMSE of 6.333, R^2 of 0.6603, RMSLE of 0.4875, and MAPE of 0.4117. The AdaBoost gave a MAE of 4.281, MSE of 5.49, RMSE of 7.356, R^2 of 0.5479, RMSLE of 1.226, and MAPE of 6.117. The K-neighbor gave a MAE of 3.757, MSE of 6.078, RMSE of 7.726, R^2 of 0.4979, RMSLE of 0.6178, and MAPE of 0.5911. The Theil-Sen gave a MAE of 4.796, MSE of 6.133, RMSE of 7.765, R^2 of 0.4987, RMSLE of 2.392, and MAPE of 21.07.

Table 3 Correlation analysis

	Open	High	Low	Close	Adjacent close	Volume
Open	1.000000	0.998812	0.998053	0.997496	0.997496	0.678341
High	0.998812	1.000000	0.997567	0.998915	0.998915	0.678466
Low	0.998053	0.997567	1.000000	0.998665	0.998665	0.680555
Close	0.997496	0.998915	0.998665	1.000000	1.000000	0.679068
Adj close	0.997496	0.998915	0.998665	1.000000	1.000000	0.679068
Volume	0.678341	0.678466	0.680555	0.679068	0.679068	1.000000

Table 4 Performance evaluation of the six regression models on Bitcoin daily price

Models	MAE	MSE	RMSE	R^2	RMSLE	MAPE
CatBoost regressor	2.592	3.585	5.854	0.7116	1.264	3.27
Gradient boosting regressor	3.049	3.981	6.196	0.6779	1.361	3.681
Extra tree	2.468	4.153	6.333	0.6603	0.4875	0.4117
AdaBoost	4.281	5.49	7.356	0.5479	1.226	6.117
K-neighbor regressor	3.757	6.078	7.726	0.4979	0.6178	0.5911
Theil-Sen regressor	4.796	6.113	7.765	0.4987	2.392	21.07

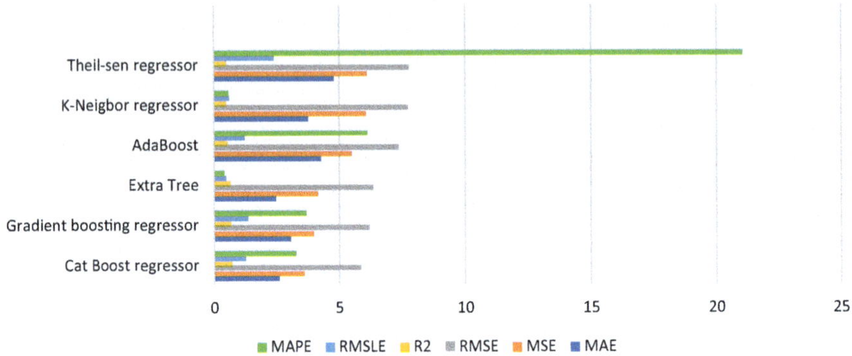

Fig. 1 Performance of the models on Bitcoin daily prices

The performance of each of the six models for the Bitcoin daily prices prediction was revealed in Fig. 1. As can be seen from Fig. 1, the Theil-Sen gave the highest MAPE, while the CatBoost gave the least MAE. The extra tree yielded the lowest MAPE.

4.3 Model Interpretation for the CatBoost Regressor

The CatBoost performance interpretation for the features is illustrated in Fig. 2. The interpretation assists the CatBoost model to analyze what CatBoost model thinks is needed and important. The CatBoost interpretation is performed based on SHAP. The plot in Fig. 2 sorts the attributes by the summation of SHAP value over all the samples and utilized SHAP values to describe the circulation of each attribute impacts has on the output of the model. The color denotes the attribute value (blue low, red high).

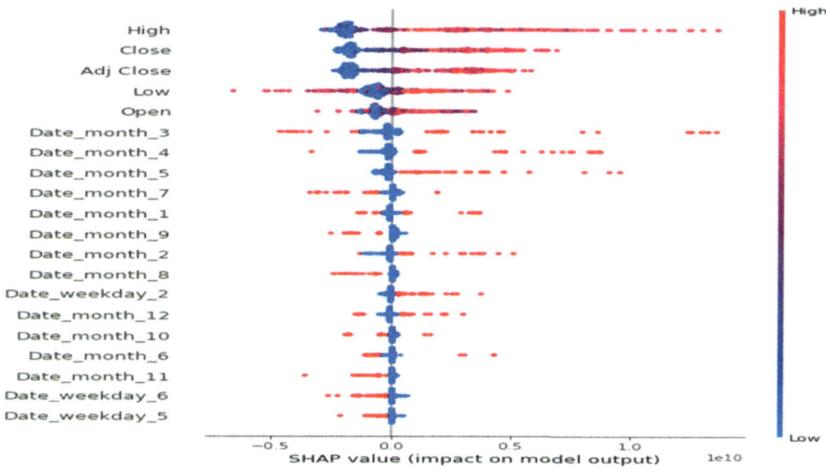

Fig. 2 CatBoost regressor model interpretation

4.4 Model Interpretation for the Gradient Boosting Regressor

The gradient boosting performance interpretation for the features is illustrated in Fig. 3. The interpretation assists the gradient boosting model to analyze what gradient boosting model thinks is needed and important. The gradient boosting interpretation is performed based on SHAP. The plot in Fig. 3 sorts the attributes by the summation of SHAP value over all the samples and utilized SHAP values to describe the circulation of each attribute impacts has on the output of the model. The color denotes the attribute value (blue low, red high).

4.5 Model Interpretation for the Extra Tree Regressor

The extra tree performance interpretation for the features is illustrated in Fig. 4. The interpretation assists the extra tree model to analyze what extra tree model thinks is needed and important. The extra tree interpretation is performed based on SHAP. The plot in Fig. 4 sorts the attributes by the summation of SHAP value over all the samples and utilized SHAP values to describe the circulation of each attribute impact has on the output of the model. The color denotes the attribute value (blue low, red high).

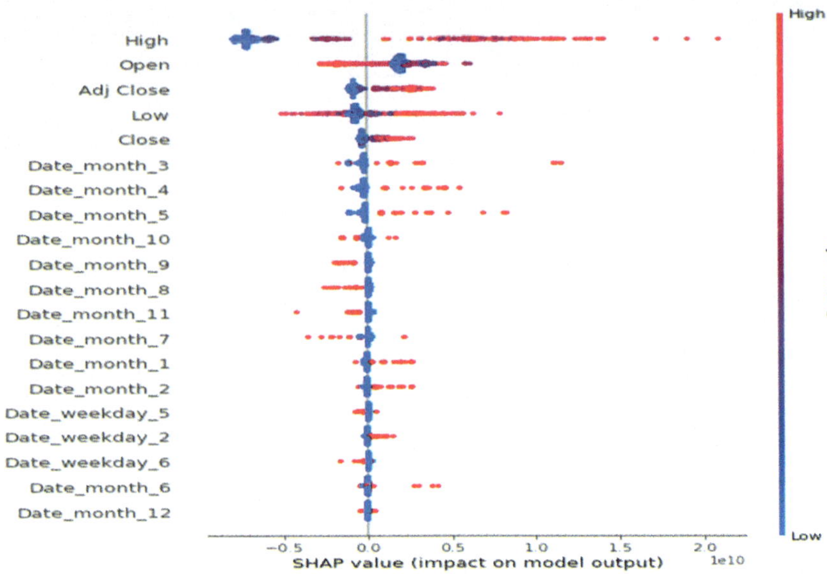

Fig. 3 Gradient boosting regressor model interpretation

4.6 *Predictive Performance of the Six Models for the Bitcoin Price in the Next Five Days*

The predictive performance for the six regressor models was carried out in order to ascertain the price of Bitcoin in the next 5 days. The experimental findings for each of the regressor is illustrated in Table 5. The CatBoost regressor revealed a MAE of 3.2843, MSE of 5.5696, RMSE of 7.4629, R^2 of 0.6248, RMSLE of 1.2071, and MAPE of 2.8631. The gradient boosting regressor generated a MAE of 3.4984, MSE of 5.4842, RMSE of 7.4055, R^2 of 0.6305, RMSLE of 1.1271, and MAPE of 2.4062. The extra tree regressor gave a MAE of 3.1447, MSE of 6.2393, RMSE of 7.8989, R^2 of 0.5797, RMSLE of 0.553, and MAPE of 0.4783. The AdaBoost gave a MAE of 4.6398, MSE of 6.7166, RMSE of 8.1955, R^2 of 0.5475, RMSLE of 0.9367, and MAPE of 1.3695. The K-neighbor regressor gave a MAE of 4.0944, MSE of 7.2222, RMSE of 8.4984, R^2 of 0.5135, RMSLE of 0.6387, and MAPE of 0.5899. The Theil-Sen regressor generated a MAE of 5.3670, MSE of 8.1152, RMSE of 9.0084, R^2 of 0.4533, RMSLE of 2.2644, and MAPE of 18.3188.

The performance of each of the six model for the Bitcoin prices prediction for the next 5 days was revealed in Fig. 5. As can be seen from Fig. 5, the Theil-Sen gave the highest MAPE; the extra tree gave the least RMSLE and also yielded the least MAPE. The gradient boost gave the least MSE.

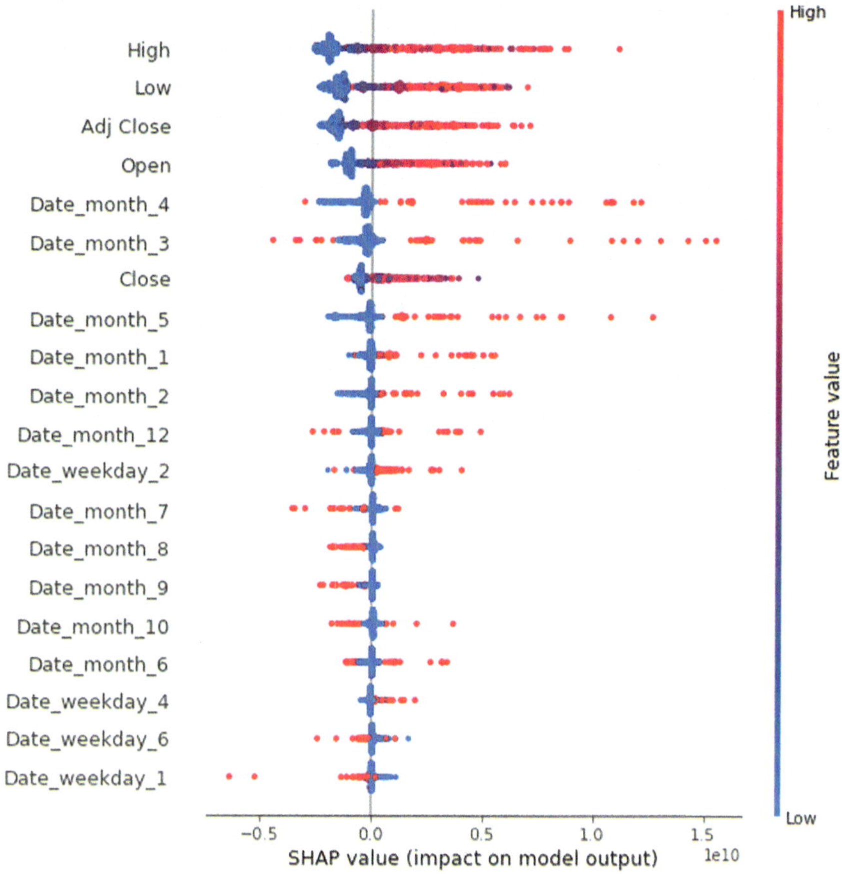

Fig. 4 Extra tree regressor model interpretation

Table 5 Performance evaluation of the six regression models on Bitcoin price in the next 5 days

Models	MAE	MSE	RMSE	R^2	RMSLE	MAPE
CatBoost regressor	3.2843	5.5696	7.4629	0.6248	1.2071	2.8631
Gradient boosting regressor	3.4984	5.4842	7.4055	0.6305	1.1271	2.4062
Extra tree	3.1447	6.2393	7.8989	0.5797	0.553	0.4783
AdaBoost	4.6398	6.7166	8.1955	0.5475	0.9367	1.3695
K-neighbor regressor	4.0944	7.2222	8.4984	0.5135	0.6387	0.5899
Theil-Sen regressor	5.3670	8.1152	9.0084	0.4533	2.2644	18.3188

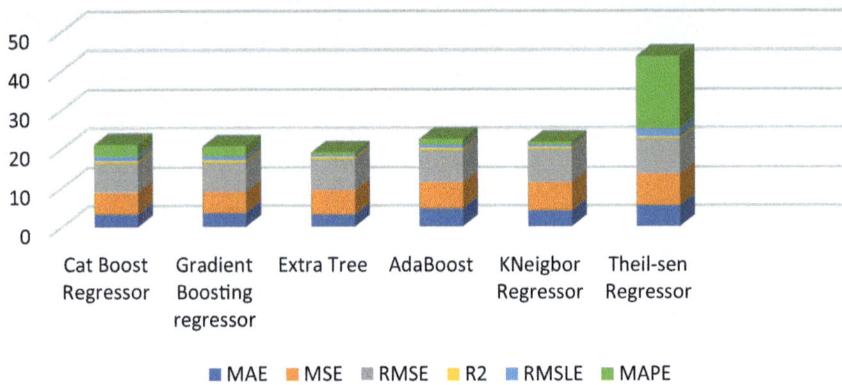

Fig. 5 Performance of the models of Bitcoin prices for the next 5 days

4.7 *Comparison with Previous Work in the Literature*

We compared the proposed models with the previous studies in the literatures. The findings of our model showed competitive results when compared with the state-of-the-art models for Bitcoin price prediction as shown in Table 6.

5 Conclusion and Future Work

This study examines the price of Bitcoin prediction using the past historical data in order to predict for the next 5 days. This research adopted the regression analysis models for the Bitcoin price prediction. The experimental analysis was performed using six regression models. The daily price prediction and the prices for the next 5 days was carried out and each model performance was evaluated in terms of different statistical metrics. The experimental results for the next 5 days revealed that the extra tree model gave a MAE of 3.1447, RMSLE of 0.553, and MAPE of 0.4783, which outperformed other regressors in terms of the MAE, RMSLE, and MAPE. The gradient boosting regressor model gave an MSE of 5.4842 and RMSE of 7.4055, which is better than other models in terms of MSE and RMSE. The Theil-Sen regressor model gave R^2 of 0.4533, which performed better than other regressor models. This study would assist investors to know the prices of Bitcoin in a day and in the next 5 days whether to invest in the stock or not. The future work direction will be to implement an autoregressive integrated moving average (ARIMA) model and deep learning models for the prediction of Bitcoin prices.

Table 6 Comparison with existing works

Authors/year	Methodology	Cryptocurrency	Forecast duration	Results
[43] 2017	BP2NN, NEAT, GABPNN	Bitcoin	One (1) day	MAPE and BPNN = 1.998 ± 0.038, GABPNN = 1.883 ± 0.066, NEAT = 2.175 ± 0.096
[44] 2018	Rprop algorithm and ANN	Bitcoin, bitcoin cash, dash	Thirty-one (31) days and 150 hours	Accuracy = 75% to 97.3% for bitcoin
[47] 2018	LSTM, AR	Bitcoin	Seventy-one (71) days	RMSE: 247.33
[48] 2018	Bidirectional LSTM, Word2vec	Bitcoin	Six (6) months	Accuracy = 50%. Precision = 60.99%
[49] 2018	ARIMA	Bitcoin	Ten (10) days	ARIMA = 90.31%. AR = 89.25%. MA = 87.58%
[53] 2019	Regression, decision tree	Bitcoin	Five (5) Days	Accuracy: Decision tree = 95.88013, regression = 97.59812
[54] 2019	ARIMA, LSTM, GRU	Bitcoin	Four hundred and ninety-two (492) days	RMSE = ARIMA = 302.53; LSTM = 603.68; GRU = 381.34
[55] 2019	ANN, LSTM	Bitcoin, Ethereum, ripple, stellar,	One (1), ten (10), twenty (20), thirty (30) days	RMSE = 53.3: One day, 68.0: Ten days, 91.41: Twenty days; 45.71: Thirty days for bitcoin
[56] 2019	SVM with PSO	Bitcoin, Ethereum, Litecoin, Nem, ripple, stellar	One (1) year	Bitcoin = 90.4, Ethereum = 97, Litecoin = 92.1, Nem = 57.8, ripple = 82.8, stellar = 64.5
[57] 2019	CNN, LSTM, GRU	Bitcoin	One (1) and three (3) months	RMSE; gold prices CNN = 201.34; LSTM = 151.67; GRU = 179.23; twitter sentiments LSTM = 32.98
[6] 2020	LSTM, GRU	Litecoin and Monero	One (1), three (3), seven (7) days	RMSE = 1 day; L = 2.2986, M;3.2715,3 days. L = 2.0327, M = 5.5005. 7 days. L = 4.5521, M = 20.2437

(continued)

Table 6 (continued)

Authors/year	Methodology	Cryptocurrency	Forecast duration	Results
Our models	**CatBoost regressor, gradient boosting regressor, extra tree regressor, AdaBoost regressor, K-neighbor regressor, and the Theil-Sen regressor**	**Bitcoin**	**One (1) day, five (5) days**	**CatBoost:MAE = 3.2843, MSE = 5.5696; GBM: MSE of 5.4842, RMSE of 7.4055; AdaBoost:R^2 =0.5797, RMSLE = 0.553, KNN:MAE = 4.0944; MSE = 7.2222; Theil-Sen = RMSLE = 2.2644, and MAPE = 18.3188**

References

1. McNally, S., Roche, J., & Caton, S. Predicting the price of bitcoin using machine learning. In *Proceedings - 26th euromicro international conference on parallel, distributed and network-based processing* (Vol. 2018, pp. 339–343). PDP 2018. https://doi.org/10.1109/PDP2018.2018.00060
2. Chen, Z., Li, C., & Sun, W. (2020). Journal of computational and applied bitcoin price prediction using machine learning: An approach to sample dimension engineering. *Journal of Computational and Applied Mathematics, 365*, 112395. https://doi.org/10.1016/j.cam.2019.112395
3. Monti, M., & Rasmussen, S. (2017). RAIN: A bio-inspired communication and data storage infrastructure. *Artificial Life, 23*(4), 552–557. https://doi.org/10.1162/ARTL_a_00247
4. Katsiampa, P. (2017). Volatility estimation for bitcoin: A comparison of GARCH models. *Economic Letters, 158*, 3–6. https://doi.org/10.1016/j.econlet.2017.06.023
5. Hua, Y. (2020). Bitcoin price prediction using ARIMA and LSTM. *E3S Web Conference, 218*(4), 396–406. https://doi.org/10.1051/e3sconf/202021801050
6. Patel, M. M., Tanwar, S., Gupta, R., & Kumar, N. (2020). A deep learning-based cryptocurrency Price prediction scheme for financial institutions. *Journal of Information Security and Applications, 55*, 102583. https://doi.org/10.1016/j.jisa.2020.102583
7. Abayomi-Zannu T.P., Odun-Ayo I., Tatama B.F., "Implementing a mobile voting system utilizing blockchain technology and two-factor authentication in Nigeria," 2020.
8. Mounika, S. (2021). Crypto-currency Price prediction using CNN and LSTM models. *International Journal for Research in Applied Science and Engineering Technology, 9*(3), 107–114. https://doi.org/10.22214/ijraset.2021.33191
9. Mai, F., Shan, Z., Bai, Q., Wang, X. S., & Chiang, R. H. L. (2018). How does social media impact bitcoin value? A test of the silent majority hypothesis. *Journal of Management Information Systems, 35*(1), 19–52. https://doi.org/10.1080/07421222.2018.1440774
10. Rogojanu, A. (2014). The issue of competing currencies. Case study – Bitcoin. *Theoretical and Applied Economics, XXI*(1), 103–114.
11. Elbahrawy, A., Alessandretti, L., Kandler, A., Pastor-Satorras, R., & Baronchelli, A. (2017). Evolutionary dynamics of the cryptocurrency market. *Royal Society Open Science, 4*(11). https://doi.org/10.1098/rsos.170623
12. Pirolo, S. L. (2016) CANDIDATE Carlotta Borelli. Department of business and management Chair of Management.

13. Ali, R., Barrdear, J., Clews, R., & Southgate, J. (2014). The economies of digital currencies. *Bank English Q Bulletin, 43*(3), 276–286.
14. Cheah, E. T., & Fry, J. (2015). Speculative bubbles in bitcoin markets? An empirical investigation into the fundamental value of bitcoin. *Economic Letters, 130*, 32–36. https://doi.org/10.1016/j.econlet.2015.02.029
15. Roche, J., & Mcnally, S. (2016). P*redicting the price of bitcoin using machine learning Sean McNally supervisor*.
16. Greaves, A., & Au, B. (2015). *Using the bitcoin transaction graph to predict the price of bitcoin* (pp. 1–8).
17. Madan, I., Saluja, S., & Zhao, A. (2015). Automated bitcoin trading via machine learning algorithms. *URL http//cs229. stanford. edu/proj2014/Isaac\% 20Madan, 20*, 1–5. [Online]. Available: http://cs229.stanford.edu/proj2014/Isaac Madan, Shaurya Saluja, Aojia Zhao,Automated Bitcoin Trading via Machine Learning Algorithms.pdf
18. Saad, M., Choi, J., Nyang, D., Kim, J., & Mohaisen, A. (2020). Toward characterizing blockchain-based cryptocurrencies for highly accurate predictions. *IEEE Systems Journal, 14*(1), 321–332. https://doi.org/10.1109/JSYST.2019.2927707
19. Ho, T. K. (1995). Random decision forests Tin Kam Ho perceptron training. In *Proceedings of 3rd international conference on document analysis and recognition* (pp. 278–282) [Online]. Available: https://ieeexplore.ieee.org/abstract/document/598994/
20. Barmaki, R. (2015). Multimodal assessment of teaching behavior in immersive rehearsal environment - TeachLivE™. In *ICMI 2015 - Proceedings of the 2015 ACM on international conference on multimodal interaction* (Vol. 139, pp. 651–655). https://doi.org/10.1145/2818346.2823306
21. Li, T. R., Chamrajnagar, A. S., Fong, X. R., Rizik, N. R., & Fu, F. (2018). Sentiment-based prediction of alternative cryptocurrency price fluctuations using gradient boosting tree model. *arXiv, 7*, 1–8. https://doi.org/10.3389/fphy.2019.00098
22. Bin Kim, Y., et al. (2016). Predicting fluctuations in cryptocurrency transactions based on user comments and replies. *PLoS One, 11*(8), 1–17. https://doi.org/10.1371/journal.pone.0161197
23. Phaladisailoed, T., & Numnonda, T. (2018). Machine learning models comparison for bitcoin price prediction. In *Proceedings of 2018 10th international conference on information technology and electrical engineering: smart technology for better society* (pp. 506–511). ICITEE 2018. https://doi.org/10.1109/ICITEED.2018.8534911
24. Sin, E., & Wang, L. (2018). Bitcoin price prediction using ensembles of neural networks. In *ICNC-FSKD 2017 - 13th international conference on natural computation fuzzy systems and knowledge discovery* (pp. 666–671). https://doi.org/10.1109/FSKD.2017.8393351
25. Georgoula, I., Pournarakis, D., Bilanakos, C., Sotiropoulos, D. N., & Giaglis, G. M. (2015). Using time-series and sentiment analysis to detect the determinants of bitcoin prices. *SSRN Electronic Journal*. https://doi.org/10.2139/ssrn.2607167
26. Tiwari, A. K., Jana, R. K., Das, D., & Roubaud, D. (2018). Informational efficiency of bitcoin—An extension. *Economic Letters, 163*, 106–109. https://doi.org/10.1016/j.econlet.2017.12.006
27. de la Horra, L. P., de la Fuente, G., & Perote, J. (2019). The drivers of bitcoin demand: A short and long-run analysis. *International Review of Financial Analysis, 62*, 21–34. https://doi.org/10.1016/j.irfa.2019.01.006
28. Wang, G. J., Xie, C., Wen, D., & Zhao, L. (2019). When bitcoin meets economic policy uncertainty (EPU): Measuring risk spillover effect from EPU to bitcoin. *Finance Research Letters, 31*, 489–497. https://doi.org/10.1016/j.frl.2018.12.028
29. Poyser, O. (2019). Exploring the dynamics of Bitcoin's price: A Bayesian structural time series approach. *Eurasian Economic Review, 9*(1).
30. Zhang, Y. J., Bouri, E., Gupta, R., & Ma, S. J. (2021). Risk spillover between bitcoin and conventional financial markets: An expectile-based approach. *The North American Journal of Economics and Finance, 55*, 101296. https://doi.org/10.1016/j.najef.2020.101296
31. Cretarola, A., & Figà-Talamanca, G. (2019). Detecting bubbles in bitcoin price dynamics via market exuberance. *Annals of Operations Research*. https://doi.org/10.1007/s10479-019-03321-z

32. Bariviera, A. F., Basgall, M. J., & Naiouf, M. (2017). *Accepted Manuscript*. https://doi.org/10.1016/j.physa.2017.04.159
33. Deokar, R. S., Dandage, P. S. M., & Jawandhiya, P. M. (2020). Design & Implementation of crypto currency prediction using machine learning approach, *4*(3), 7–12.
34. Ji, S., Kim, J., & Im, H. (2019). A comparative study of bitcoin price prediction using deep learning. *Mathematics, 7*(10). https://doi.org/10.3390/math7100898
35. Corbet, S., Lucey, B., Urquhart, A., & Yarovaya, L. (2019). Cryptocurrencies as a financial asset: A systematic analysis. *International Review of Financial Analysis, 62*, 182–199. https://doi.org/10.1016/j.irfa.2018.09.003
36. Alessandretti, L., Elbahrawy, A., Aiello, L. M., & Baronchelli, A. (2018). Anticipating cryptocurrency prices using machine learning. *arXiv, 2018*.
37. Awotunde, J. B., Ogundokun, R. O., Jimoh, R. G., & Misra, S. (2021). *Machine learning algorithm for cryptocurrencies Price prediction, artificial*. Springer.
38. Jang, H., & Lee, J. (2017). An empirical study on modeling and prediction of bitcoin prices with bayesian neural networks based on blockchain information. *IEEE Access, 6*, 5427–5437. https://doi.org/10.1109/ACCESS.2017.2779181
39. Nakano, M., Takahashi, A., & Takahashi, S. (2018). Bitcoin technical trading with artificial neural network. *Physica A: Statistical Mechanics and its Applications, 510*, 587–609. https://doi.org/10.1016/j.physa.2018.07.017
40. Huisu, J., Lee, J., Ko, H., & Lee, W. (2018). Predicting bitcoin prices by using rolling window LSTM model. *Data Mining and Knowledge Discovery*, 9. [Online]. Available: https://doi.org/10.475/123_4.
41. Shintate, T., & Pichl, L. (2019). Trend prediction classification for high frequency bitcoin time series with deep learning. *Journal of Risk and Financial Management, 12*(1), 17. https://doi.org/10.3390/jrfm12010017
42. Mallqui, D. C. A., & Fernandes, R. A. S. (2019). Predicting the direction, maximum, minimum and closing prices of daily bitcoin exchange rate using machine learning techniques. *Applied Soft Computing - Journal, 75*, 596–606. https://doi.org/10.1016/j.asoc.2018.11.038
43. Radityo, A. (2017). *Prediction of bitcoin exchange rate to American Dollar using artificial neural network methods*.
44. Almasri, E., & Arslan, E. (2018). Predicting cryptocurrencies prices with neural networks. In *2018 6th international conference on control engineering and information technology* (pp. 1–5). CEIT 2018. https://doi.org/10.1109/CEIT.2018.8751939
45. Jain, A., Tripathi, S., Dhardwivedi, H., & Saxena, P. (2018). Forecasting price of cryptocurrencies using tweets sentiment analysis. In *2018 11th international conference on contemporary computing* (pp. 2–4). IC3 2018. https://doi.org/10.1109/IC3.2018.8530659
46. Wimalagunaratne, M., & Poravi, G. (2018). A predictive model for the global cryptocurrency market: A holistic approach to predicting cryptocurrency prices. In *Proceedings - International conference on intelligent systems, modelling and simulation* (pp. 78–83). ISMS, 2018. https://doi.org/10.1109/ISMS.2018.00024
47. Wu, C. H., Lu, C. C., Ma, Y. F., & Lu, R. S. (2019). A new forecasting framework for bitcoin price with LSTM. In *IEEE international conference on data mining work* (Vol. 2018, pp. 168–175). ICDMW. https://doi.org/10.1109/ICDMW.2018.00032
48. Mohanty, P., Patel, D., Patel, P., & Roy, S. (2018). Predicting fluctuations in cryptocurrencies' price using users' comments and real-time prices. In *2018 7th international conference on reliability, infocom technologies and optimization: Trends and future directions* (pp. 477–482). ICRITO 2018. https://doi.org/10.1109/ICRITO.2018.8748792
49. Roy, S., Nanjiba, S., & Chakrabarty, A. (2019). Bitcoin price forecasting using time series analysis. In *2018 21st international conference of computer and information technology* (pp. 1–5). ICCIT 2018. https://doi.org/10.1109/ICCITECHN.2018.8631923
50. Hashish, I. A., Forni, F., Andreotti, G., Facchinetti, T., & Darjani, S. (2019). A hybrid model for bitcoin prices prediction using hidden Markov models and optimized LSTM networks. In *IEEE international conference on emerging technologies and factory automation* (pp. 721–728). ETFA, 2019. https://doi.org/10.1109/ETFA.2019.8869094

51. Wirawan, I. M., Widiyaningtyas, T., & Hasan, M. M. (2019). Short Term Prediction on Bitcoin Price Using ARIMA Method. In *Proceedings – 2019 International Seminar on Application for Technology of Information and Communication: Industry 4.0: Retrospect, Prospect, and Challenges* (pp. 260–265). iSemantic 2019. https://doi.org/10.1109/ISEMANTIC.2019.8884257
52. Anupriya, & Garg, S. (2018). Autoregressive integrated moving average model based prediction of bitcoin close price. In *Proceedings of the international conference on smart systems and inventive technology* (pp. 473–478). ICSSIT 2018. https://doi.org/10.1109/ICSSIT.2018.8748423
53. Rathan, K., Sai, S. V., & Manikanta, T. S. (2019). Crypto-currency price prediction using decision tree and regression techniques. In *Proceedings of the international conference on trends in electronics and informatics* (pp. 190–194). ICOEI 2019. https://doi.org/10.1109/icoei.2019.8862585
54. Yamak, P. T., Yujian, L., & Gadosey, P. K. (2019). A comparison between ARIMA, LSTM, and GRU for time series forecasting. In *ACM international conference proceedings series* (pp. 49–55). https://doi.org/10.1145/3377713.3377722
55. Zhengyang, W., Xingzhou, L., Jinjin, R., & Jiaqing, K. (2019). Prediction of cryptocurrency price dynamics with multiple machine learning techniques. In *ACM international conference proceedings series* (pp. 15–19). https://doi.org/10.1145/3340997.3341008
56. Hitam, N. A., Ismail, A. R., & Saeed, F. (2019). An optimized support vector machine (SVM) based on particle swarm optimization (PSO) for cryptocurrency forecasting. *Procedia Computer Science, 163*, 427–433. https://doi.org/10.1016/j.procs.2019.12.125
57. Aggarwal, A., Gupta, I., Garg, N., & Goel, A. (2019). Deep learning approach to determine the impact of socio economic factors on bitcoin price prediction. In *2019 12th international conference on contemporary computing* (pp. 1–5). IC3 2019. https://doi.org/10.1109/IC3.2019.8844928
58. Misra, S. (2021). *A step by step guide for choosing project topics and writing research papers in ICT related disciplines* (Vol. 1350). Springer International Publishing.
59. Bitcoin Data from 2014 to 2020 | Kaggle. https://www.kaggle.com/khalilbrick/bitcoin-data-from-2014-to-2020. Accessed 03 Apr 2021.
60. Ahmad, T., & Aziz, M. N. (2019). Data preprocessing and feature selection for machine learning intrusion detection systems. *ICIC Express Letters, 13*(2), 93–101. https://doi.org/10.24507/icicel.13.02.93
61. Saranya, C., & Manikandan, G. (2013). A study on normalization techniques for privacy preserving data mining. *International Journal of Engineering & Technology, 5*(3), 2701–2704.
62. Patro, S. G. K., & Sahu, K. K. (2015). Normalization: A preprocessing stage. *Iarjset*, 20–22. https://doi.org/10.17148/iarjset.2015.2305
63. Prokhorenkova, L., Gusev, G., Vorobev, A., Dorogush, A. V., & Gulin, A. (2018). Catboost: Unbiased boosting with categorical features. *Advances in Neural Information Processing Systems, 2018*, 6638–6648.
64. Ridgeway, G. (2007). Generalized Boosted Models: A guide to the gbm package. *Compute, 1*(4), 1–12. [Online]. Available: http://cran.r-project.org/web/packages/gbm/vignettes/gbm.pdf
65. Ferov, M., & Modrý, M. (2016). Enhancing LambdaMART using oblivious trees. [Online]. Available: http://arxiv.org/abs/1609.05610.
66. Kang, P., Lin, Z., Teng, S., Zhang, G., Guo, L., & Zhang, W. (2019). Catboost-based framework with additional user information for social media popularity prediction. In *MM 2019 – Proceedings of the 27th ACM international conference on multimedia* (pp. 2677–2681). https://doi.org/10.1145/3343031.3356060
67. Jha, S. B., Babiceanu, R. F., Pandey, V., & Jha, R. K. (2020). Housing market prediction problem using different machine learning algorithms: A case study. *arXiv*.
68. Bishop, C. M. (2006). Pattern Recognition and Machine Learning Springer Mathematical notation Ni. *Springer-Verlag New York, Inc.,, Secaucus, NJ, USA*, p. 9, [Online]. Available: http://cds.cern.ch/record/998831/files/9780387310732_TOC.pdf.

69. sklearn.ensemble.GradientBoostingClassifier — scikit-learn 0.24.1 documentation. https://scikit-learn.org/stable/modules/generated/sklearn.ensemble.GradientBoostingClassifier.html. Accessed 03 Apr 2021.
70. Keprate, A., & Ratnayake, R. M. C. (2018). Using gradient boosting regressor to predict stress intensity factor of a crack propagating in small bore piping. *IEEE International Conference on Industrial Engineering and Engineering Management, 2017*, 1331–1336. https://doi.org/10.1109/IEEM.2017.8290109
71. A Gentle Introduction to the Gradient Boosting Algorithm for Machine Learning. https://machinelearningmastery.com/gentle-introduction-gradient-boosting-algorithm-machine-learning/. Accessed 03 Apr 2021.
72. Bouktif, S., Fiaz, A., Ouni, A., & Serhani, M. A. (2018). Optimal deep learning LSTM model for electric load forecasting using feature selection and genetic algorithm: Comparison with machine learning approaches. *Energies, 11*(7). https://doi.org/10.3390/en11071636
73. Gkerekos, C., Lazakis, I., & Theotokatos, G. (2019). Machine learning models for predicting ship main engine fuel oil consumption: A comparative study. *Ocean Enginerring, 188*, 106282. https://doi.org/10.1016/j.oceaneng.2019.106282
74. Schapire, R. E. (2003). *The boosting approach to machine learning: An overview* (pp. 149–171). https://doi.org/10.1007/978-0-387-21579-2_9
75. Drucker, H. (1997) Improving regressors using boosting techniques. *14th international conference on machine learning*, (pp. 107–115). [Online]. Available: http://www.researchgate.net/publication/2424244_Improving_Regressors_using_Boosting_Techniques/file/3deec51ae736538cec.pdf%5Cnhttp://citeseerx.ist.psu.edu/viewdoc/summary?doi=10.1.1.31.314.
76. Saheed, Y. K., Oladele, T. O., Akanni, A. O., & Ibrahim, W. M. (2018). Student performance prediction based on data mining classification techniques. *Nigerian Journal of Technology, 37*(4), 1087. https://doi.org/10.4314/njt.v37i4.31
77. Liu, H., Tian, H. Q., Li, Y. F., & Zhang, L. (2015). Comparison of four Adaboost algorithm based artificial neural networks in wind speed predictions. *Energy Conversion and Management, 92*, 67–81. https://doi.org/10.1016/j.enconman.2014.12.053
78. Guo, L., Ge, P. S., Zhang, M. H., Li, L. H., & Zhao, Y. B. (2012). Pedestrian detection for intelligent transportation systems combining AdaBoost algorithm and support vector machine. *Expert Systems with Applications, 39*(4), 4274–4286. https://doi.org/10.1016/j.eswa.2011.09.106
79. Imandoust, S. B., & Bolandraftar, M. (2013). Application of K-nearest neighbor (KNN) approach for predicting economic events : Theoretical background. *International Journal of Engineering Research and Applications, 3*(5), 605–610.
80. Saheed, Y. K., & Hamza-usman, F. E. (2020). Feature selection with IG-R for improving performance of intrusion detection system, *12*(3), 338–344.
81. Oladejo, A. K., Oladele, T. O., & Saheed, Y. K. (2018). Comparative evaluation of linear support vector machine and K-nearest neighbour algorithm using microarray data on leukemia cancer dataset, *11*(2), 1–10.
82. Borhani, S., Kilmarx, J., Saffo, D., Ng, L., Abiri, R., & Zhao, X. (2019). Optimizing prediction model for a noninvasive brain-computer Interface platform using channel selection, classification, and regression. *IEEE Journal of Biomedical and Health Informatics, 23*(6), 2475–2482. https://doi.org/10.1109/JBHI.2019.2892379
83. Uludag, K., & Korcak, O. (2017). Energy and rate modeling of data download over LTE with respect to received signal characteristics. In *2017 27th international telecommunication networks and application conference* (pp. 1–6). ITNAC 2017. https://doi.org/10.1109/ATNAC.2017.8215395

84. Chai, T., & Draxler, R. R. (2014). Root mean square error (RMSE) or mean absolute error (MAE)? -arguments against avoiding RMSE in the literature. *Geoscientific Model Development, 7*(3), 1247–1250. https://doi.org/10.5194/gmd-7-1247-2014
85. Miles, J. (2014). R Squared, Adjusted R Squared. *Wiley StatsRef: Statistics Reference Online, 2*, 2–4. https://doi.org/10.1002/9781118445112.stat06627
86. Chen, P., Hsieh, H., Su, K., Sigalingging, X. K., Chen, Y., & Leu, J. (2020). Predicting station level demand in a bike-sharing system using recurrent neural networks. *IET Intelligent Transport Systems, 14*(6), 554–561. https://doi.org/10.1049/iet-its.2019.0007

Adoption of Blockchain Technology in the Indian Business Market: Obstacles and Opportunities

Ratnesh Litoriya, Abhishek Arora, Raddhant Bajaj, and Abhik Gulati

1 Introduction

Blockchain is a technology that allows you to store decentralized and distributed records of digital transactions. Blockchain was first time introduced in 2009 concerning Bitcoin [1]. Although its use is varied to different industries, the procedure to be followed is the same (first, the transactions are communicated to multiple nodes on the network. Then, these nodes verify the transactions and turn them into blocks). Every single block in the chain is corroborated by hash (a coded unique value calculated based on the block's number and contents). Block includes a reference to the previous block explicitly so that blocks are linked to each other. This chain of blocks acts as a track record of multiple transactions, referred to as a ledger [2].

By the above procedure way, all nodes rectify that the keys used are correct, which results in maintaining secrecy, which further ensures that the Bitcoins are transferred with utmost anonymity (free from intruders). Under this technology, once the transactions become part of the block contents and are added to the chain, they become confirmed and verified; adding blocks into the chain is not easy. It is essential to mine it first, which requires calculating its hash by solving a single mathematical calculation which requires expert knowledge and a reasonable amount of time.

R. Litoriya (✉)
Medi-Caps University, Indore, Madhya Pradesh, India

A. Arora · R. Bajaj
Symbiosis Centre for Management Studies, Noida, Uttar Pradesh, India

A. Gulati
Jaypee University of Engineering and Technology, Guna, Madhya Pradesh, India

S. Misra, A. Kumar Tyagi (eds.), *Blockchain Applications in the Smart Era*,
EAI/Springer Innovations in Communication and Computing,
https://doi.org/10.1007/978-3-030-89546-4_11

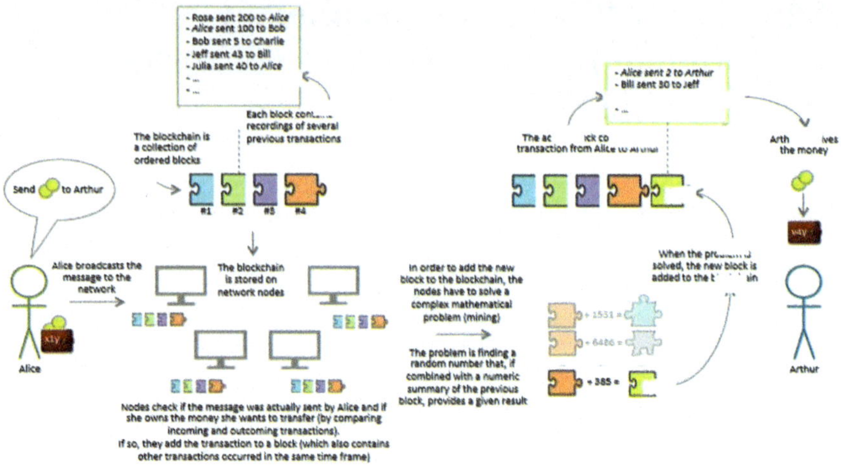

Fig. 1 Schematic transaction of Bitcoin through blockchain

In case of some problems, the hash adjusts itself periodically to adapt to the network's processing capabilities. As the power of networking is increasing day by day, the difficulty of the problems is growing simultaneously. Modifying a block directly leads to changes in its hash, which results in breaking the chain as the next block would fail, and as we know, repairing it is a big task.

Various issues have come under the limelight after the societies started implementing digitization in their processes. Money and market, supported by money, were found to have the most significant challenges directly concerned with confidence and security in transactions. At first, when digital money was introduced, it was based on a single centralized host, probably to develop customers' trust by ensuring non-fraudulent activities. But despite significant advances in technologies like centralization, cryptography, and anonymity, preventing fraud is still a back-breaking job. Bitcoin, which is mostly known by all, is the only digital currency that functions as a decentralized, anonymous system, which is proved by various tests conducted by researchers in the past.

Blockchain technology works on a sophisticated concept of chains of blocks, evolving technology at an immense potential with no less integrated risk. In their paper, Sachin Kamble et al. [3] advance the literature on blockchain technology and its adoption in the supply chain by developing and statistically validating a model for understanding the user perceptions on blockchain adoption. A conceptual framework is also proposed by Janssen et al. [4] for the adoption of blockchain technology considering the multifaceted relationships amid market, institutional, and technical factors. Figure 1 illustrates a sample transaction of Bitcoin on the blockchain. This revolutionary technology has already made significant inroads into a variety of industries, for instance, real estate, healthcare, crowdfunding, education, supply chain, etc., and soon, it would have a significant impact on the whole economy surrounding mobile apps as well [5–10]. Its decentralized architecture provides some fantastic benefits that practically all industries can take advantage of. The world of mobile apps will be able to take advantage of these to correct some shortcomings that have become evident in recent years. Blockchain technology focuses on creating a world without intermediaries. This technology's various developments are being discussed by discussing its implementation in multiple industries worldwide where its cases are met.

This chapter aimed to recognize (RQ1) various industries in which blockchain tech is being used (which is discussed with the help of past use cases) and (RQ2) the multiple obstacles due to which blockchain implementations are still weak in India.

The rest of the paper is organized as follows. Section 2 elaborates the different variants of blockchain with a comparative analysis. Section 2.3 discusses the characteristics of the blockchain, followed by the finalization of research questions. Existing use cases of blockchain and its deployment are discussed in Sect. 3. Section 4 consists of the research methodology used in this study. Statistical methods used in the study are described in Sect. 5. Discussion against each RQ is presented in Sect. 6. The subsequent section discusses applying technology acceptance model. Section 7 talks about the limitation of this work, and the last section are for conclusions and future directions.

2 Variants of Blockchain Technology

At present, businesses are at a point where different blockchain technology variants exist. Each variant serves its specific purpose and solves a particular set of problems for a business transaction environment. Various organizations are using these variants in their day-to-day business to leverage the maximum benefits they can. A brief detail of each variant is explained in the subsections given below.

2.1 Public Blockchains

Like cloud computing, public blockchains can be accessed and updated by anyone registered to the network. This type of blockchain is based on a distributed decentralized architecture, which results in no central authority. Everyone can navigate and verify transactions in the database concerning the predetermined set of norms. Famous examples are Bitcoin and Ethereum. These networks are generally based on consensus protocols.

Table 1 Comparison table of different types of blockchain

Property	Blockchain types		
	Public	Consortium	Public
Process	Permissionless	Process	Permissionless
Consensus	All users	Consensus	All users
Immutability	Nearly impossible to tamper	Immutability	Nearly impossible to tamper
Centralized	No	Centralized	No
Efficiency	Low	Efficiency	Low
Read permission	Public	Read permission	Public

2.2 *Private Blockchains vs. Consortium Blockchains*

Private blockchains work on strict control; it does not incorporate decentralization over the access to the network and better performances. This type of blockchain is suitable for industrial and business tools. The Web is accessible only to verified and registered members of a group, or administrators can edit an organization and transactions. Table 1 explains the difference between various types of blockchains. Hyperledger and Ripple are examples of popular private blockchains that many companies use, unlike public blockchains that operate on an advance agreement. Permissioned blockchains are distinguished in private blockchains and consortium blockchains. At the top, a central authority can modify the blocks and retain complete control of the data records or ledger. Conversely, blockchain consortiums are based on predefined mechanisms in which predefined nodes control the whole database .

2.3 *Characteristics of Blockchain*

These are the characteristics that differentiate blockchain from various other technologies, and these are at the core of blockchain implementation.

- **Fast Transactions**

In blockchains, transactions work on a peer-to-peer system involving very few or no intermediaries, eliminating the double-spending problem while giving reliable information. Transactions are executed simultaneously at both sides of the payment, and digital signatures are enabled, thus increasing the possibility to avoid fraudulent activities by these companies which can gain trustworthiness and accountability.

- **Resiliency**

Resiliency is the key to preserving data in a securable form on a server or storage system, making them 24/7 available to users, even in any failure.

- **Traceability**

Companies lacking tools to manage supply chains effectively have difficulties in handling demand and supply changes. With various variables involved in the supply chain, firms have to face a high uncertainty level and rely on reliable data. Blockchain technology can ensure product origin's exact traceability and help if there is any default in the supply chain.

- **Cost-Effectiveness**

The blockchain's main aim is to make a world free of intermediaries, leading to cost-effective transactions on the part of both consumers and service providers.

- **Automation**

The programmable nature of blockchain allows setting up and updating actions, events, and payments.

After its introduction into the market, which was around 10 years ago, many industries start grabbing the opportunities by implementing it into various sectors beyond cryptocurrency, such as banking, insurance, energy, supply chain, etc. After considering all past studies and relevant bodies of knowledge in this context, we decided to discuss various issues and use cases that have emerged in recent years to picture this technology better. We prepared two different research questions.

RQ1 – The various industries in which blockchain technology is being used (which is discussed with the help of past use cases).

RQ2 – What are the various obstacles due to which blockchain implementations are still deficient in India?

3 Current Blockchain Adoption in Varying Domains

Under this section, we will discuss various domains where blockchain is extensively used worldwide with the help of use cases discussed in past research. Upadhyay [11] presents a systematic review of the scholarly research work published on the blockchain concerning its challenges, potential opportunities, and applications. Discussions for individual domains are elaborated in subsequent subsections.

3.1 Healthcare Domain

The use of blockchain is increasing in the healthcare sector faster, leveraging blockchain technology in healthcare operations to help organizations achieve lower transaction processing costs. Zhang et al. [12] show that blockchain can help the pharmaceutical supply chain by storing and accessing medical products and other relevant information. Healthcare data can be securely and privately managed with blockchain's help, ensuring anonymity and integrity on the part of service providers during patients' lifetime [13–15]. A software made of blockchain is presented by [16, 17] to manage electronic medical records so that data transfer and handling become more secure and straightforward, resulting in transparency to a low level of third-party interference. Sloane and Perepa [18] give information about MedRec (a live prototype where patients and doctors can access data related to diagnosing reports, treatments, test reports, and prescriptions in one place). Older adults are vulnerable to various diseases. Mostly they are living in isolation; special and secure health vigilance and alert systems can make their life risk-free [19–22]. ARIA (a prototype that combines radiation, medical, and surgical oncology information and helps clinicians manage different kinds of medical data, develop oncology-specific plans, and monitor radiation dose received by patients) is also discussed. A demonstration of a lightweight backup and efficient recovery scheme is given, which helps collect personal records and provides notification to patients in real time in case of emergency [23, 24].

3.2 Government Sector

In recent years, most governments have successfully set up their services in online modes to provide transparency and ease of access to citizens. This has led to maintaining records of high-quality data that is confidential but of significant importance to the government bodies and everyday people. Blockchain technology can further help governments to build innovative applications to handle information transactions securely. Abayomi-Zannu et al. [25] developed a framework for mobile voting that utilizes a two-factor authentication to validate the vote-casters and BT to securely store the votes. Zhang et al. [26] present a chain code that can make the voting process transparent and ensure that participants cannot manipulate the votes counting process. Taylor et al. [27] discusses BC systems, which identify data's credibility and accuracy during the digital transformation from people to the government resource directory and ensure no data breach. Svein and Arild [28] state that blockchain tools can support digital infrastructure by maintaining secrecy in the country's interests, such as storing digital IDs of citizens, secure documents, handling, armed forces operations, etc.

3.3 Supply Chain Management

Various studies are carried out to model the supply chain's trust deficits and analyzed decentralization in it [29, 30]. Under the supply chain [31], the best use case we found was that BC could help find the exact origin of the product [32] irrespective of whether it is connected to the network. Counterfeit drug detection is also implemented through blockchain to ensure a genuine supply of medicines

[33]. Dubey et al. [34] proposed a theoretical model to comprehend how blockchain technology can influence operational supply chain transparency (OSTC) and swift-trust among different actors engaged in disaster relief operations. Further, many use cases like ownership management software are discussed, which helps prevent any type of mismanagement. At the same time, the product reaches the end of the supply chain, and it can help the consumers track the origin of the product after purchasing or acquiring it [35]. Aitzhan et al. [36] discuss how blockchain can maintain secrecy in logistics processes with a higher threat, for example, cash, gold, etc. This can be done by ensuring that there are no problematic parts. This directly leads to building trust in the logistics service provider. Konstantinidis et al. [23] present an everledger that can ensure quality and accuracy in the food supply chain by ensuring speedy delivery and exact traceability due to low shelf life.

3.4 Energy Domain

Energy systems are also evolving around a more decentralized model. A software system is made of blockchain technology presented, which can manage monthly transactions between consumers or pro-consumers, be it B2B or B2C [37]. These are the consumers connected with the local energy grid; this will lower the cost as there will be no third-party interference of companies like the bill desk. Efficient waste management techniques are now available, which ultimately conserve energy and nature at large [38–42]. Suhaliana et al. [43] presented a use case where companies have implemented the technology at local energy grids, which resulted in allowing the distribution, metering, and billing of electricity to be regulated by the consumer itself rather than relying on third parties rendering their services; further, this case can help companies to save up on cost. Mathiyazhagan et al. [44] tried to present a benchmarking framework for ranking the Green supply chain management implementation drivers. Teufel et al. [45] discusses blockchain-based, intelligent systems developed to calculate energy consumption by plants connected to energy grids successfully. This system is more suitable in cases where there are chances of manipulation tampering.

3.5 Education Sector

Most of the articles talked about certification management as a whole, while some present a prototype to regulate the procedure of storing, issuing, and sharing certificates—further, Alammary et al. [17] discussed how blockchain tools help save and share information related to teaching outcomes students have achieved in their courses. Software development paradigm, which always faces challenges of selecting the appropriate development model, can also be transformed by introducing blockchain in the software process [46–48]. Authors in [49] present different use cases of applications that can manage fees and credit transfer processes. Lilian Yedigaryan [50] focuses on acquiring consent from legal guardians in matters related to data. Furthermore, Chen et al.[51] in their article concerned with enhancing the efficiency and transparency of competitions help, which will help manage competitions. Bartolomé et al. [52] talk about a blockchain tool that helps conduct examinations by securing the audit processes, sharing, and checking exam papers.

3.6 Insurance Sector

The insurance industry is in dire need of technological revolutions that could reduce overhead costs and time wastage. It is difficult to tell whether blockchain technology could overcome regulatory obstacles to become the insurance industry's future. Still, opportunities are endless for established companies as well as startups in this sector.

(A) ***Fraud Detection and Risk Prevention*** – Using Distributed Ledger Technology (DLT), insurers can record transactions and claims for better collaboration and check suspicious behavior. Many companies invest in public and private domain companies that predict fraudulent activities, but creating pan-industry fraud prev. is usually hampered by constraints around sharing personal information between different companies, while the introduction to the blockchain-based initiative would give insurers the following benefits:

- Prevention of multiple claims booking and secondly ensuring that processing of claims for the same accident is not being done
- Providing ownership beforehand with the help of digital certificates signed by both insurer
- Ensure eliminating diversion of premium, in case of unlicensed brokers selling insurance

(B) ***Property and Casualty Insurance*** – Dealing with P&C claims is subject to errors. It requires a large amount of manual data entry processes while simultaneously maintaining coordination between different parties to an agreement. Smart contracts derived out of blockchain technology can convert old-school paper contracts into coded and encrypted documents. This helps in automated claims processing for all parties involved in the agreement by collecting real-time casualty data and calculating claims accordingly.

(C) ***Healthcare Insurance*** – Currently, the healthcare insurance industry faces two significant problems: transferring patient's sensitive medical records and the second being old-school back-end infrastructure for claims processing. Using blockchain technology, patient's privacy can be maintained while, at the same time, smooth transfer of information is possible on a case-by-case basis.

(D) ***Reinsurance*** – Reinsurance can help in cases where a usually large number of claims are needed to process at once during times of calamities. To mitigate risk, different insurers involve more than two reinsurers at least. While in agreements, each risk in a contract must be individually underwritten, made worse by a highly inefficient reinsurance process. Using blockchain technology, the insurer and reinsurer of any company can have detailed information about insured assets, including info about claims processed, past accidents, premium rates, asset value, ongoing condition, etc.

3.7 Finance and Banking

Combining various MCDM methods is applied to calculate the efficiency of different public sector banks in India [53]. Xiongfeng et al. [54] prove that implementation of blockchain technology has a constructive impact on enhancing the asset turnover rate and minimizing the sales expense rate. Blockchain is presently being tested in many firms like Goldman Sachs, J.P. Morgan, and other banking institutions that have set up labs that collaborate with other blockchain sites. Ripple, a blockchain platform, is being used to run Standard Chartered's first cross-border transactions through the platform [55]. Awotunde et al. [56] applied a machine learning algorithm for the price prediction of cryptocurrencies. The platform took about 10 seconds to complete the transaction that would generally take 2 days to complete via present banking infrastructure. Besides, blockchain use will enable banks to facilitate overseas transactions by collecting nodes in a blockchain instead of having a central institution [57, 58]. Blockchain has disrupted the banking sector; as highlighted by IBM, "In 4 years, 66% of banks would have a commercialized blockchain" [59]. The technology can be applied to various areas such as clearing and settlement of financial assets and collateralization of financial derivatives to minimize costs and risks. It has gathered a lot of attention in major IT companies' eyes: Microsoft Azure (Azure, 2016) and IBM (IBM, 2016) will commence offering blockchain-as-a-service.

4 Research Methodology

Here, we will explain in detail the research method and the methodology implemented in our study. We will first describe our research approach and then the research design.

4.1 Research Approach

In this research, we have used a mix of both qualitative and quantitative strategies. This method was needed because we set two research:

RQ1– The various industries in which blockchain technology is being used (which is discussed with the help of past use cases).

RQ2– What are multiple obstacles due to which blockchain implementations are still weak in India?

Different types of research and their conduction is widely reported in pieces of literature [60]. The research approach is shown in Fig. 2. We are using a quantitative method in our research to measure the RQ2. This tries to showcase issues and challenges related to the implementation of blockchain technology, specifically in India. It is vital to measure the obstacles aspect because we need to understand why the lagging or slow adoption of blockchain in Indian industries refrains them from implementing technology in their businesses.

To measure RQ1 (which focuses on reviewing the past literature and finding out various applications of the technology currently being used and the recent use cases in different industries such as banking supply chain, etc.), we reviewed approximately 45 literature papers to derive the above information.

Many of the databases were searched by us, such as ScienceDirect, IEEE, ACM Digital Library, Springer, etc., to find out literature concerning our research questions. Out of the databases we referred to, approximately 112 papers were derived based on their titles. Still, multiple repetitions were found in the research contents. Hence, after the exclusion process, all the papers were now carefully inspected by us, which resulted in a selection of 55 articles; the reason for excluding 57 papers at this stage was the non-relevancy to domains we were interested in (e.g., technicality and practical models). After this, we searched for duplicates. This round resulted in selecting the final 44 papers, which we carefully reviewed to determine the benefits and use cases to support our study.

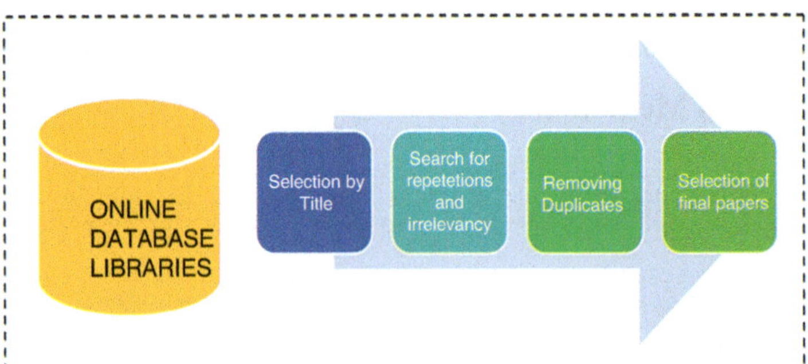

Fig. 2 Research approach

4.2 Research Design

This section discusses the type of research, data source, sampling method, methods of data collection, and data analysis methods, describing them with their pros and cons.

So, the type of research which we have used is conceptual. For the subtype, we are using a cross-sectional study for our research. As we know, a cross-sectional study analyzes data from the population or its representative subset at a specific point in time, that is, cross-sectional data. We have made a sample out of the population described in the sampling strategy section. Hence, our research becomes a descriptive cross-sectional study.

4.3 Questionnaires

These are used because they are reliable methods to collect data from many persons efficiently and conveniently quickly. The format of the questionnaire is shown in Fig. 3.

It becomes beneficial in the case of big projects. But we also faced one disadvantage: their structured format, which restricts more critical and in-depth information about our topic. So, we felt that some of the aspects of our research were left uncovered.

4.4 Sampling Strategy and Survey Locations

For this study, we have employed a stratified random sampling strategy. With this method, every person in any group has an equally likely chance of being chosen. We applied the technique during our questionnaire's circulation as we distributed it randomly among people belonging to various categories in terms of age groups, occupations, etc. So, we collected data from a sample of 41 respondents. The medium of circulation of the questionnaire was through links shared on WhatsApp, emails, Instagram DMs, etc. Survey locations belong to an area where work is proposed to gather information for research. Our study's survey locations are generally some cities of India, for which we did a questionnaire-type survey for the study.

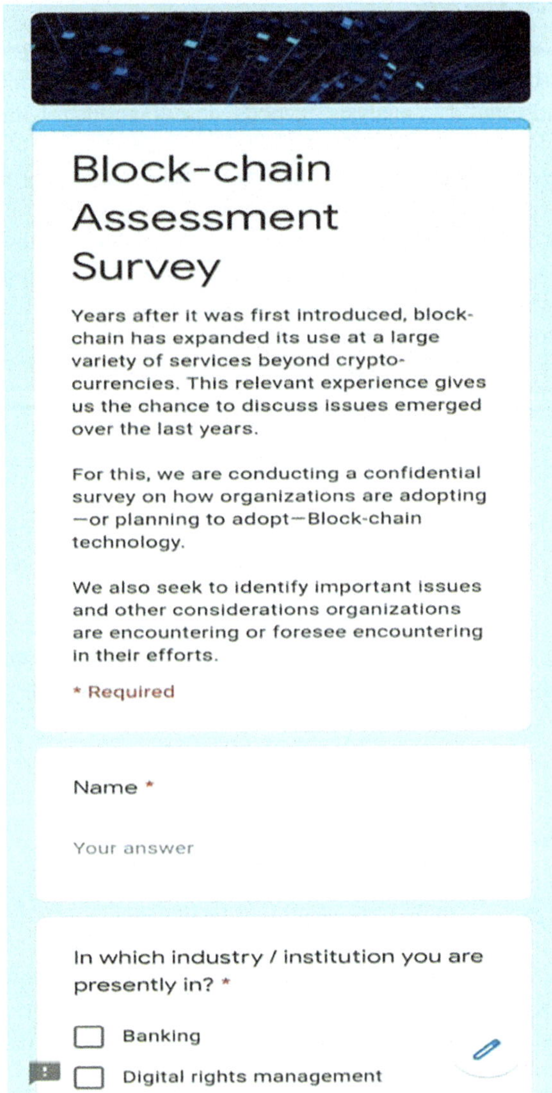

Fig. 3 Snapshot of the questionnaire used in the research study

5 Statistical Methods Used in the Study

Bar Graph – This chart pictorifies categorical data into bars with height according to the respective values. Vertically or horizontally, the graph can be arranged accordingly as per the requirement. A vertical chart can also be called a line graph.

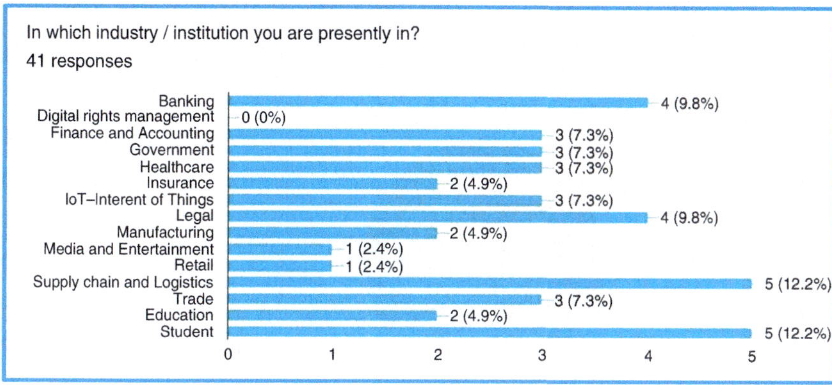

Fig. 4 Participation of various domain in the survey

In our study, the bar graph is used to represent the data collected from survey respondents.

Pie Chart – It's a circular graphic, proportionately divided according to the given data. In this chart, the arc length is proportional to the data it is representing. In our study, the pie chart describes information like gender, qualification, cadre, and industry type from where the respondent is associated. The statistical data analysis is performed and represented as different graphs, as shown in Figs. 4, 5, 6, 7, 8, 9, 10, 11, 12, and 13.

5.1 *Data Interpretations (Survey)*

Most respondents worked in the supply chain, legal, and banking sectors.

Most respondents were from the engineering and management background and knew blockchain tech.

Most respondents responded negatively when asked whether they had implemented blockchain.

Most respondents felt that they might have blockchain-based applications in the next year or so.

Most organizations haven't established a dedicated group for a blockchain initiative.

A total of 73.2% of respondents stated that their organization had not developed a budget for a blockchain initiative.

The knowledge sample had about blockchain technology (1 – lowest, 5 – highest).

Most professionals felt that blockchain had high adoption cost (1– high cost, 5 – low cost).

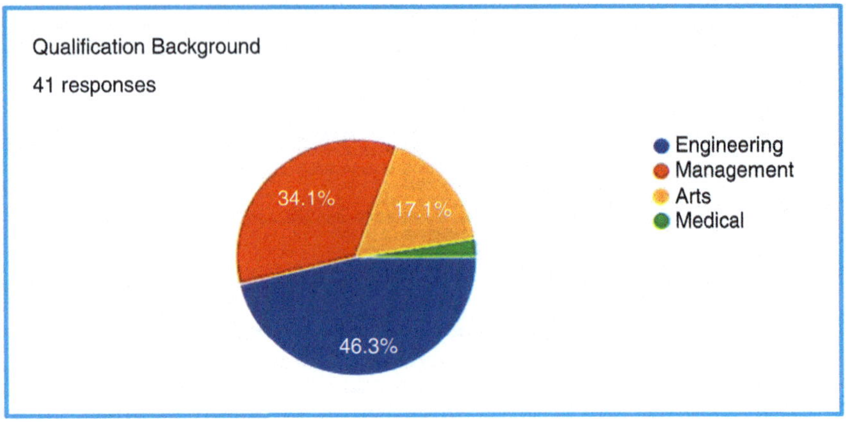

Fig. 5 Pie chart representing qualification background

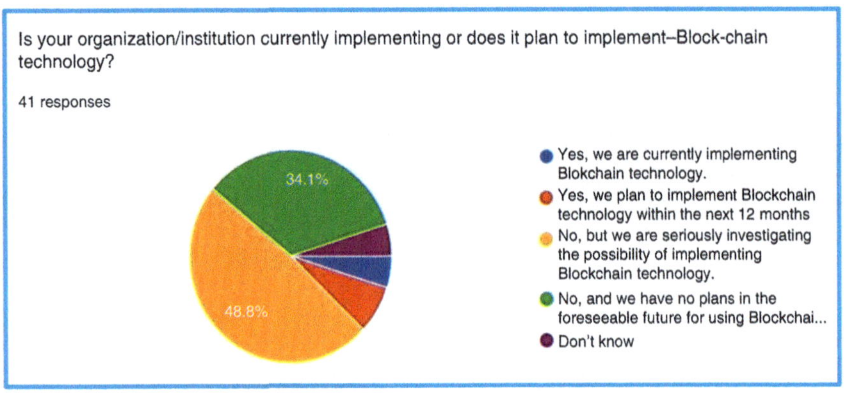

Fig. 6 Pie chart representing the usage/possible usage of blockchain

Most professionals found the implementation of blockchain tech very challenging (1 – very challenging, 5 – not challenging).

A total of 87.8% of respondents responded negatively when asked whether their organization was making revenue from blockchain-based products/services.

6 Discussions

RQ1 The research brought to the notice that blockchain can be the base technology for the birth of the latest groundbreaking apps. However, this section lists some essential tools that blockchain can make a significant impact on various businesses.

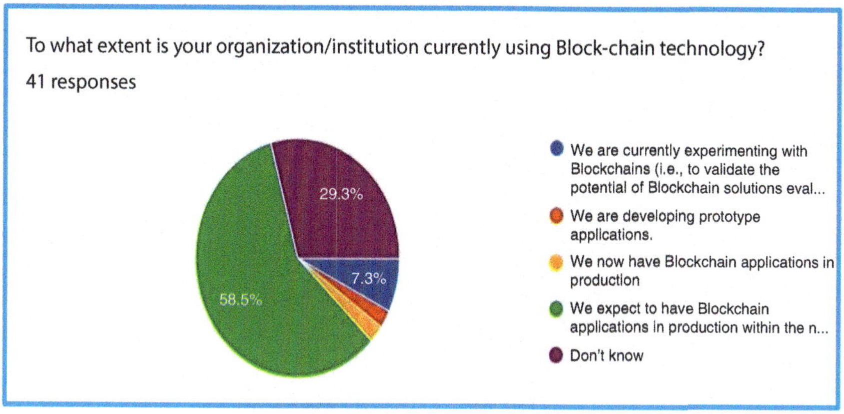

Fig. 7 Pie chart representing the usage extent

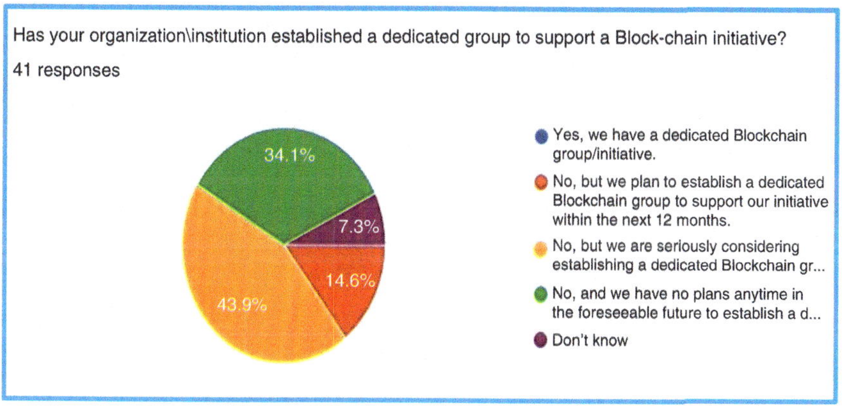

Fig. 8 Pie chart representing the formation of a dedicated group for a blockchain initiative

- **Smart Contracts (SC):** A SC is a computer-based protocol that automatically executes the terms of the agreement [61]. After being long proposed, this tool can now turn into reality using blockchain. The SC is a code that would be run by itself as terms are fulfilled. These days, many SC development platforms are coming to significance, which could help SC become enriched with more functionalities. With SCs, real-time fulfillment of the contract is possible.
- **Distributed Ledger Technology (DLT):** DLT gives decentralization in its application and provides a new data storage paradigm. It can offer cheaper, faster, more secure, and decentralized storage by spreading the data across various nodes-many industries. They are attempting to adopt this technology to cut down the cost by increasing efficiency, resulting in better time management.
- **Identity Management (IM):** Blockchain could prove to be a reliable tool for IM. As people are always asked to share information that could be used to personally

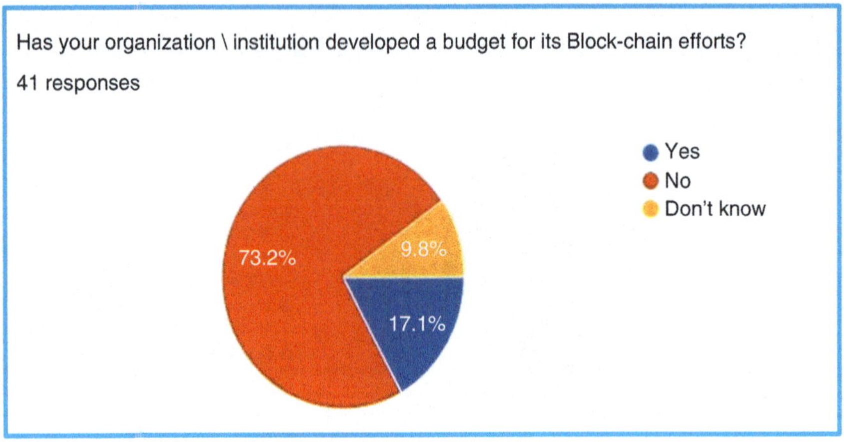

Fig. 9 Pie chart representing budgetary arrangements

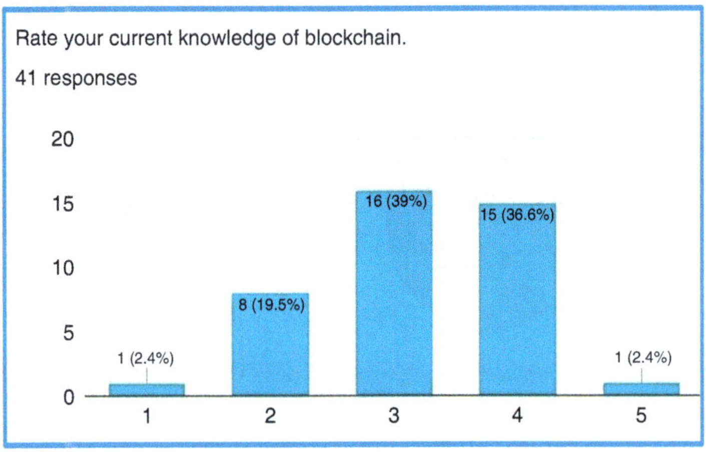

Fig. 10 Bar graph representing blockchain awareness rating

identify them to interact with websites or try to do business with various companies, they are at risk of being defrauded. Blockchain is the underlying tech for IM via decentralized networks. Svein Ølnes [62] discusses that IM with blockchain can provide every citizen with a valid digital identity, increasing efficiency and speed of processes and providing government approval authority. It can aid privacy risks as users can provide their personal information on a case-by-case basis.

RQ2 After discussing blockchain applications, it is equally important to look at the challenges of identifying its business value truly. This tech is faced with two types of challenges – technological challenges and adoption-based challenges.

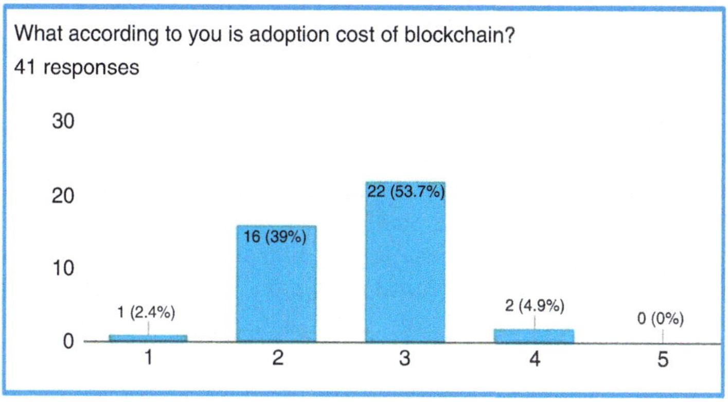

Fig. 11 Bar graph representing blockchain adoption cost awareness

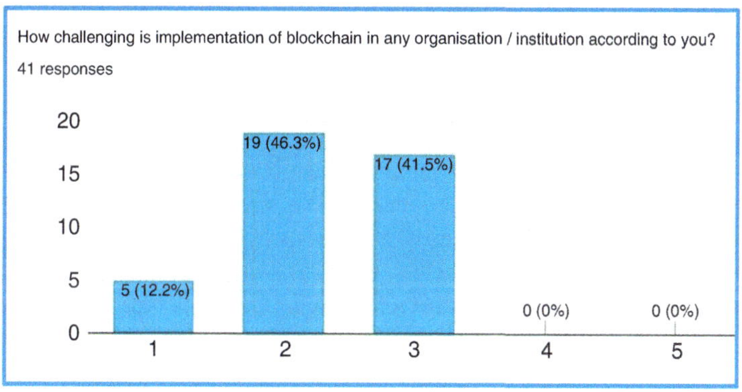

Fig. 12 Bar graph representing challenges in implementing blockchain

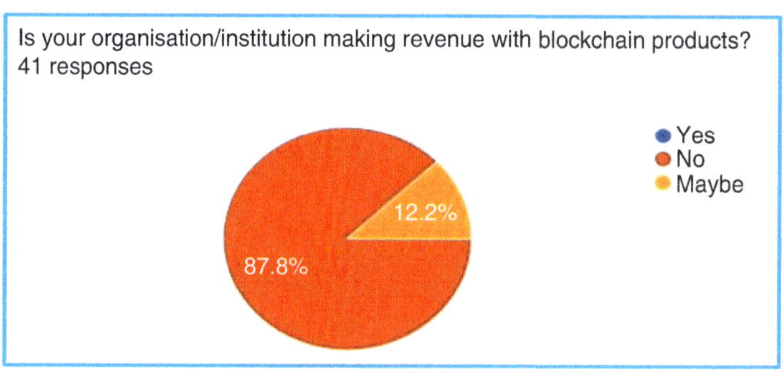

Fig. 13 Pie chart representing blockchain revenue information

6.1 Technological Challenges

- **Security:** Although blockchain is employed in many industries, some security issues remain to be solved. In [43], the "transaction manipulation" attack is described. It takes place if the attacker changes the unique transaction ID before the transaction is verified. Thus, cryptocurrencies are getting transferred to the attacker's account. In [58], the author suggests that blockchain nodes are too risky as the keys could be stolen, leading to malicious transactions. Shae and Tsai [63] describe the initial critical issue proof of the labor framework faces when miners attempt to take control of quite 50% of the network's computational power to prevent transactions from getting verified 51% attack.
- **Privacy:** Privacy is one of the main issues that need to be addressed [64, 65]. Tsai et al. [57] suggest that the feature of cryptographic keys could not ascertain the user's anonymity. Big data analytics of knowledge sets across the Web have compromised 60% of the blockchain users' identities. Moreover, in [66], the authors suggest that the banking systems must provide high privacy compared to current blockchain tech's low privacy level.
- **Latency:** The most prominent limitation of the consensus algorithm is that every node processes each transaction. Decentralization, the most innovative characteristic of blockchain tech, leads to latency issues. Marco and Lakhani [67] state that the Bitcoin blockchain's transactional speed is seven transactions/second. Inside an economic system where speedy executions at high rates are mandatory, the peer-to-peer network that blockchain provides is not a good idea to implement.
- **Computational Cost:** The most challenging issue in the blockchain-based initiative is the specialized hardware often needed to perform transaction verification through the blockchain platform [68, 69]. This means that higher energy consumption results in higher computational costs. Due to this mining power concentration, every potential application based on a blockchain tech should be subjected to further study and research.

6.2 Adoption-Based Challenges

- **Lack of Awareness and Understanding:** The principal issue here in India is the lack of awareness and understanding of how this technology works, because of the low level of technological infrastructure. This further dampens investment and exploration of ideas. But India is constantly coping with this challenge by opening up a tech market for investments. India comes second worldwide in terms of tech. Startup Indian who want to excel with this technology could set aside an internal team focused on understanding the technology, its impact, and usage areas.
- **Culture:** In India, most people still trust the traditional way of doing things due to economic backwardness and low education. Blockchain technology

focuses on developing confidence and authority in a decentralized network instead of a centralized institution/network. Blockchain is about 80% business process change and 20% technology implementations. This means that a creative approach is needed to understand the upcoming opportunities and predict how things will change.

- **Cost and Efficiency:** The execution speed of P2P transactions depends on the type of blockchain. Aggregate execution cost is high since every node is working on the same problem on its copy of the data to find the solution first. So before implementing blockchain, applications need to be carefully thought through its cost, which depends directly upon its scale of operations and its desired budget for blockchain.
- **Regulation and Governance:** Advances in technology have always kept the regulations struggling to remain relevant. The most tricky challenge of the blockchain approach turned out to be its original motivations, i.e., lack of oversight.

Centralized systems "act as shock absorbers in times of crisis" regardless of their drawbacks [70]. This is because these decentralized networks are often much less resilient to shocks impacting users directly, unless careful thought is given to their design.

7 Applying Technology Acceptance Model

The TAM (technology acceptance model) is a theory formulated for IT systems to adopt the technology. A typical TAM framework is shown in Fig. 14.

Mainly two factors affect a person's decision about how and when they will use the given technology, notably:

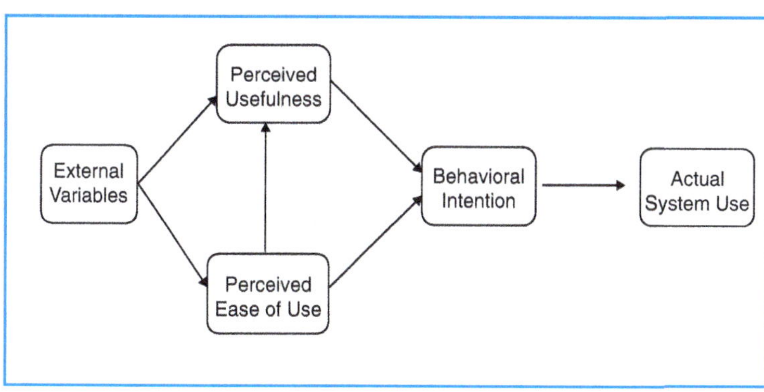

Fig. 14 TAM framework

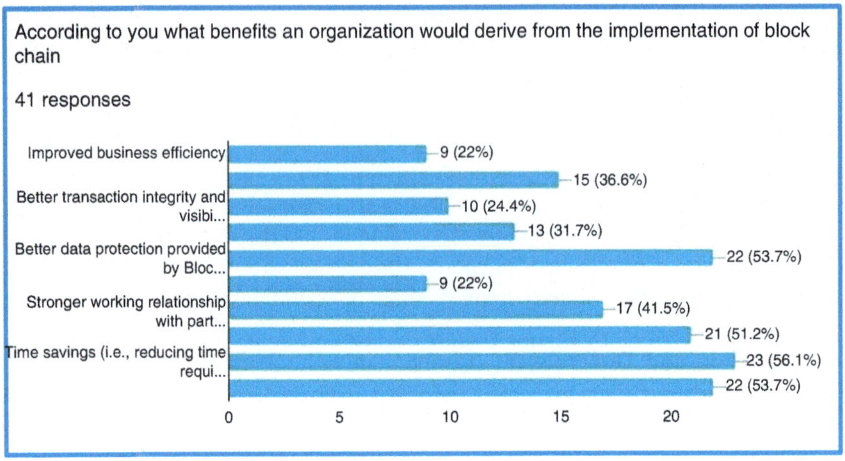

Fig. 15 Benefit analysis graph

(1) **Perceived usefulness (PU)** – It is the measure to which someone perceives that technology to be valuable enough to improve their job performance.
(2) **Perceived ease-of-use (PEOU)** – It's the measure to which a person believes the technology to be straightforward enough to use so that the barriers are crossed easily. If it is easy to use and the interface is simple, nobody will feature a negative attitude.

External variables like social influence affect the intention of people to use technology. However, the perception may alter w.r.t age and gender since everyone seems to have different experiences.

According to our survey, "time-saving" was the most significant benefit, with 56% of respondents feeling the same, followed by "better data protection" and "reduction of risk" with a vote share of about 54% (refer to Fig. 15).

According to the survey result, "lack of experts skilled in blockchain technology" was the biggest adoption challenge for about 73% of respondents. In comparison, "lack of understanding and awareness of what blockchain can do/is good for" and "lack of industry standards and regulatory constraints" was the next significant challenge, with about 66% of respondents facing them (refer to Fig. 16).

Perceived ease-of-use is very low, while perceived usefulness is moderate as compared to the former. Assuming external variables such as social influence is constant on everyone, we can say that behavioral intention to use the technology will be low to moderate.

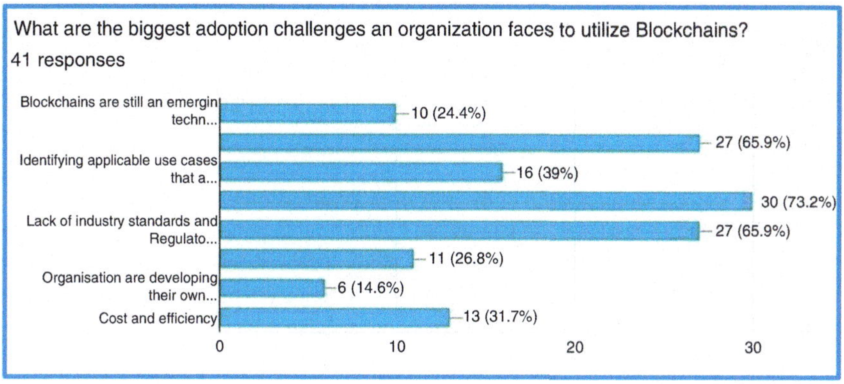

Fig. 16 Clustered graph representing blockchain adoption challenges

8 Limitations of Research

The following are the limitations of the research study, which are observed during the work.

(1) **Lack of previous studies in the research area –** Literature review helps identify the scope of findings that have been done in the research area. In this research, it was complicated to find a significant number of studies that were presenting use cases of blockchain in various industries
(2) **Scope of discussions –** Since this study was conducted only by a group of students, it may not generalize the finding to the best possible degree because of poor past experiences with research and literature in this specific area.
(3) **Sample size –** The study was conducted by analyzing the data collected from approx. 40 respondents. But doing this research on a broader basis of a particular country could confirm the findings and maybe successfully provide a significant contribution to this field of study.
(4) **Respondent fatigue –** A well-documented phenomenon occurs when survey participants become tired of filling surveys and the response quality starts to deteriorate. There are many reasons which can cause such a phenomenon. Still, in our case, it was probably the length of the questionnaire, which resulted in lowering down the concertation while answering.

9 Conclusions and Future Line

Blockchain has garnered unparalleled attention from the industry and is taken into account as a breakthrough technology. However, it would be risky if a company decides to implement the technology just because it is in the spotlight, without thinking about whether it is developed enough for daily use. This chapter presents a summary of potential applications and uses cases/applications of blockchain to help companies eliminate this risk. The chapter also discussed various technology-based and adoption-based challenges faced by blockchain initiatives in India. Results reveal that 56% of respondents agree that "time-saving" is the most significant benefit and "lack of experts skilled in blockchain technology" is the biggest adoption challenge with 73% response. Some possible future directions are also discussed below. The authors extracted and analyzed 45 original papers from various scientific databases using the systematic mapping study process. Blockchain technology continues to disrupt multiple business sectors at an increasing pace. Thus, authors believe that further critical research is required to take advantage of its capabilities and understand its restrictions when applied on a vast scale.

The following research could be undertaken further – blockchain and its amalgamation with various emerging technologies with IoT devices, artificial intelligence, big data analytics, machine learning, cloud computing, and its challenges associated with them.

References

1. Satoshi, N. (2008). Bitcoin: A peer-to-peer electronic cash system
2. Prateek, P., & Ratnesh, L. (2020). Promoting trustless computation through blockchain technology. *National Academy Science Letters*. https://doi.org/10.1007/s40009-020-00978-0
3. Kamble, S., Gunasekaran, A., & Arha, H. (2019). Understanding the blockchain technology adoption in supply chains-Indian context. *International Journal of Production Research, 57*, 2009–2033. https://doi.org/10.1080/00207543.2018.1518610
4. Janssen, M., Weerakkody, V., Ismagilova, E., Sivarajah, U., & Irani, Z. (2020). A framework for analysing blockchain technology adoption: Integrating institutional, market and technical factors. *International Journal of Information Management, 50*, 302–309. https://doi.org/10.1016/j.ijinfomgt.2019.08.012
5. Pandey, M., Litoriya, R., & Pandey, P. (2019). Application of fuzzy DEMATEL approach in analyzing mobile app issues. *Programming and Computer Software, 45*. https://doi.org/10.1134/S0361768819050050
6. Pandey, M., Litoriya, R., & Pandey, P. (2020). Applicability of machine learning methods on mobile app effort estimation: Validation and performance evaluation. *International Journal of Software Engineering and Knowledge Engineering, 30*, 23–41. https://doi.org/10.1142/S0218194020500023
7. Pandey, M., Litoriya, R., & Pandey, P. (2019). Application of fuzzy Dematel approach in analyzing mobile app issues. *Programming and Computer Software, 45*, 268–287. https://doi.org/10.1134/S0361768819050050
8. Pandey, M., Litoriya, R., & Pandey, P. (2019). Perception-based classification of mobile apps: A critical review. In A. K. Luhach, K. B. G. Hawari, I. C. Mihai, P.-A. Hsiung, & R. B. Mishra (Eds.), *Smart computational strategies: Theoretical and practical aspects* (pp. 121–133). Springer Singapore. https://doi.org/10.1007/978-981-13-6295-8_11
9. Pandey, M., Litoriya, R., & Pandey, P. (2019). Identifying causal relationships in mobile app issues: An interval type-2 fuzzy DEMATEL approach. *Wireless Personal Communications, 108*, 683–710.
10. Bhattacharya, P., Tanwar, S., Shah, R., Ladha, A. Mobile edge computing-enabled blockchain framework—A survey. Presented at the (2020). https://doi.org/10.1007/978-3-030-29407-6_57.

11. Upadhyay, N. (2020). Demystifying blockchain: A critical analysis of challenges, applications and opportunities. *International Journal of Information Management, 54*, 102120. https://doi.org/10.1016/j.ijinfomgt.2020.102120
12. Zhang, J., Xue, N., & Huang, X. (2016). A secure system for pervasive social network-based healthcare. *IEEE Access., 4*, 9239–9250. https://doi.org/10.1109/ACCESS.2016.2645904
13. Mettler, M. (2016). Blockchain technology in healthcare: The revolution starts here. In *2016 IEEE 18th International Conference on e-Health Networking, Applications and Services (Healthcom)* (pp. 1–3). IEEE. https://doi.org/10.1109/HealthCom.2016.7749510
14. Prateek, P., & Ratnesh, L. (2020). Securing and authenticating healthcare records through blockchain technology. *Cryptologia., 44*, 341–356. https://doi.org/10.1080/01611194.2019.1706060
15. Pandey, P., & Litoriya, R. (2020). Implementing healthcare services on a large scale: Challenges and remedies based on blockchain technology. *Health Policy and Technology., 9*, 69–78. https://doi.org/10.1016/j.hlpt.2020.01.004
16. Wijaya, D. A., Liu, J. K., Suwarsono, D. A., & Zhang, P. (2017). A new blockchain-based value-added tax system. In T. Okamoto, Y. Yu, M. H. Au, & Y. Li (Eds.), *Provable security* (pp. 471–486). Springer. https://doi.org/10.1007/978-3-319-68637-0_28
17. Alammary, A., Alhazmi, S., Almasri, M., & Gillani, S. (2019). Blockchain-based applications in education: A systematic review. *Applied Sciences, 9*, 2400. https://doi.org/10.3390/app9122400
18. Brakeville, S., & Perepa, B. Blockchain basics: Introduction to distributed ledgers, https://developer.ibm.com/technologies/blockchain/tutorials/cl-blockchain-basics-intro-bluemix-trs/, last accessed 2020/07/21.
19. Pandey, P., & Litoriya, R. (2019). Elderly care through unusual behavior detection: A disaster management approach using IoT and intelligence. *IBM Journal of Research and Development, 64*, 15:1–15:11. https://doi.org/10.1147/JRD.2019.2947018
20. Pandey, P., & Litoriya, R. (2020). Ensuring elderly well being during COVID-19 by using IoT. *Disaster Medicine and Public Health Preparedness*, 1–10. https://doi.org/10.1017/dmp.2020.390
21. Pandey, P., & Litoriya, R. (2019). An activity vigilance system for elderly based on fuzzy probability transformations. *Journal of Intelligent Fuzzy Systems, 36*, 2481–2494. https://doi.org/10.3233/JIFS-181146
22. Pandey, P., & Litoriya, R. (2020). An IoT assisted system for generating emergency alerts using routine analysis. *Wireless Personal Communications, 112*, 607–630. https://doi.org/10.1007/s11277-020-07064-0
23. Konstantinidis, I., Siaminos, G., Timplalexis, C., Zervas, P., Peristeras, V., Decker, S. Blockchain for business applications: A systematic literature review. Presented at the (2018). https://doi.org/10.1007/978-3-319-93931-5_28.
24. Casino, F., Dasaklis, T. K., & Patsakis, C. (2019). A systematic literature review of blockchain-based applications: Current status, classification and open issues. *Telematics and Informatics, 36*, 55–81. https://doi.org/10.1016/j.tele.2018.11.006
25. Abayomi-Zannu, T.P., Odun-Ayo, I., Tatama, B.F., Misra, S. Implementing a mobile voting system utilizing blockchain technology and two-factor authentication in Nigeria. Presented at the (2020). https://doi.org/10.1007/978-981-15-3369-3_63.
26. Zhang, S., Wang, L., & Xiong, H. (2019). Chaintegrity: Blockchain-enabled large-scale e-voting system with robustness and universal verifiability. *International Journal of Information Security*. https://doi.org/10.1007/s10207-019-00465-8
27. Taylor, P. J., Dargahi, T., Dehghantanha, A., Parizi, R. M., & Choo, K.-K. R. (2020). A systematic literature review of blockchain cyber security. *Digital Communications and Networks., 6*, 147–156. https://doi.org/10.1016/j.dcan.2019.01.005
28. Ølnes, S., Jansen, A. Blockchain technology as s support infrastructure in e-government. Presented at the (2017). https://doi.org/10.1007/978-3-319-64677-0_18.

29. Heredia-Roldán, M.-J., Gibaja-Romero, D.-E., Martínez-Flores, J.-L., & Caballero-Morales, S.-O. (2019). The impact of trust in the strategic decisions of a decentralized supply chain. *Opsearch, 56*, 757–779. https://doi.org/10.1007/s12597-019-00377-0
30. Baidya, A. (2019). Stochastic supply chain, transportation models: Implementations and benefits. *Opsearch, 56*, 432–476. https://doi.org/10.1007/s12597-019-00370-7
31. Saberi, S., Kouhizadeh, M., Sarkis, J., & Shen, L. (2019). Blockchain technology and its relationships to sustainable supply chain management. *International Journal of Production Research, 57*, 2117–2135. https://doi.org/10.1080/00207543.2018.1533261
32. Paul Brody. (2017). How blockchain is revolutionizing supply chain management
33. Pandey, P., & Litoriya, R. (2020). Securing E-health networks from counterfeit medicine penetration using Blockchain. *Wireless Personal Communications*. https://doi.org/10.1007/s11277-020-07041-7
34. Dubey, R., Gunasekaran, A., Bryde, D. J., Dwivedi, Y. K., & Papadopoulos, T. (2020). Blockchain technology for enhancing swift-trust, collaboration and resilience within a humanitarian supply chain setting. *International Journal of Production Research, 58*, 3381–3398. https://doi.org/10.1080/00207543.2020.1722860
35. Toyoda, K., Mathiopoulos, P. T., Sasase, I., & Ohtsuki, T. (2017). A novel blockchain-based product ownership management system (POMS) for anti-counterfeits in the post supply chain. *IEEE Access., 5*, 17465–17477. https://doi.org/10.1109/ACCESS.2017.2720760
36. Aitzhan, N. Z., & Svetinovic, D. (2018). Security and privacy in decentralized energy trading through multi-signatures. *Blockchain and Anonymous Messaging Streams. IEEE Transactions on Dependable and Secure Computing., 15*, 840–852. https://doi.org/10.1109/TDSC.2016.2616861
37. Imbault, F., Swiatek, M., de Beaufort, R., & Plana, R. (2017). The green blockchain: Managing decentralized energy production and consumption. In *2017 IEEE International Conference on Environment and Electrical Engineering and 2017 IEEE Industrial and Commercial Power Systems Europe (EEEIC/I&CPS Europe)* (pp. 1–5). IEEE. https://doi.org/10.1109/EEEIC.2017.7977613
38. Sharma, N., Litoriya, R., & Sharma, A. (2021). Application and analysis of K-means algorithms on a decision support framework for municipal solid waste management. In A.E. Hassanien, R. Bhatnagar, & A. Darwish (Ed.), *Advanced machine learning technologies and applications. AMLTA 2020. Advances in intelligent systems and computing* (pp. 267–276). Springer. https://doi.org/10.1007/978-981-15-3383-9_24
39. Sharma, N., Litoriya, R., & Sharma, D. (2021). Forecasting the most predictable municipal solid wastes for improving the quality of waste management system in urban & rural areas of India. *Mukt Shabd Journal., 10*, 403–408.
40. Sharma, N., Litoriya, R., & Sharma, D. (2019). An analytical study on the importance of data mining for designing a decision support system. *Journal Of Harmonized Research (JOHR)., 7*, 44–48. https://doi.org/10.30876/JOHR.7.2.2019.44-48
41. Sharma, N., Litoriya, R., Pratap Singh, H., & Sharma, D. (2021). Modern approach for the significance role of decision support system in solid waste management system (SWMS). In D. Goyal, V. E. Bălaş, A. Mukherjee, C. Hugo, V. de Albuquerque, & A. K. Gupta (Eds.), *Algorithms for intelligent systems* (pp. 619–628). Springer. https://doi.org/10.1007/978-981-15-4936-6_67
42. Sharma, N., Litoriya, R., Sharma, D., & Singh, H. P. (2019). Designing a decision support framework for municipal solid waste management. *International Journal on Emerging Technologies, 10*, 374–379.
43. Suhaliana bt Abd Halim, N., Rahman, M.A., Azad, S., Kabir, M.N. Blockchain security hole: Issues and solutions. Presented at the (2018). https://doi.org/10.1007/978-3-319-59427-9_76.
44. Mathiyazhagan, K., Datta, U., Bhadauria, R., Singla, A., & Krishnamoorthi, S. (2018). Identification and prioritization of motivational factors for the green supply chain management adoption: Case from Indian construction industries. *Opsearch, 55*, 202–219. https://doi.org/10.1007/s12597-017-0316-7

45. Teufel, B., Sentic, A., & Barmet, M. (2019). Blockchain energy: Blockchain in future energy systems. *Journal of Electronic Science and Technology., 17*, 100011. https://doi.org/10.1016/j.jnlest.2020.100011
46. Pandey, P., & Litoriya, R. (2020). Software process selection system based on multicriteria decision making. *Journal of Software: Evolution and Process.*https://doi.org/10.1002/smr.2305
47. Pandey, P., & Litoriya, R. (2020). Fuzzy AHP based identification model for efficient application development. *Journal of Intelligent Fuzzy Systems, 38*, 3359–3370. https://doi.org/10.3233/JIFS-190508
48. Prateek, P., & Ratnesh, L. (2020). Fuzzy cognitive mapping analysis to recommend machine learning based effort estimation technique for web applications. *International Journal of Fuzzy Systems, 22*, 1212–1223. https://doi.org/10.1007/s40815-020-00815-y
49. Grech, L., & Camilleri, A.F. Blockchain in education. (2017). https://doi.org/10.2760/60649.
50. Lilian Yedigaryan: Blockchain applications in education: Use cases, https://nooor.io/blog/blockchain-in-education/, last accessed 2020/05/22.
51. Chen, G., Xu, B., Lu, M., & Chen, N.-S. (2018). Exploring blockchain technology and its potential applications for education. *Smart Learning Environments., 5*, 1. https://doi.org/10.1186/s40561-017-0050-x
52. Bartolomé, A., Adell, J., & Castañeda, L. (2017). Block chain in education: Introduction and critical review of the state of the art. *EDUTEC. Revista Electr—nica de Tecnolog'a Educativa., 61*, 1–14.
53. Guru, S., & Mahalik, D. K. (2019). A comparative study on performance measurement of Indian public sector banks using AHP-TOPSIS and AHP-grey relational analysis. *Opsearch, 56*, 1213–1239. https://doi.org/10.1007/s12597-019-00411-1
54. Pan, X., Pan, X., Song, M., Ai, B., & Ming, Y. (2020). Blockchain technology and enterprise operational capabilities: An empirical test. *International Journal of Information Management, 52*, 101946. https://doi.org/10.1016/j.ijinfomgt.2019.05.002
55. Khullar, S.: Utilising blockchain for cross-border payments: Implications for India. (2018).
56. Awotunde, J.B., Ogundokun, R.O., Jimoh, R.G., Misra, S., Aro, T.O. Machine learning algorithm for cryptocurrencies price prediction. Presented at the (2021). https://doi.org/10.1007/978-3-030-72236-4_17.
57. Tsai, W.-T., Blower, R., Zhu, Y., & Yu, L. (2016). A system view of financial blockchains. In *2016 IEEE Symposium on Service-Oriented System Engineering (SOSE)* (pp. 450–457). IEEE. https://doi.org/10.1109/SOSE.2016.66
58. Bhardwaj, S., Kaushik, M. Blockchain—Technology to drive the future. Presented at the (2018). https://doi.org/10.1007/978-981-10-5547-8_28.
59. Guo, Y., & Liang, C. (2016). Blockchain application and outlook in the banking industry. *Financial Innovation., 2*, 24. https://doi.org/10.1186/s40854-016-0034-9
60. Misra, S. (2021). A step by step guide for choosing project topics and writing research papers in ICT related disciplines. In: Communications in Computer and Information Science. pp. 727–744 https://doi.org/10.1007/978-3-030-69143-1_55.
61. Gopie, N. What are smart contracts on blockchain?, https://www.ibm.com/blogs/blockchain/2018/07/what-are-smart-contracts-on-blockchain/, last accessed 2020/05/22.
62. Ølnes, S. Beyond bitcoin enabling smart government using blockchain technology. Presented at the (2016). https://doi.org/10.1007/978-3-319-44421-5_20.
63. Shae, Z., Tsai, J.J.P. (2017). On the Design of a Blockchain Platform for Clinical Trial and Precision Medicine. In: 2017 IEEE 37th International Conference on Distributed Computing Systems (ICDCS). pp. 1972–1980
64. Litoriya, R., Gulati, A., Yadav, M., Ghosh, R. S., & Pandey, P. (2021). Social, ethical, and regulatory issues of fog computing in healthcare 4.0 applications. In *Fog computing for healthcare 4.0 environments* (pp. 593–609). Springer. https://doi.org/10.1007/978-3-030-46197-3_23

65. Pandey, P., Litoriya, R. Legal/regulatory issues for MMBD in IoT BT – Multimedia big data computing for IoT applications: Concepts, paradigms and solutions. Presented at the (2020). https://doi.org/10.1007/978-981-13-8759-3_14.
66. Kabra, N., Bhattacharya, P., Tanwar, S., & Tyagi, S. (2020). MudraChain: Blockchain-based framework for automated cheque clearance in financial institutions. *Future Generation Computer Systems, 102*, 574–587. https://doi.org/10.1016/j.future.2019.08.035
67. Marco Iansiti, Lakhani, K.R. The truth about blockchain (2017).
68. Buterin, V. (2014). Ethereum White Paper. Etherum White Paper. 1–36
69. Pandey, P., & Litoriya, R. (2021). Technology intervention for preventing COVID-19 outbreak. *Information Technology & People. ahead-of-p*. https://doi.org/10.1108/ITP-05-2020-0298
70. Mengelkamp, E., Notheisen, B., Beer, C., Dauer, D., & Weinhardt, C. (2018). A blockchain-based smart grid: Towards sustainable local energy markets. *Computer Science - Research and Development, 33*, 207–214. https://doi.org/10.1007/s00450-017-0360-9

Machine Learning, IoT, and Blockchain Integration for Improving Process Management Application Security

Roseline Oluwaseun Ogundokun, Michael Olaolu Arowolo, Sanjay Misra, and Joseph Bamidele Awotunde

Abbreviations

Measures	Percentage Values
ML	Machine Learning
BT	Blockchain Technology
TEL	Traditional Ensemble Learning
IEL	Improved Ensemble Learning
ROC	Receiver Operating Characteristics
TP	True Positive
TN	True Negative
FP	False Positive
FN	False Negative
FPR	False Positive Rate
CNN	Convolutional Neural Network
LSTM	Long Short-Term Memory

R. O. Ogundokun (✉) · M. O. Arowolo
Department of Computer Science, Landmark University, Omu Aran, Nigeria
e-mail: ogundokun.roseline@lmu.edu.ng

S. Misra
Department of Computer Science and Communication, Østfold University College (HIOF), Halden, Norway

J. B. Awotunde
Department of Computer Science, University of Ilorin, Ilorin, Nigeria

S. Misra, A. Kumar Tyagi (eds.), *Blockchain Applications in the Smart Era*, EAI/Springer Innovations in Communication and Computing,
https://doi.org/10.1007/978-3-030-89546-4_12

1 Introduction

Information and communication technology (ICT) is rising at a high rate. Advances in semiconductor gadgets and networking systems allow the Internet to connect with a multitude of devices. These devices allow communication from machine to machine (MTM) and from machine to person (MTP). Many words, like IoT, Internet of Everything (IoE), Internet of Vehicles (IoV), Internet of Medical Things (IoMT), Internet of Battlefield Things (IoBT), etc., can refer to such a pattern. These gadgets generally have beams that can identify data from the corporeal setting. The data observed is thereby processed for analysis and retrieval by different applications in consolidated cloud storage [1]. The knowledge in the centralized cloud is prone to different types of attacks. In 2008, Satoshi Nakamoto laid the groundwork for blockchain technologies by proposing a compromise to anonymous individuals for decentralized confidence [2]. In the following years, Bitcoin, the first decentralized digital currency, influenced financial institutions, and a wide variety of cryptocurrencies joined the sector. Digital cryptocurrency is the bulk in blockchain implementations, where consumers share monetary capital via the transparent system with each other [3]. The core features of blockchain technologies are allowing decentralized trust via a consensus protocol and distributed storage through a tamper-proof ledger. These characteristics will benefit from any program that requires multiple parties because it allows open interactions without involving a trustworthy third party. IoT implementations consist of various partners in the field of smart cities and supply chain management, where the blockchain platform can be used to reinforce the trust between the institutions and organizations concerned [4]. While the system has been available for almost a decade, only in the last two years have its technological underpinnings been made simpler. On the one hand, IoT technology developers are well aware of the limits and capabilities of current IoT systems and technologies while on the other hand, developers and proponents of blockchain understand the functional specifics and feasibility of the blockchain implementations on various classes of computing and storage systems. We see a divide between the two groups, and this gap must be bridged to completely leverage blockchain technology resources beyond cryptocurrency and FinTech applications [3].

IoT has been fetched to existence to bridge the distance between the network and the corporeal world by lending perceptual senses to objects to help serve us [4]. To create an increasingly cognitive framework for the distribution of personalized resources to consumers and optimize their overall experience, IoT has been integrated into different industries. A well-correlated and concentrated activity and administration are required because of numerous issues as discovered in pieces of literature relevant to IoT. Therefore, blockchain originated into existence and the combination aims to encourage, control, and facilitate P2P connectivity because the combination will eliminate technological bottlenecks and inefficiencies [5]. In essence, blockchain is a shared network where both parties retain a copy of each transaction [6]. The transfers are straightforward and it is easy to spot any changes in them. Take the case of a smart city, where parking spaces are revealed to drivers instantaneously. When an unoccupied carport is identified by sensors, they refresh the consolidated database. Without revealing this slot to others, it is possible for a system manager who operates this account to claim a parking spot for himself. The credibility of the data from the sensor is undermined in this situation. An interconnected system blockchain network aims to avoid the utilization of a third party and, thus, to guarantee that the sensor's instantaneous data will enter any node in the system beyond unspecified change [7]. Furthermore, BT helps IoT machines connect and instantly make decisions. There are several benefits of distributing the IoT link, for instance, decreased prices connected with managing a single IoT transaction archive, together with increased safety and confidentiality, removing the necessity for a third party. Its debris, however, indistinct whence these characteristics could be applied in the IoT system. This is primarily attributed to the computing capacity, power, and storage constraints for IoT computers. The IoT has a big safety dilemma. To map sensor information and avoid replication with other malicious data, blockchain is used. Thus, radars can share data with the help of BT rather than exploring an intermediary to create confidence. The combination of BT and IoT allows independence and facilitates P2P communication because the combination will eliminate technological bottlenecks and inefficiencies. The cost of IoT implementation may be greatly decreased due to the lack of any other mediator. The blend of IoT and BT is very suitable for commercial applications and price effectiveness.

Blockchain technology could be exploited by IoT in four aspects, specifically confidence building, cost discount, rapid data sharing, and scaled security [8]. This would result in the development of innovative commercial activities for value, environment optimization, risk mitigation, capital freeing, lower transaction costs, processing speed, security provision, confidence, certification confirmation, the integrity of the architecture, anti-counterfeiting, prognostics, distant facilities, and micro-dealings [9]. Devices will exploit smart contracts to allow message transfers and then model agreements between two parties, allowing smart devices to operate autonomously lacking dominant experts [10]. Nevertheless, the key challenge with the BT is the problem of scalability and extreme transaction performance due to the billions of network devices that produce a massive volume of data on an ongoing basis. Amalgam IoT project [11] uses PoW blockchains and byzantine fault resistant to overcome this. BT's key advantages are dissemination, which is a mutual organization of archives between partakers of a business link; authorized, where all affiliate has entree liberties; and confidentiality, where all network members need consensus [10]. Although blockchain is a safer way to create IoT apps, it is not necessarily the appropriate concept to utilize. It should only be done where the design of the applications requires the attributes it offer since it can solitarily acquire extra costs to do it aimlessly. Therefore, before evaluating the architecture basis, the framework must be thoroughly analyzed. Blockchain can only be used if those variables are taken into accounts, such as decentralized structure, P2P system, civic and consecutive business tracking, micro-transactions, and computing power.

This study, therefore, employed the combination of BT and ML to protect network communications and manage datasets to overwhelm the counterfeit dataset. To bring about and evaluate the gathered dataset, big data procedures were employed. Likewise, the fault diagnosis forecast aspect was evaluated on the predictive ML approach proposed which is the improved ensemble learning (IEL) classification ML technique. The system was implemented using the traditional ensemble learning (TEL) and the improved ensemble learning (IEL), and performance matrices like accuracy, precision, sensitivity, and false positive rate (FPR) were used to evaluate the system performance.

The remaining part of the article is structured as follows: section two discussed the state-of-the-art pieces of literature on BT and IoT systems. The related works in this field are likewise discussed. Section 3 discussed the materials in terms of datasets and methods in terms of ML algorithms used in accomplishing the implementation of the system. Section 4 discussed the results discovered and the interpretation of the finding were also discussed. The article was concluded in section 5 and here future works were suggested.

2 Literature Review

The integration of blockchain and IoT literature adds encryption and privacy solutions. Kshetri [12] validates blockchain applications for IoT stability. CIoTA [13] is helped by Tomer et al. to classify irregularities in IoT applications. In tandem with the extensible Markov model (EMM), CIoTA implements the principles of blockchain to detect malicious activities. Dorri et al. [14] highlight the weaknesses in existing protection and privacy policies and leads to LSB, a lightweight and flexible IoT security and privacy blockchain. The lightweight protocols from LSB reduce the cost of bandwidth and measurement. Danzi et al. [15] analyze the overhead connectivity of IoT blockchain synchronization protocols and demonstrate the demands for uplink and downlink bandwidth. A testing platform to handle and deploy blockchain networks for transactive IoT implementations was proposed by PlaTIBART [16]. Shafagh et al. [17] are developing a distributed data management system using the blockchain for IoT applications. Shafagh et al. [17] guarantee that the IoT data control remains with the stakeholders. These papers discuss some of the problems outlined in section 5, but the architectural specifics and performance consequences, especially for resource-constrained IoT platforms, are not addressed. In the literature, the benefits and difficulties of applying blockchain for the IoT are discussed. Huckle et al. [18] address blockchain technologies for monetizing IoT applications, but the problems are not defined in their work. Huh et al. [19] illustrate how blockchain can be used using smart contracts to store sensor data. Canoscenti et al. [20] discuss blockchain usage cases and highlight the transparent integrity, confidentiality, and adaptability challenges while storing IoT data in a decentralized network. The developers of [15] assess the overhead correspondence for the IoT of blockchain synchronization. In comparison to [19, 20], our work focuses on architectural problems and performance consequences for data management, monetization, stability, and privacy using blockchain for IoT. Current methods are largely clustered in IoT applications, posing many security issues such as solitary points of catastrophe, reliance, and safety. Additionally, it restricts their scalability and alerts the necessity for a distributed IoT belief system. Beyond the necessity of dominant authority, blockchain will provide trust by cryptographic techniques. Numerous blockchain-built IoT technologies have recently acquired popularity because of their ability to improve safety and confidentiality. A new report by Juniper Research [21] forecasts that by reducing retailers' prices, simplifying regulatory enforcement, and tackling bribery, a mix of IoT and BT in the food business will lay aside billions of dollars. Food business leaders such as Carrefour, Nestle, and Cermaq have already begun employing hyperledger cloth, an IBM-designed blockchain technology [22–24].

Few researchers have work in the field of BT and IoT, some of them have adopted ML techniques to predict bitcoin prices such as cryptocurrency while others have used ML techniques to solve one challenge or the others encountered in the field of BT and IoT. Some of such researches are discussed as follows:

Liu, Richard Yu, Li, Ji, and Leung [25] examine a variety of difficulties and benefits associated with merging blockchain with machine learning. They also talk about topics and difficulties in a larger sense. The authors likewise present a comprehensive examination of combining BT with AI for the exchange of information and networking arrangements, as well as a comparison of prevailing BT and ML technologies [25].

Mohandas, Dhanaraj, and Gao [26] suggested an artificial neural network-based solution for energy-effective illumination in smart cities. The concept was put into action in a domestic setting and verified in various settings and periods. The suggested model uses data from multiple sensors, a neural network, and a fuzzy logic regulator to make effective decisions for demand-built usage and generated power conservation.

Ferrag et al. [27] published a report that summarized previous surveys on blockchain technology and its use in the Internet of Things. A Categorization of a blockchain menace method that incorporates the BT system's susceptibilities as well as loopholes in blockchain protocols in IoT systems was considered by the authors. The most up-to-date techniques in the field of safety and confidentiality-conserving BT are examined, alongside assessment done concerning intricacy, exchange of information overhead, restrictions, safety aims, and performance.

Awotunde, Ogundokun, Misra, Adeniyi, and Sharma [28] postulated the utilization of an ML technique to construct a system for the stock and cryptocurrency price forecasting by employing technical indicators that are of utmost significance for investigation on market inclinations. The authors made use of the long short-term memory (LSTM) ML technique to build the cryptocurrency price forecasting system. The system had an accuracy of 67.43%.

Researchers have been addressing several concerns, obstacles, and many types of difficulties related to BT and their incorporation with IoT, ML, and cloud computing (CC) structures in recent years [29]. B-IoT: blockchain technology for

IoT in intelligent transportation systems, [30]) explored B-IoT, or BT for IoT, smart transportation schemes, process management systems, and the likes. Few of such researches are discussed as follows:

Polina et al. [31] addressed how to decentralize and speed health-care research by combining BT and immediate generation AI pieces of machinery. Khan, Asif, Ahmad, Alharbi, and Aljuaid [32] explored smart grid safety concerns and their implementation in health care for long-term sustainability. A cutting-edge network for BT medical applications was demonstrated, along with a mobile application dubbed HDG for medical record computerization. Bibri [33] presented a background of sensor data used for imminent IoT smart city sustainability. The scholars researched existing material to explain current big data applications and to investigate prospects for enhancing the information scenery of justifiable smart municipalities. The scholars similarly concentrated on precise pieces of machinery, submissions, and further theoretical investigative study, as well as innovative uses of these technologies. Yli-Huumo, Deokyoon, Choi, Park, and Smolander [34] conducted a review of current blockchain research with the goal of better understanding the technology's present study subjects, problems, and future developments. Following a survey of 40 published publications, the authors concluded that blockchains' principal limitations are privacy and security problems [34]. Park, Jin, and Park [35] investigated the acceptance of blockchain technology and protection challenges in CC settings. The researchers observed investigation on protection issues, for instance, threats, and analyzed the block structure. Li et al. [36] examined BT safety by examining dangers and actual-world assaults. The study examines the actual-world assaults on BTs and the susceptibilities that were employed. The scholars similarly covered safety resolutions and scholarly accomplishments, along with imminent investigative prospects. Joshi, Han, and Wang [37] examined the safety and confidentiality problems of BT, as well as some potential resolutions for further protected settings. The scholars focused on the problems and prospects of consensus algorithms concerning protection and confidentiality. In addition, the investigators analyzed potential BT developments. Numerous researchers have looked at various BT applications. Panarello, Tapas, Merlino, Longo, and Puliafito [38], for example, investigated a variety of application areas and classified use patterns as gadgets management or data administration, expansion resolutions, or issues connected to incorporating BT into IoT. Banarjee et al. [39] investigated blockchain-IoT safety and discovered a few concerns with IoT datasets that need to be addressed. The authors investigated prevailing IoT protection techniques concerning intrusion discovery and avoidance, application categorization, system organization classification, prognostic protection, and self-recovering, as well as the blockchain's probable applicability in securing the distribution of data in IoT records [40]. Ferrag et al. [27] published a report that summarized previous surveys on blockchain technology and its use in the Internet of Things. A Categorization of a blockchain menace method that incorporates the BT system's susceptibilities as well as loopholes in blockchain protocols in IoT systems was considered by the authors. The prevailing techniques in the field of protection and privacy-conserving BT are examined, with a side-by-side assessment done concerning complexity, communication overhead, restrictions, protection aims, and performance.

Sharma, You, Jayakody, Reina, and Choo [41] also discuss neural blockchain-built drone hoarding for edge UAV systems. Li et al. looked at some of the issues and solutions to safeguarding data privacy using federated learning [42]. Researches by Bhattacharya, Kaluri, Singh, Alazah and Tariq [43]; Gadakallu et al., [44]; Lightweight Proof of Game (LPoG) [45] describe few additional relevant pieces of research on deep learning and the BT [43–45].

3 Materials and Method

In this section, the dataset used for the implementation of the system was discussed here. The proposed system flow and the performance of the system implemented were discussed here as well.

3.1 Dataset

The IoT blockchain dataset was used to test the KNN classifiers' performance. The information was obtained from the Mendeley database repository. There are 17 characteristics and 81 occurrences in all. The dataset can be found using this link: https://data.mendeley.com/datasets/rxsdfg8ct9/1. DOI: 10.17632/rxsdfg8ct9.1

Bio studies: supporting data is http://www.ebi.ac.uk/biostudies/studies/S-EPMC6412473?xr=true. Figure 1 shows the implemented sample dataset attributes.

3.2 Proposed System

The proposed system used for the implementation and test in this study was the IoT-blockchain dataset. The dataset consists of 17 attributes with 81 instances initially uploaded into the system after which the training and testing phases were performed on the dataset. The testing dataset was passed into the ensemble learning ML technique and the confusion matrix was gotten. Thereafter the testing dataset was secondly passed into the improved ensemble learning ML technique which also computed the confusion matrix and the result was derived as well. Finally, the performance of the system was evaluated using the confusion matrix values, and the values of both EL and IEL were compared together to know the best ML algorithm out of the two classifiers. Figure 2 shows the block flow diagram of the proposed system.

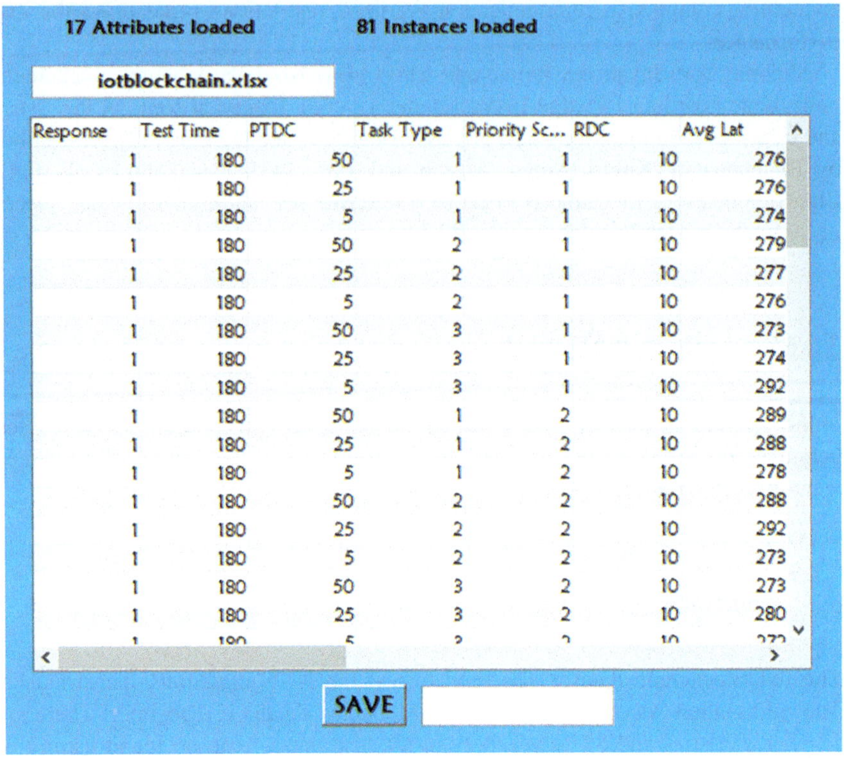

Fig. 1 Dataset attributes uploaded

3.3 *Performance Evaluation*

For the performance evaluation of the system that was implemented, a confusion matrix was used. The matrices used to evaluate the system are accuracy, detection rate, and false positive rate [43]. A classifier successfully and efficiently gives high accuracy, precision, and sensitivity as well as a low false positive rate (FPR), as may be determined. The following is the formula for the four metrics seen in Eqs. 1, 2, 3, and 4:

$$\text{Accuracy} : \frac{TP + TN}{TP + FP + FN + TN} \tag{1}$$

$$\text{Precision} : \frac{TP}{(TP + FP)} \tag{2}$$

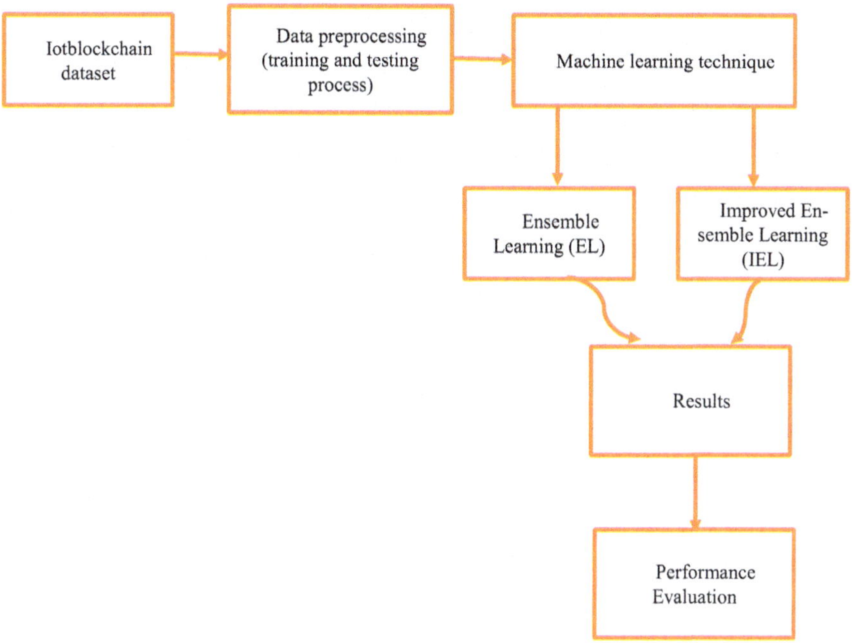

Fig. 2 Proposed system block diagram

$$\text{Sensitivity} : \frac{TP}{(TP + FN)} \tag{3}$$

$$\text{FPR} : \frac{FP}{FP + TN} \tag{4}$$

4 Results and Discussion

The results from the implementation of the system are discussed in this section. Figure 3 shows the scattered plots diagram of the data for the IEL ML classification technique. Figure 4 displays the confusion matrix for the EL ML classification technique, whereas Fig. 5 displays the confusion matrix for the IEL ML classification technique. Figure 6 shows the receiver operating characteristics (ROC) curve for the IEL classification ML technique.

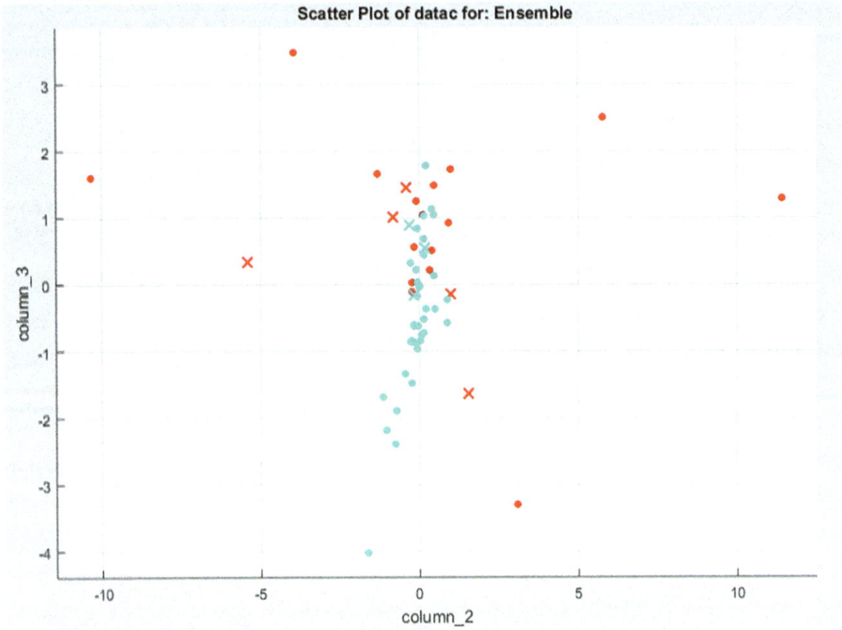

Fig. 3 Scattered plots diagram for IEL

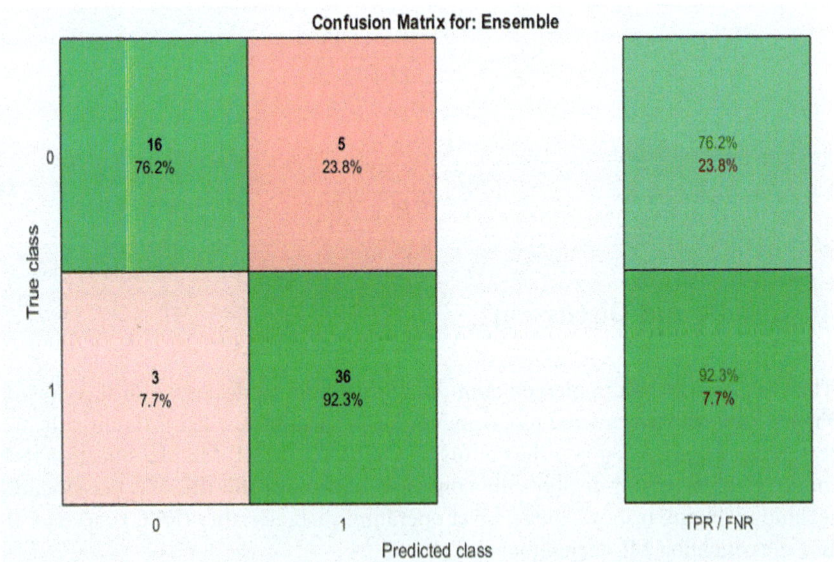

Fig. 4 Confusion matrix for EL

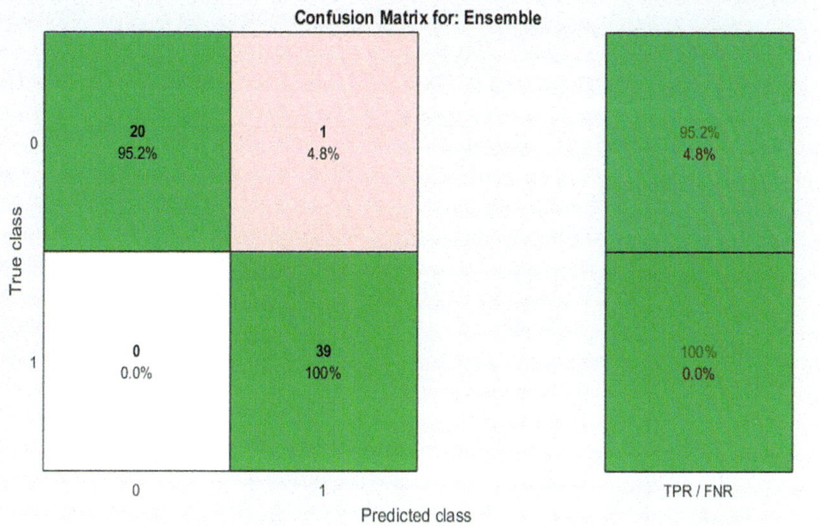

Fig. 5 Confusion matrix for IEL

4.1 Discussion

Table 1 shows the confusion matrix for the classifiers (TEL and IEL). It is stated that once the ROC curve is 1, the ML classifier is capable of flawlessly distinguishing between all the positive and negative class points appropriately that is once the arc is nearer to the uppermost left angle, it specifies an improved performance [47]. In this study, the ROC value of the IEL classification ML technique had a ROC value of 1 as shown in Fig. 6, and this implies that the IEL classifier performed better than the TEL classifier. Table 2 displays the performance metrics employed to assess the technique for the TEL classifier, and Table 3 displays the performance metrics employed to assess the technique for the IEL classifier.

Figure 7 shows the comparative analysis between the two implemented classification ML algorithms. It was demonstrated on the graph that the improved ensemble learning classifier system performance surpassed that of the traditional ensemble learning ML classification technique with a 98.3% accuracy over 86.7% for TEL, the precision of 97.5% over that of TEL of 87.8% precision, 100% sensitivity over 92.3% for TEL, and 4.8% FPR for IEL over 23.8% for TEL. It is deduced that the higher the accuracy, sensitivity, and precision and the lower the FPR, the better the performance of the system in the detection and recognition process in ML procedure. Figure 8 displays the comparative analysis of the projected technique with the existing ones, and it was discovered that the proposed system performance surpassed that of the state-of-the-arts with an accuracy of 98.3 over that of Lahmiri and Bekiros [45] having an accuracy of 50.35%; Wu, Lu, Ma, and Lu [46] having 49.08% accuracy; and Alonso-Monsalve, Suarez-Cetrulo, Cervantes, and Quintana [47] having an accuracy of 50.67%.

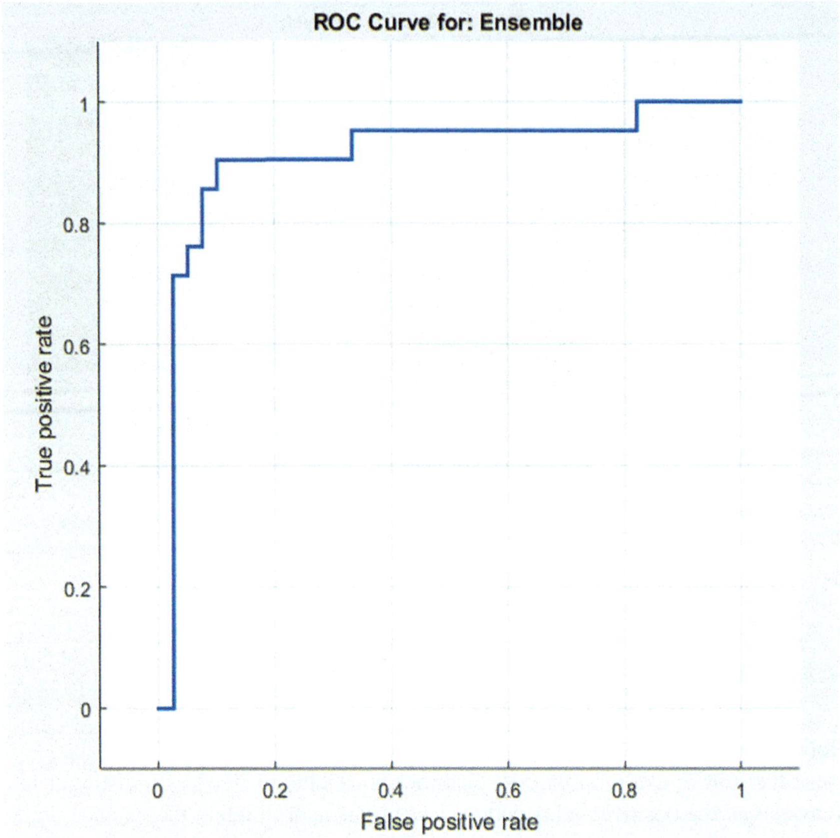

Fig. 6 ROC curve for the improved ensemble learning

Table 1 Confusion matrix for the classifiers

Classifier	TP	TN	FP	FN
Ensemble	36	16	5	3
Improved ensemble	39	20	1	0

Table 2 Performance metrics for TEL

Measures	Percentage values
Accuracy	86.7
Precision	87.8
Sensitivity	92.3
False positive rate	23.8

Table 3 Performance metrics for IEL

Measures	Percentage values
Accuracy	98.3
Precision	97.5
Sensitivity	100
False positive rate	4.8

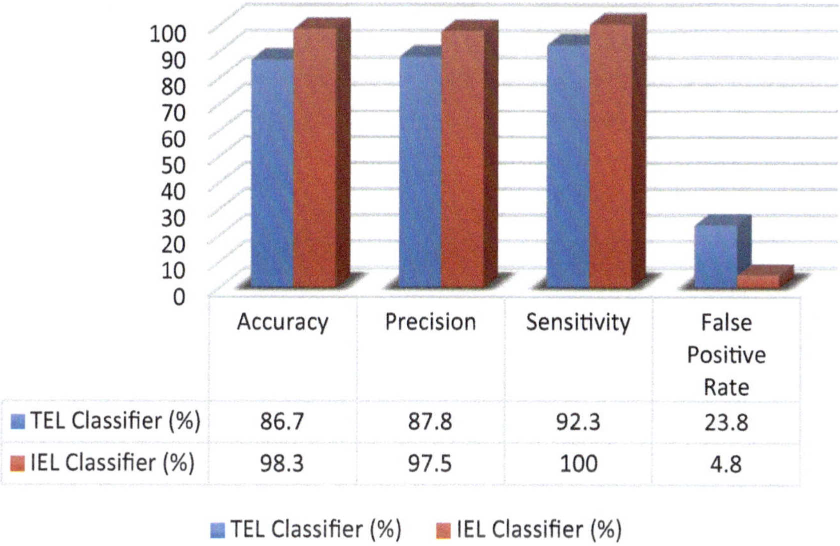

Fig. 7 Comparative analysis between the two proposed classifiers

5 Conclusion

The recent resolutions to IoT protection and confidentiality using ML techniques, BT approaches, and a combination of both have been discussed in this article. The study also sought to offer an ML menace method for IoT founded on past researches to better comprehend the protection and confidentiality challenges in an ML approach. Currently, combining ML techniques with BC approaches to accomplish IoT protection and confidentiality is a novel subject that has to be explored further. This study, therefore, employed the combination of BT and ML to protect network communications and manage datasets to overwhelm the counterfeit dataset. To bring about and evaluate the gathered dataset, big data procedures were employed. Likewise, the fault diagnosis forecast aspect was evaluated on the predictive ML approach proposed which is the improved ensemble learning (IEL) classification ML technique. The study also demonstrated that the proposed system performance when compared with the existing ones outperformed that of the state-of-the-arts with an accuracy of 98.3 over that of Lahmiri and Bekiros [45] having an accuracy of 50.35%; Wu, Lu, Ma, and Lu [46] having 49.08% accuracy; and Alonso-

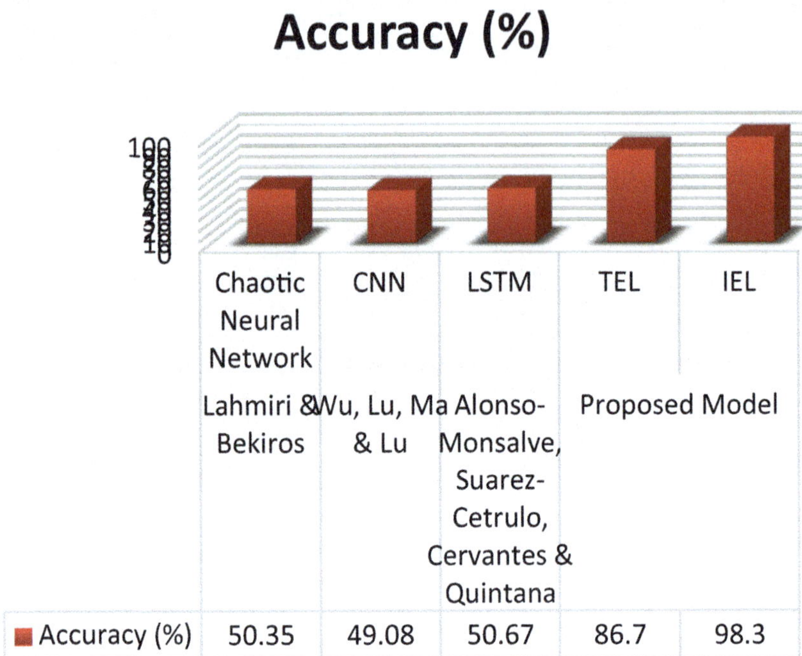

Fig. 8 Comparative analysis with state-of-the-arts

Monsalve, Suarez-Cetrulo, Cervantes, and Quintana [47] having an accuracy of 50.67%. The ROC of the system is discovered to be 1, sensitivity is 100%, and the FPR was 4.8%, which indicated that the proposed system performance is excellent, and it is recommended that the proposed classifier can be adopted for detection and recognition of security threats or attack in a particular network and can be used for improving process management application security. In the future, a researcher creates a privacy-preserving IoT framework that allows for data exchange and analysis while maintaining privacy.

References

1. Shahidinejad, A., Ghobaei-Arani, M., & Masdari, M. (2021). Resource provisioning using workload clustering in cloud computing environment: A hybrid approach. *Cluster Computing, 24*(1), 319–342.
2. Nakamoto, S. (2019). *Bitcoin: A peer-to-peer electronic cash system*. Manubot.
3. Ramachandran, G. S., & Krishnamachari, B. (2018). Blockchain for the IoT: Opportunities and challenges. arXiv preprint arXiv:1805.02818.
4. Abayomi-Zannu, T. P., Odun-Ayo, I., Tatama, B. F., & Misra, S. (2020). Implementing a mobile voting system utilizing blockchain technology and two-factor authentication in Nigeria. In *Proceedings of first international conference on computing, communications, and cyber-security (IC4S 2019)* (pp. 857–872). Springer.

5. Thakore, R., Vaghashiya, R., Patel, C., & Doshi, N. (2019). Blockchain-based IoT: A survey. *Procedia Computer Science, 155*, 704–709.
6. Cho, S., & Lee, S. (2019). Survey on the application of blockchain to IoT. In *2019 International Conference on Electronics, Information, and Communication (ICEIC)* (pp. 1–2). IEEE.
7. Awotunde, J. B., Ogundokun, R. O., Jimoh, R. G., Misra, S., & Aro, T. O. (2021). Machine learning algorithm for cryptocurrencies price prediction. In S. Misra & A. Kumar Tyagi (Eds.), *Artificial intelligence for cyber security: Methods, issues, and possible horizons or opportunities* (Studies in computational intelligence) (Vol. 972). Springer. https://doi.org/10.1007/978-3-030-72236-4_17
8. Gopal, S. *Blockchain for the Internet of Things, Tata Consultancy Services White Paper.* URL: https://www.tcs.com/blockchain-for- io
9. URL: https://www.transparencymarketresearch.com/blockchain-technology-market.html [visited: 10/02/2019]
10. IBM Research Editorial Staff, 'What is blockchain?', 2018, URL: https://www.ibm.com/downloads/cas/K54GJQJY [visited: 10/02/2019]
11. URL: https://blockgeeks.com/guides/what-is-blockchain-technology/ [visited: 10/02/2019]
12. Kshetri, N. (2017). Can blockchain strengthen the internet of things? *IT Professional, 19*(4),
13. Golomb, T., Mirsky, Y., & Elovici, Y. (2018). CIoTA: Collaborative IoT anomaly detection via blockchain. arXiv preprint arXiv:1803.03807.
14. Dorri, A., Kanhere, S. S., Jurdak, R., & Gauravaram, P. (2019). LSB: A Lightweight Scalable Blockchain for IoT security and anonymity. *Journal of Parallel and Distributed Computing, 134*, 180–197.
15. Danzi, P., Kalor, A. E., Stefanovic, C., & Popovski, P. (2018). Analysis of the communication traffic for blockchain synchronization of IoT devices. In *2018 IEEE International Conference on Communications (ICC)* (pp. 1–7). IEEE.
16. Walker, M. A., Dubey, A., Laszka, A., & Schmidt, D. C. (2017). Platibart: A platform for transactive IoT blockchain applications with repeatable testing. In *Proceedings of the 4th workshop on middleware and applications for the Internet of Things* (pp. 17–22).
17. Shafagh, H., Burkhalter, L., Hithnawi, A., & Duquennoy, S. (2017). Towards blockchain-based auditable storage and sharing of IoT data. In *Proceedings of 2017 on cloud computing security workshop* (pp. 45–50).
18. Huckle, S., Bhattacharya, R., White, M., & Beloff, N. (2016). Internet of things, blockchain, and shared economy applications. *Procedia Computer Science, 98*, 461–466.
19. Huh, S., Cho, S., & Kim, S. (2017). Managing IoT devices using a blockchain platform. In *2017 19th international conference on advanced communication technology (ICACT)* (pp. 464–467). IEEE.
20. Conoscenti, M., Vetro, A., & De Martin, J. C. (2016). Blockchain for the Internet of Things: A systematic literature review. In *2016 IEEE/ACS 13th International Conference of Computer Systems and Applications (AICCSA)* (pp. 1–6). IEEE.
21. Juniperresearch.com. Blockchain to Save the Food Industry $31 Billion by 2024. https://www.juniperresearch.com/press/pressreleases/blockchain-to-save-the-food-industry-%2431-billion-b (2019). Accessed 4 Dec 2019
22. Carrefour Group. Carrefour launches Europe's first food blockchain. https://www.carrefour.com/current-news/carrefour-launches-europes-first-food-blockchain (2018). Accessed 4 Dec 2019.
23. Cermaq.com. Cermaq | Cermaq contributes to traceability with blockchain. https://www.cermaq.com/wps/wcm/connect/cermaq/news/mynewsdesk-press-release-2945012/ (2019). Accessed 4 Dec 2019
24. Nestle´ Global. Nestle´ breaks new ground with an open blockchain pilot. https://www.nestle.com/media/pressreleases/allpressreleases/nestle-open-blockchain-pilot (2019). Accessed 4 Dec 2019
25. Liu, Y., Richard Yu, F., Li, X., Ji, H., & Leung, V. C. M. (2020a). Blockchain and machine learning for communications and networking systems. *IEEE Communications Surveys & Tutorials*.

26. Mohandas, P., Dhanaraj, J. S. A., & Gao, X. Z. (2019). Artificial neural network-based smart and energy-efficient street lighting system: A case study for a residential area in Hosur. *Sustainable Cities and Society, 48*, 101499.
27. Ferrag, M. A., Derdour, M., Mukherjee, M., Derhab, A., Maglaras, L., & Janicke, H. (2018). Blockchain technologies for the internet of things: Research issues and challenges. *IEEE Internet of Things Journal, 6*(2), 2188–2204.
28. Awotunde, J. B., Ogundokun, R. O., Misra, S., Adeniyi, E. A., & Sharma, M. M. (2021). Blockchain-based framework for secure transaction in mobile banking platform. *Advances in Intelligent Systems and Computing, AIST, 1375*, 525–534.
29. Conoscenti, M., Vetro, A., & Martin, J. C. D. (2016). Blockchain for the internet of things: A systematic literature review. In *Proceeding of the 3rd international conference on computer systems and applications*, (pp. 1–6).
30. B-IoT: Blockchain technology for IoT in intelligent transportation systems. (Accessed on 2019) http://iot.ed.ac.uk/projects/b-iot/
31. Polina, M., Lucy, O., Yury, Y., Alex, O., Alex, B., Pavel, P., et al. (2018). Converging blockchain and next-generation artificial intelligence technologies to decentralize and accelerate biomedical research and healthcare. *Oncotarget, 9*(5), 5665–5681.
32. Khan, F. A., Asif, M., Ahmad, A., Alharbi, M., & Aljuaid, H. (2020). Blockchain technology, improvement suggestions, security challenges on the smart grid, and its application in healthcare for sustainable development. *Sustainable Cities and Society, 102018.*
33. Bibri, S. E. (2018). The IoT for smart sustainable cities of the future: An analytical framework for sensor-based big data applications for environmental sustainability. *Sustainable Cities and Society, 38*, 230–253.
34. Yli-Huumo, J., Deokyoon, K., Choi, S., Park, S., & Smolander, K. (2016a). Where is current research on Blockchain technology? —A systematic review. *PloS, 11*(10), 1–27.
35. Park, J., & Jin, & Park, J. H. (2017). Blockchain security in cloud computing: Use cases, challenges, and solutions. *Symmetry, 9*(8), 164–177.
36. Li, X., Jiang, P., Chen, T., Luo, X., & Wen, Q. (2016). A survey on the security of Blockchain systems. *Future Generation Computer Systems, 9604*, 106–125.
37. Joshi, A. P., Han, M., & Wang, Y. (2018). A survey on security and privacy issues of Blockchain technology. *Mathematical Foundations of Computing, 1*(2), 121–147.
38. Panarello, A., Tapas, N., Merlino, G., Longo, F., & Puliafito, A. (2018). Blockchain and IoT integration: A systematic survey. *Sensors, 18*(8), 1–37.
39. Banerjee, M., Lee, J., & Choo, K. K. R. (2018). A Blockchain future for the internet of things security: A position paper. *Digital Communications and Networks, 4*(3), 149–160.
40. Halim, N. S. A., Rahman, M. A., Azad, S., & Kabir, M. N. (2017). Blockchain security hole: Issues and solutions. In *Proceedings of the international conference of reliable information and communication technology*, (pp. 739–746).
41. Sharma, V., You, I., Jayakody, D. N. K., Reina, D. G., & Choo, K.-K. R. (2019). Neural blockchain-based ultrareliable caching for edge-enabled UAV networks. *IEEE Transactions on Industrial Informatics, 15*(10), 5723–5736.
42. Li, Z., Sharma, V., & Mohanty, S. P. (2020). Preserving data privacy via federated learning: Challenges and solutions. *IEEE Consumer Electronics Magazine, 9*(3), 8–16.
43. Li, Y., & Chen, Z. (2018). Performance evaluation of machine learning methods for breast cancer prediction. *Applied and Computational Mathematics, 7*(4), 212–216.
44. Yang, S., & Berdine, G. (2017). The receiver operating characteristic (ROC) curve. *The Southwest Respiratory and Critical Care Chronicles, 5*(19), 34–36.
45. Lahmiri, S., & Bekiros, S. (2019). Cryptocurrency forecasting with deep learning chaotic neural networks. *Chaos, Solitons & Fractals, 118*, 35–40.
46. Wu, C. H., Lu, C. C., Ma, Y. F., & Lu, R. S. (2018, November). A new forecasting framework for bitcoin price with LSTM. In *2018 IEEE International Conference on Data Mining Workshops (ICDMW)* (pp. 168–175). IEEE.
47. Alonso-Monsalve, S., Suárez-Cetrulo, A. L., Cervantes, A., & Quintana, D. (2020). Convolution on neural networks for high-frequency trend prediction of cryptocurrency exchange rates using technical indicators. *Expert Systems with Applications, 149*, 113250.

A Blockchain-Powered Energy Monitoring System

A. Swain, K. P. Swain, G. Palai, and M. N. Mohanty

1 Introduction

Smart grid technology is emerging at a rapid pace by leveraging various IoT technologies that increase the efficiency and accuracy of the systems as well as reduce the requirement of manpower. It adopts intelligent systems for energy production and consumption that can monitor and communicate with each other. For efficient utilization of energy, an energy monitoring system is used, in which a system of computer-aided tools is providing real-time digitally maintained energy records used by operators of utility grids to monitor, control, and optimize the performance of the generation and transmission system [1]. It provides real-time metering, utility bill tracking, energy audits, and carbon and sustainability reporting, in order to analyze the trend in energy consumption and identify cost-saving opportunities. Due to the increasing demand for energy consumption, it is necessary to come up with energy systems that can manage individual energy transactions between the consumers and the utility provider without any loss of information. To make information available instantly and securely to authorized users, these energy records can be shared through Energy Internet that integrates power technology, electronic technology, information technology, and intelligent

A. Swain
Department of Electrical Engineering, CET, Systems Engineer (C1) Tata Consultancy Service, Bhubaneswar, Odisha, India

K. P. Swain (✉) · G. Palai
Department of Electronics and Communication Engineering, GITA, Bhubaneswar, Odisha, India

M. N. Mohanty
Department of ECE, I.T.E.R, SoA University, Bhubaneswar, Odisha, India

S. Misra, A. Kumar Tyagi (eds.), *Blockchain Applications in the Smart Era*, EAI/Springer Innovations in Communication and Computing,
https://doi.org/10.1007/978-3-030-89546-4_13

management technology to achieve energy exchange more efficiently and cleanly with the least wastage.

However, there are many issues associated with the current system of energy monitoring and billing, of which some major ones are centralized control, tampering of data, the involvement of intermediaries, high transaction costs, privacy issues, consumer issues, the manual intervention of billing process, overdue bill payment, power theft, and lack of interoperability. The involvement of several processes, intermediaries, and third parties, such as banks, regulators, price reporters, exchanges, and logistics in typical energy commodity transactions, leads to an increase in operational and transactional cost, which sometimes paves way for erroneous transactions, intentionally or otherwise. Due to a centralized control, the entire process is controlled by one central entity, and thus, in case of data tampering or loss of data due to lack of security, the data cannot be retrieved, since there is only a single copy of the records.

With the emergence of modern power systems and new age technologies, the traditional energy markets are transforming into advanced markets that are adopting various mechanisms and business models to curb the issues associated with centralized systems and encourage distributed clean energy trading. At present, blockchain technology is emerging in various fields, including financial services, health care, cybersecurity, supply chain management, banking, etc., and it is growing at a rapid rate with each passing day. It has been a major area of research as it provides numerous benefits that can help overcome the challenges faced by centralized systems. Its application in the energy sector has increased over the recent years due to its distributed ledger technology and decentralized control over energy systems. The main advantage of blockchain-based energy systems is that it eliminates the involvement of intermediaries and third parties, such as banks, regulators, exchanges, logistics, price reporters, etc., that can lead to high transactional costs and delay in payment settlements. Due to the absence of central authority, the faster and cheaper transaction is achieved in the blockchain. Along with that, the personal information and identity of blockchain users are also protected, as these are not linked to the transactions.

1.1 Blockchain Technology

Blockchain is a highly secure and immutable distributed ledger system that holds transactional records in the form of blocks, as shown in Fig. 1. These blocks are created with every transaction and are verified and validated by network engineers before adding them to the existing ledger. They are linked to each other using cryptography, that is, a reference to the previous block's hash is present in the current block, which is time-stamped with a unique cryptographic hash [2]. Blockchain is also known as a consensually shared database that stores transactional records on the Internet, thus cutting the costs of buying servers. It provides a high level of transparency and security by providing a copy of the ledger to all the

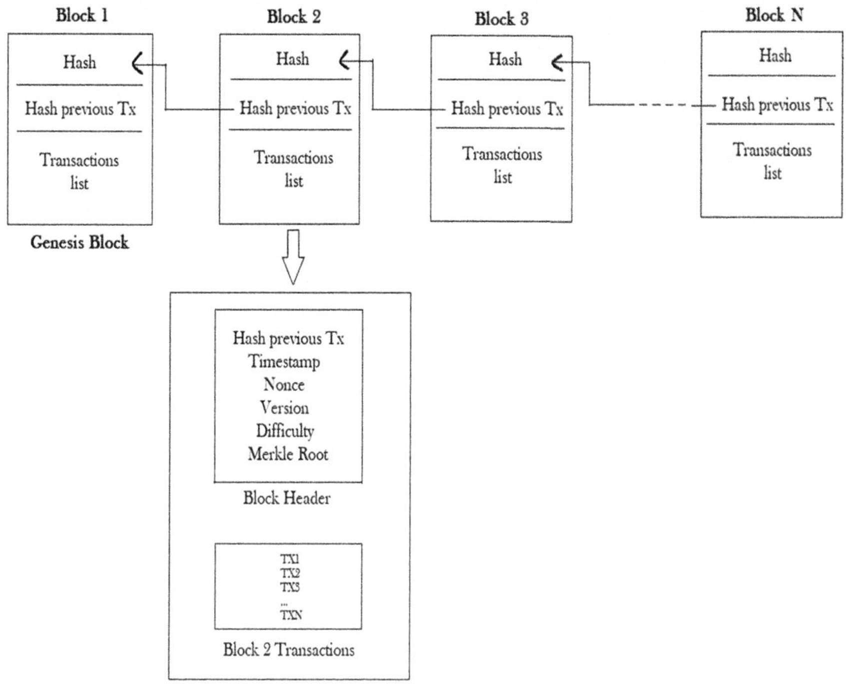

Fig. 1 Structure of a blockchain

nodes and achieving consensus among participating computers within a peer-to-peer network. Each participating node, also known as "miner," validates the transaction using certain mathematical calculations that are based on predefined rules. A block is considered verified when a consensus is achieved among at least 51% of network participants. This consensus protocol establishes system reliability and trust between unknown peers in a distributed computing environment. Therefore, in case of any malicious activity, that is, if someone tries to tamper the data in one node, the alternate transaction copies in other nodes will identify the invalid ledger and discard it and provide that node with a new copy of the ledger. Hence, the public distributed ledger is regularly updated at the same time in all the participating nodes in a blockchain network.

In Fig. 1, the "nonce" represents the resulting number that the miners are required to obtain by solving the consensus algorithm, and the term "difficulty" is the measure of the difficulty level for mining a particular block that represents the amount of computational power required to validate the block. This will make the network more secure against malicious attacks. The Merkle root is the hash of all the hashes of the transactions present in a particular block.

In a blockchain, each block is wrapped with a 256-bit hashing algorithm-based encryption, which is very difficult to decipher. Even if a particular block is altered, it will lead to an invalid ledger that will automatically get discarded, since modifying a single block will also modify the hash of that block and all subsequent blocks. Thus, it requires a huge amount of computing power to decrypt the blocks in a blockchain.

Blockchain is of three types – private blockchain, public blockchain, and consortium blockchain. The private blockchain is a permissible blockchain, in which transactions are private and are only available to authorized participants who are given access to join the network. They behave the same as centralized systems. Hyperledger is an example of a private blockchain. A public blockchain is a decentralized blockchain, also known as a distributed ledger system, which ensures a high level of transparency by providing a replica of the ledger to each blockchain node and perform verification and validation of data by a consensus mechanism. Examples of public blockchain are Bitcoin and Ethereum. Consortium blockchain behaves like a combination of both public and private blockchain as they provide only a set of pre-authorized nodes with high computational capabilities to allow them to access the ledger of transactional records and solve the proof-of-work consensus algorithm. This blockchain can enable faster transactions by reducing the overall power consumption. Corda, Quorum, and Hyperledger are few examples of consortium blockchain.

1.2 Ethereum

Ethereum is the first decentralized, open-source, public blockchain-based platform that supports smart contract [3]. It was released in 2015 by Vitalik Buterin, a cryptocurrency researcher. It enables smart contracts and distributed applications or DApps to be built and deployed on top of it, without any third-party interference. Smart contracts, which form the core of blockchain applications, are computer codes consisting of some predefined instructions that are implemented only upon meeting a particular set of conditions or certain specified actions. They are immutable, that is, once a smart contract is executed on a blockchain, its code cannot be updated like a normal application [4]. It uses Ether cryptocurrency for trading and running smart contracts for DApps. It assigns a cost called "gas" (unit – Gwei) for every execution of the task, which is a much smaller unit of Ether (1 ETH = 10^{-9} Gwei), in order to drive the competition for transactions to be added successfully to the blockchain. The price of the cryptocurrency can also be predicted by LSTM (Long Short-Term Memory) algorithm of machine learning technology [5].

The Ethash algorithm-based encryption, which is a cryptographic hash function, is used to encrypt blocks in the Ethereum blockchain. It takes input data and yields an alphanumeric output of 64 characters or 256 bits. Every user in a blockchain network has two cryptographic keys to secure identities and to encrypt and decrypt the data – a public-key, which is like an email address of the user, and a private key, which is like a password or passkey used to sign the transactions and is always kept undisclosed with the user. In the blockchain, when the transactions contained in a transaction pool are considered verified and validated, a new block is added to the existing blockchain and the miner who solves the proof-of-work (PoW) consensus algorithm gets the reward in the form of Ethers coins for mining the block.

The rest of the chapter is organized as follows. Section 2 summarizes the related literature review. Section 3 describes the structure of the proposed work. Sections 4 and 5 shows the implementation and analysis of the proposed model. Finally, Sect. 6 concludes the work.

2 Literature Review

In the smart grid domain, blockchain technology delivers excellent features like distributed ledger facilities, smart contracts, compromise mechanisms for a secure transaction. The advent of blockchain expertise in the energy segment is the key market transformation that is implemented in [6, 7]. In these, the application of blockchain technology in the energy sector is discoursed by comparing it with the existing technology in South Africa and Russia. The references [8–11] highlight the employment of Bitcoin in the distributed energy system such as energy operating conditions, monitoring, sharing, trading, etc.

A systematic review on Blockchain technology over the energy sector is thoroughly discussed [12] by considering over 140 research articles and startups related to the same. In this, it is found that blockchain technology is one of the promising technologies that can be implemented in the energy sector as it exhibits a tamper-proof, secure, and transparent system that leads to a novel business solution. It also focuses on the challenges and market barriers of the technology while implementing it in the real world. Another review is proposed in the reference [13] where the integration of blockchain technology is discussed in the energy sector by considering the related works of literature from 2014 to 2020 available in the Scopus database. In this assessment, the authors have pointed out the rapid progress of the technology in the energy sector especially in the fossil energy sector where blockchain technology can solve the bottleneck problem of existing technology. The use of cryptocurrency in the energy sector is realized [14] with some proof of works using a technical example. The reference [15] highlights some possibilities of blockchain technology in the electricity market with a review of all prospects and threats. Primarily, this work focuses on the decentralized energy sector with all sustainable configurations by using Bitcoin.

A trusted communication between the electric grid and the consumer is projected in reference [16] where the billing system is implemented via smart contact through blockchain technology. It also proposes a mathematical model to schedule the use of electric appliances for the optimum uses; according to the price, distributed energy system, and smart contract to monitor the trading of the electricity transaction system. A flexible, decentralized smart grid system is proposed in reference [17] using blockchain technology where both Bitcoin and Ethereum are implemented. A ledger is utilized for every transaction for each demand response in a very trusted manner using blockchain technology. A billing system for the electrical network is proposed [18] by using blockchain technology and IoT devices where trust and privacy are maintained. The smart contract using Blockchain technology for securing the Internet and IoT with challenges and the research associated with this technology are revealed in the references [19, 20]. IoT data sets that are meant for the research fraternity are addressed in [21] by analyzing the IoT-based research papers published in English since Jan 2016. Here, seven IoT-related research potential threats are highlighted pertaining to IoT security. The performances of IoT are analyzed [22] by adapting blockchain technology in which, the several ways of integration of blockchain with IoT, challenges, benefits are discoursed by comparing the existing blockchain-IoT platform. An assessment is also realized for the shake of the practicability of blockchain nodes with IoT devices. The integration of Blockchain technology in smart logistics with regard to the information, transport, finance, and management are discussed in [23]. The blockchain technology-based smart grid system is reviewed in the reference [24, 25] where different parameters are thoroughly discussed. It also focuses on a variety of blockchain technology applications for smart grid and other industrial applications with a standard procedure for future expansions. In addition to IoT, a combination of blockchain and big data is realized in [26] for smart city projects to reduce the carbon emission intended for a green environment.

A Blockchain technology-based smart grid system is demonstrated [27] where the database is implemented more securely. In this, a customer can monitor the uses of power on a real-time basis. Blockchain technology-based smart grid is discussed in [28] for the protection of data flow in the different layer network where a consumer can sell their individual renewable energy back to the grid. A peer-to-peer electricity exchange network is discussed in [29] with proof of benefits to attain minimum power fluctuation without any intermediate party. The reference [30] proposes a blockchain-managed smart grid system where all the transactions are managed with a smart contract. This technology ensures that each transaction must be an immutability transaction with a history of transaction which is beneficial for audit. Secure communication is achieved [31] between a service provider and a user having a smart meter through blockchain technology under a keyless signature decentralized scheme. Furthermore, blockchain technology-based cybersecurity architecture is proposed in [32], where all the cloud computing technology is implemented to improve the efficiency of the smart metering system. The integration of blockchain technology in smart logistics with regard to information, transport, finance, and management are discussed in [33]. A case study of a smart model related to blockchain for the supply chain is proposed in [34] for storing the entire transaction. The case study has been carried out in Nigeria to implement mobile voting system which averts illegible person from casting their votes using blockchain technology [35]. The complete work along with literature is well presented for reader's interest as in [36].

Keeping the view of the above literature, we propose in this work, a blockchain-powered energy monitoring system that provides the solution to the issues associated with a centralized system and addresses major challenges in utility grids such as delayed electricity bill payment, the manual intervention of the process, unethical practices in a centralized system and illegal tapping of power in transmission lines. Furthermore, the blockchain technology is implemented in Advanced Metering Infrastructure (AMI) to monitor the energy usage and send data to a decentralized Web application that is developed using an Ethereum smart contract for supervising and automating the billing process.

3 Proposed Structure

In order to overcome the challenges of a centralized system, this paper presents a blockchain-powered energy monitoring system in which Blockchain technology is integrated with advanced metering infrastructure, which, in general, comprises various software and hardware equipment, that gather energy usage information from the smart meter and transfer it to the control center equipment via a two-way communication network in near real time as shown in Fig. 2.

Initially, a smart meter will be installed in every house registered on a given network, which will send meter readings to a blockchain at regular intervals. A decentralized Web application or DApp will be used that records the cost of energy usage of each house in terms of Ether for households to pay their monthly electricity bills. This Web application is powered by a smart contract that fetches unit usage of meter from the blockchain and secures the transactions by permitting genuine and accurate data transfers among the smart meters and supervisory nodes and reports if any illegal and malicious tampering of data has occurred. Once the required amount is paid, the smart contract will verify and record the transaction on a payment blockchain and send a permit signal to the smart meter for the next month. If a household fails to pay the bill in the required time, the smart contract will not send a permit to the smart meter, resulting in the meter automatically tripping the connection of the house. Hence, the entire process is automated without the need

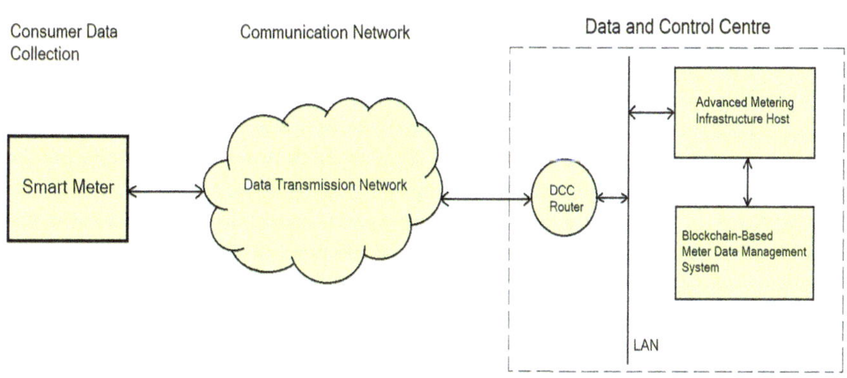

Fig. 2 Advanced metering infrastructure

for any manual intervention. This process of automated and regulated billing is illustrated in Fig. 3.

4 Implementation

The prototype for a blockchain-powered energy monitoring system is implemented by developing a Web application that can be used for enabling protected transactions of Ethers and storing energy records securely. Blockchains, like general Web servers, execute an application code and host a database. Generally, with a Web application, we can access a web page on our Web browser. All the front-end codes (written in HTML, CSS, JavaScript) for the website are stored on a central Web server, which interacts with the back end (written in PHP, JavaScript, or python), which then interacts with a database in order to render the web page. However, a blockchain application works differently. We can access our blockchain Web app with an Ethereum browser that interacts with a normal front-end website, but instead of interacting with a back-end Web server, this website will interact directly with the blockchain. Hence, the blockchain acts as a back end that hosts all the codes and data for our decentralized system.

4.1 Model of Blockchain Web Application

The blockchain Web application has a user interface for users to interact with a smart contract and a blockchain that stores information for peer-to-peer payments. For building the user interface, a client-side website is created using HTML5 and JavaScript along with web3.js and react.js libraries. The web3.js libraries are used to connect to the Ethereum network and interact with a smart contract through an inter-process communication (IPC) connection. The Ethereum smart contract is developed using Solidity programming language and deployed to the blockchain. The model of this Web application is illustrated in Fig. 4.

This application is implemented on Intel Core i5-7200U 2.50GHz Processor with 8 GB RAM and Windows 10 Pro 64-bit operating system. To develop and run dynamic Web applications, Node.js is used, which is a JS runtime environment on Chrome used to execute JavaScript codes outside a Web browser. The codes for smart contracts and client-side websites are written on Sublime Text. To build, test, and deploy the Web-based DApp, the Truffle framework is used, which provides multiple functionalities. It compiles the smart contract codes to byte codes that can be run on Ethereum Virtual Machine. It enables automated testing of the smart contract using Mocha and Chai testing frameworks. It helps in connecting to both public and private blockchain networks to run the applications. The Truffle ecosystem also provides a personal Ethereum blockchain known as "Ganache," where smart contracts are deployed. Ganache blockchain can be used instead of

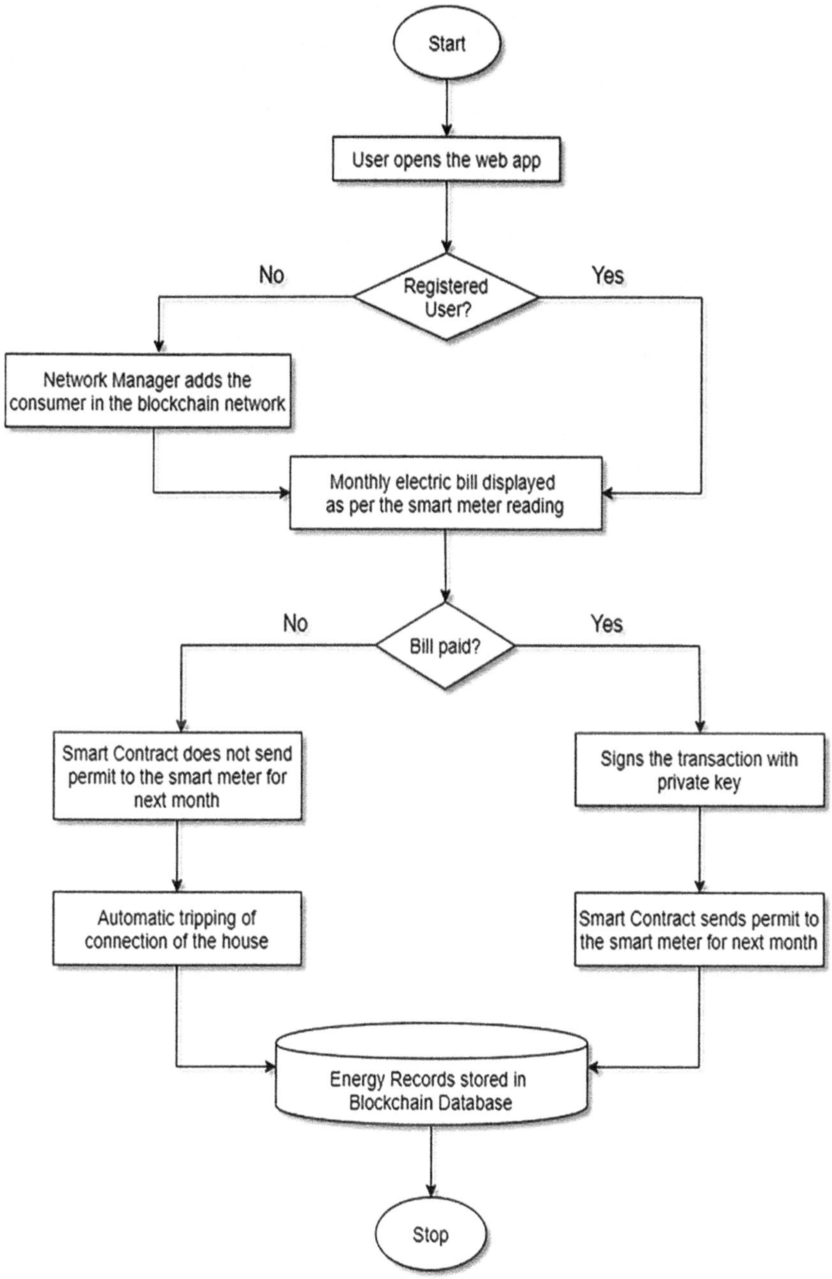

Fig. 3 Workflow overview

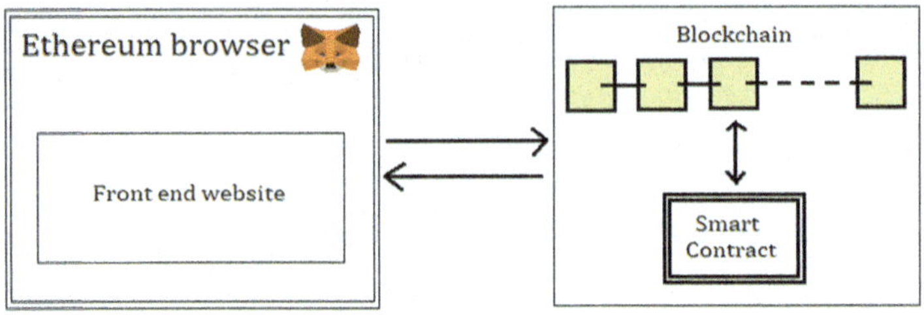

Fig. 4 Web app model

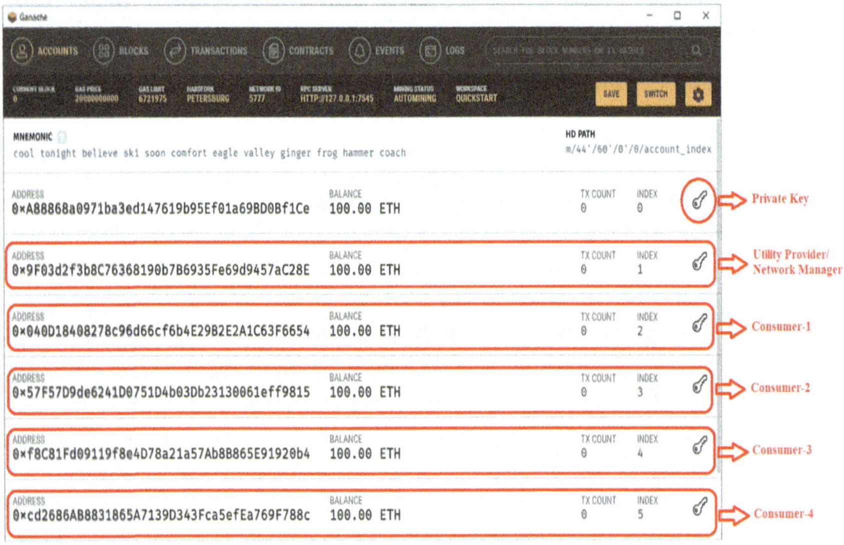

Fig. 5 Ganache console

a real Ethereum blockchain as it behaves the same as a public blockchain but won't cost us any real money. It enables us to develop, deploy, and test our DApps in a safe and deterministic environment. The user interface of Ganache is shown in Fig. 5. Its UI has a list of user accounts with different public and private keys, and each account is credited with a balance of 100 Ethers.

In order to interact with Ganache blockchain from the Web app, we need to use MetaMask browser extension, which converts regular Web browser (e.g., Chrome, Firefox) into Ethereum blockchain browser and allows us to run Ethereum DApps without having to download and run an entire Ethereum node consisting of more than 10GB blockchain on to our computers. To connect our Web app with Ganache, we need to import all Ganache accounts into the MetaMask wallet using the private key of each account and change the default main Ethernet network of MetaMask to Localhost 7545. The Ether balance of each user is now visible in the MetaMask wallet of each account.

Using this model, a decentralized application is built that allows registered consumers to pay electricity bills on the website with Ethereum cryptocurrency and uploads energy transaction data on the Ganache blockchain. It also allows new houses to be added to the group of registered consumers in the power grid network through the Web app.

4.2 Front End of Web Application

The front end of our Web app is the user interface through which consumers can directly pay their monthly electricity bills online. The front end, also known as the client-side application, is shown in Fig. 6. It consists of two sections, the admin's section and the consumer's section. The admin's section has a field for the admin to add new consumers to the network by entering their usernames and the bill amount as per their smart meter reading. The monthly bill amount will be in Ethers. It is assumed that the smart meter reading will be regularly updated on the web page by the AMI. However, in this prototype, we will manually add the readings on this Web app. These readings are then uploaded on the blockchain along with the information of new consumers in an encrypted manner. The consumer's section contains the bill payment status of all the consumers as per their blockchain transaction records. It includes the usernames of consumers, their monthly bill amount in Ether, and their public-keys in the "account number" section. It also contains the link in the form of

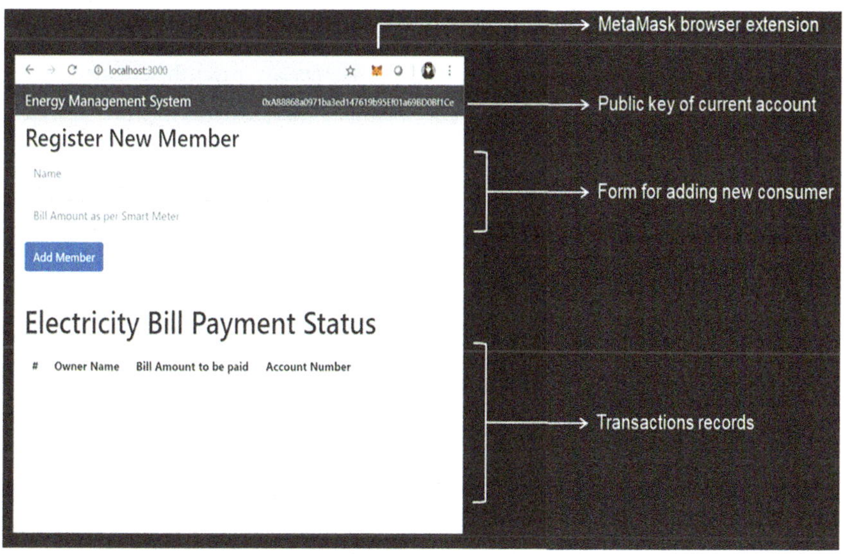

Fig. 6 Client-side web page in MetaMask browser

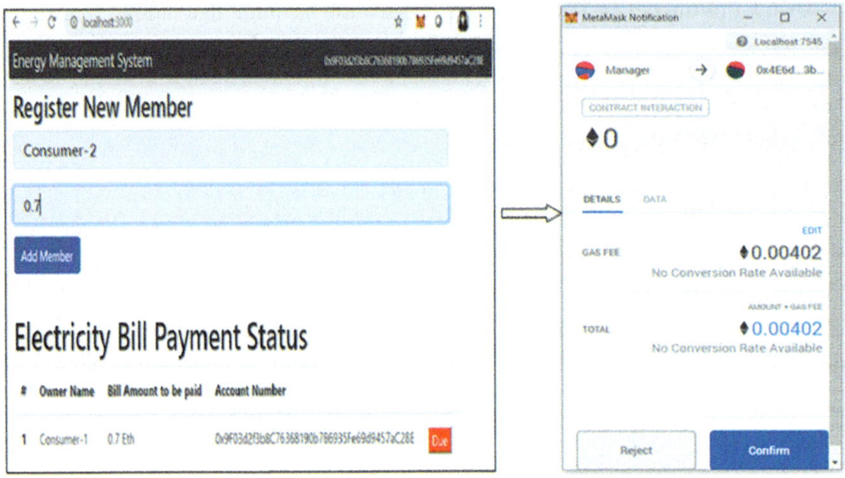

Fig. 7 MetaMask confirmation for adding new participants to the network

a button to pay their respective bills online, which will be added once a consumer record is created.

4.3 Role of Admin

The admin is an independent entity whose role is to only add new consumers in the blockchain network, who are willing to purchase energy from that particular grid network. To add new consumers, the admin has to fill the "Name" field and the "Bill Amount as per Smart Meter" field and click on the "Add member" button on the webpage, after which a MetaMask confirmation will be generated for signing the transaction initiated due to the interaction with a smart contract, as shown in Fig. 7. Thus, a new block will be created and added to the Ganache blockchain, and a new transaction record will be created in the consumer's section, as shown in Fig. 7. In this record, the account number is the public-key of the user, and the red "Due" button indicates that the bill has not been paid yet. Currently, this record has the public-key of admin in the account number section, since Consumer-1 is newly added to the Ethereum network. This key will change into the consumer's public-key only after the payment of the first month's electricity bill, which indicates the transfer of ownership of energy from the admin to the consumer.

The smart contract is written in a way that restricts the admin to pay the consumer's bill. If the admin tries to pay any consumer's bill, the smart contract will throw a transaction error in the MetaMask notification, which means the transaction will not be successful. This can help in checking unethical selling of power units by the electricity board.

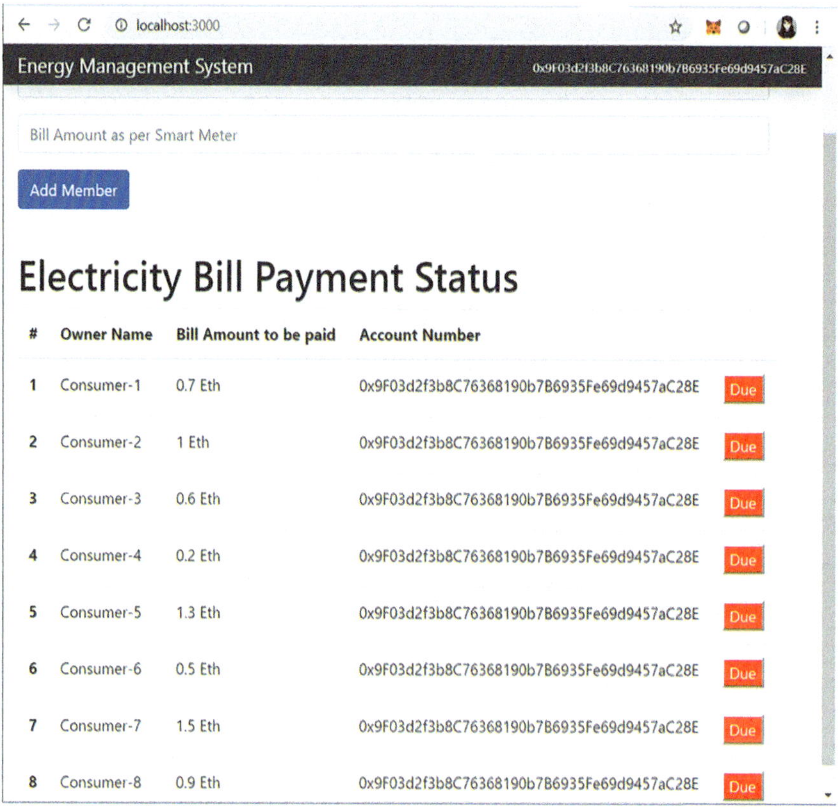

Fig. 8 Consumer's payment status

4.4 Role of Consumer

The consumer's section on the web page is a list of energy records of all the registered consumers, which contain only the username, the bill amount as per the smart meter reading, and the public-key of consumers, as shown in Fig. 8. This ensures that the personal information and identity of users are hidden. In these records, a payment link is also present in the form of "Due" button, which redirects to the MetaMask wallet once it is clicked by the respective consumers to pay their bills.

It is assumed that the energy usage of a house is constantly monitored by the AMI and the bill amounts in the record are updated at regular intervals by the blockchain. It is also assumed that each consumer record is marked "Due" at the end of the month, indicating that the monthly electricity bill is due the current month and the payment portal will remain active till the end of the month, exceeding which the smart contract will not send any permit to the smart meter that will eventually trip

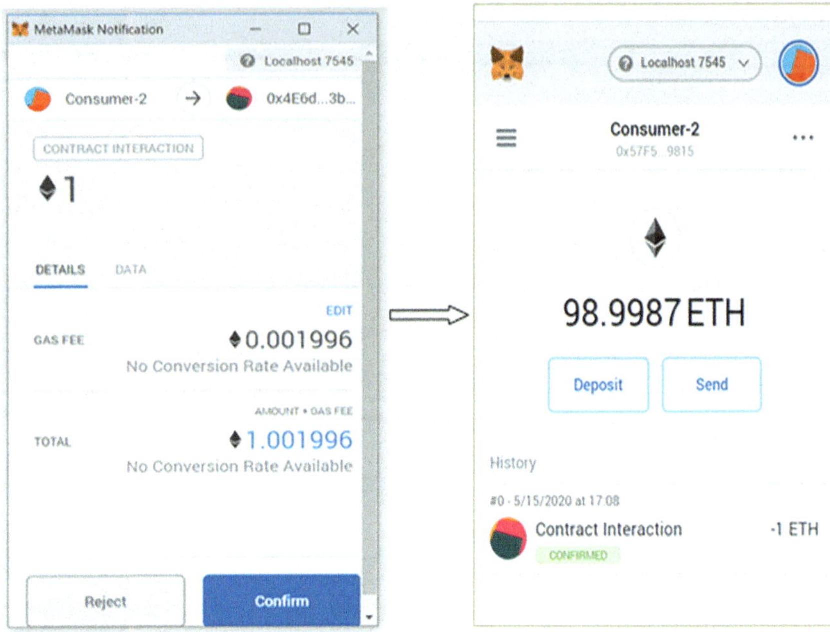

Fig. 9 MetaMask confirmation for consumers

the power flow to the house. On clicking the respective "Due" button, a MetaMask confirmation will be generated, where the consumers have to sign their transactions using their private key in order to complete the payment process and create a block, as shown in Fig. 9.

The MetaMask notification contains a detailed summary of a transaction, which contains the Ether value of bill amount along with the gas fee, and it is generated by the smart contract to let the users sign or confirm their transactions. As soon as the contract interaction is confirmed in MetaMask, a block will be added to the Ganache blockchain, and the red "Due" button automatically changes to the green "Bill paid" mark, as seen in Fig. 10, which signifies successful transaction of Ethers from the consumer's account to admin's account.

At the end of the month, the client-side UI will appear as shown in Fig. 10. The consumers who have paid their electricity bills are now easily distinguishable from the ones who have not and will get a permit for the next month in their smart meter from the smart contract. Consumers who have passed the due will get their connection automatically tripped by the smart meter as it would not receive any permit from the smart contract. This way, our Web application is able to monitor all the bill payments from the consumers by listing out the users who have paid and who have not, respectively, at the end of the month.

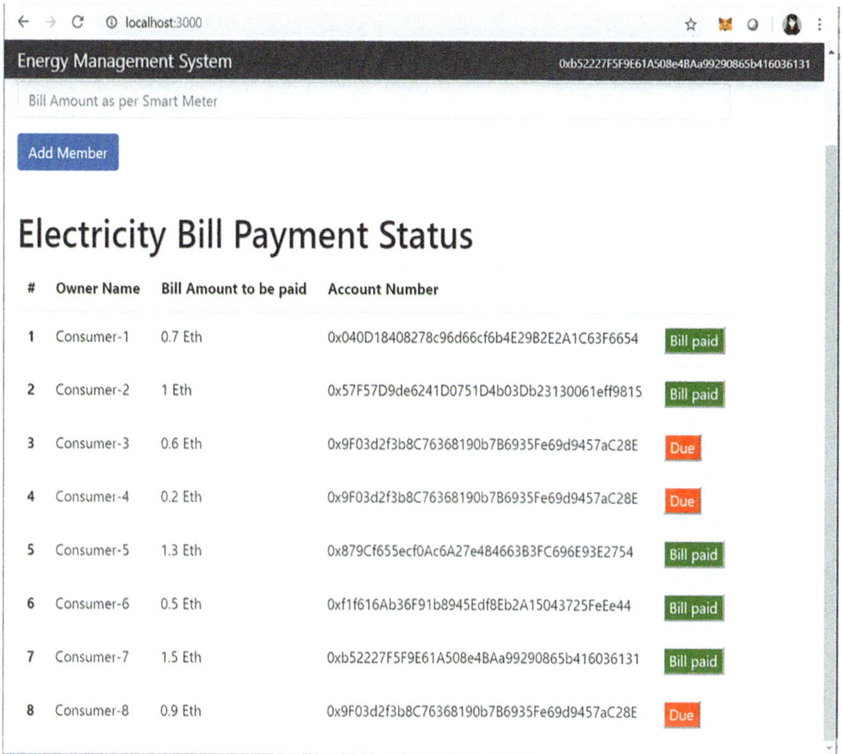

Fig. 10 Resulting database after bill payment

4.5 Creation of Blocks in Ganache

In the Web app, a transaction is initiated either when the admin adds new consumers to the blockchain network or the consumers pay their monthly bills. Blocks are created and mined every time a transaction is initiated and added to the Ganache blockchain, as shown in Fig. 11. Each transaction block is verified and validated during mining, which requires a certain amount of gas fee to complete the process. Thus, the blocks contain different time stamps on which they are added to the ledger, along with the amount of gas fee spent during the process of mining.

In the Ganache console, we can find the list of blockchain transactions in the transaction section of Ganache, which is shown in Fig. 12.

All the transaction details stored in the blocks are encrypted using the 256-bit hash algorithm known as Ethash, which is very difficult to crack. This will prevent data tampering due to cyberattacks and also protects the personal information of the user. The transactional information of each user is converted into a transaction hash, as shown in Fig. 12. Along with transaction hash, the gas amount and the public-key of the sender and receiver are also present in the block.

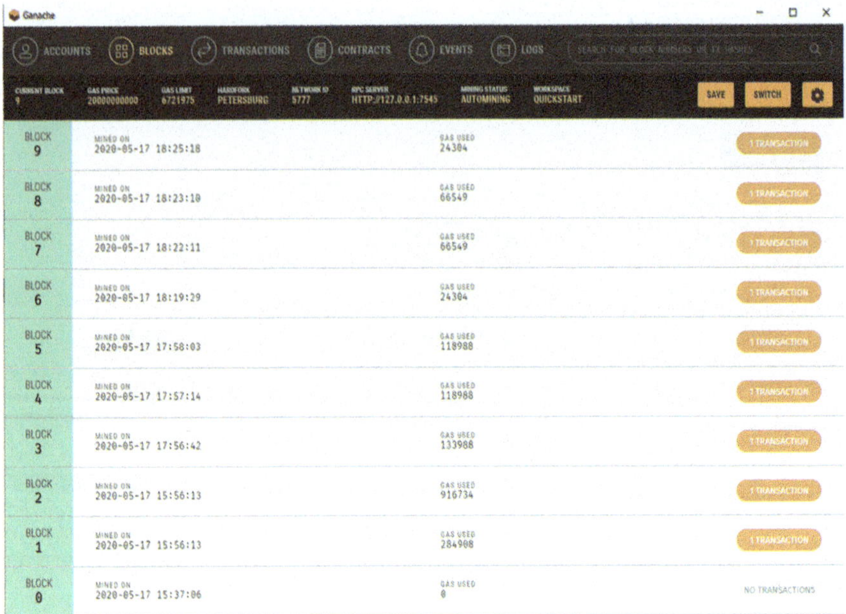

Fig. 11 Blocks in Ganache

Fig. 12 Transactions in Ganache

5 Analysis

It is evident from the proposed system of a blockchain application that it is capable of executing the smart contract functionality in automated billing and storing transactional data in the blockchain. This prototype can be used in large-scale applications for monitoring energy usage in real time and tracking the electricity billing process. Along with ease in payment process through the Web app, this will also ensure payment settlements in real time through consensus among all the nodes in the network. The automated tripping of electricity connections in a house by the smart contract will prevent manual interventions and other problems arising due to delays in bill payments. In our DApp prototype, the Web app is hosted on the "LocalHost 7545" network since we are implementing our prototype on the Ganache blockchain, which is present in our local machine. We can also implement the proposed system on a public blockchain network such as the "Kovan Test Network," where we have to pay with real Ethereum cryptocurrency. The households can pay directly through the Web app if they are connected to a given public blockchain network for energy transactions of the power grid.

The power generated in power stations passes through large and complex networks, such as transformers, overhead lines, cables, and other equipment, and reaches the end users. It is a fact that the unit of electric energy generated by the power station does not match with the units distributed to the consumers. Some percentages of the units are lost in the distribution network. This difference in the generated and distributed units is known as transmission and distribution loss. Transmission and distribution losses are the amounts that are not paid for by users. Distribution lines account for nearly 15% loss of the total power produced. However, in the case of illegal tapping of power, the sum of electricity usage by individual houses and power generated from transformers/power plants after eliminating the losses will not be equal, i.e., losses will be greater than 15% of the total power generated. Therefore, in order to detect power theft, transformers can be connected to the blockchain on which their electricity transactions are uploaded regularly. So, if there is an illegal tapping at any instant, then the net power available to the consumers will be less than 0.85 times the total power generated by power plants. This change will be evidently identified by the blockchain networks due to the transparency in the transactions of all the consumers.

6 Conclusion

In this work, a system is proposed successfully by using blockchain as a medium to reserve energy transactions and user accounts. With the help of the Internet and the encryption technique in blockchain, the proposed system provides a better solution for a smart grid system that is decentralized, interoperable, and secure. As seen in the proposed system, the decentralized application offers numerous benefits over centralized systems, such as immutability, security, transparency, reduction in transactional cost, faster transactions and prevention of unethical access, data tampering, and single point of failure. Due to the abovesaid advantages, the corresponding institutions will find easy and cost-effective solutions to implement in their system. The projected blockchain system will be highly favorable for both the customers and service providers to make profitability in the business.

Currently, there is a high potential for blockchain-based systems and applications in the energy sector. It aims to provide the solution for real-time energy monitoring and faster, secure, and efficient energy transactions. However, it faces major obstacles, which have prevented its optimal usage. Blockchain technology is still in its nascent stage of development, and applications are still in the pilot stage, due to limited research work. Lack of guidelines, laws, and regulations is also there. Future work can be done to integrate a blockchain-based energy monitoring system with the current system and include incentives for users. Possible long-term consequences are the creation of an Internet of Value and decentralized society and organizations. This could be a future energy market with a decentralized and unified energy grid, efficient and secure P2P transaction platform, interchangeable roles between prosumers and consumers, automated billing and logistics, energy flexibility, and competitive and cost-efficient market.

References

1. Dhananjayan, R., & Shanthi, E. (2014). Smart energy meter with instant billing and payment. *International Journal of Innovative Research in Computer and Communication Engineering, 2*, 1397–1403.
2. Nakamoto, S. (2008). *Bitcoin: A peer-to-peer electronic cash system*. www.bitcoin.org
3. Zhang, Y., Chu, J., Chan, S., & Chan, B. (2019). The generalised hyperbolic distribution and its subclass in the analysis of a new era of cryptocurrencies: Ethereum and its financial risk. *Physica A: Statistical Mechanics and its Applications, 526*, 120900.
4. Münsing, E., Mather, J., & Moura, S. (2017). Blockchains for decentralized optimization of energy resources in microgrid networks. *IEEE Control Technology and Applications (CCTA)*, 27–30.
5. Abayomi-Zannu, T. P., Odun-Ayo, I., Tatama, B. F., & Misra, S. (2019). Implementing a mobile voting system utilizing blockchain technology and two-factor authentication in Nigeria. *IC4S, 121*, 857–872.
6. Mollah, M. B., Zhao, J., Niyato, D., Lam, K. Y., Zhang, X., Ghias, A. M. Y. M., Koh, L. H., & Yang, L. (2021). Blockchain for future smart grid: A comprehensive survey. *IEEE Internet of Things Journal, 8*, 18–43.
7. Brilliantova, V., & Thurner, T. W. (2019). Blockchain and the future of energy. *Technology in Society, 57*, 38–45.
8. Kumar, N. M. (2018). Blockchain: Enabling wide range of services in distributed energy system. *Beni-Suef University Journal of Basic and Applied Sciences, 7*, 701–704.
9. Dimitriou, T., & Mohammed, A. (2020). Fair and privacy-respecting bitcoin payments for smart grid data. *IEEE Internet of Things Journal, 7*, 10401–10417.
10. Ghorbanian, M., Dolatabadi, S. H., Siano, P., Kouveliotis-Lysikatos, I., & Hatziargyriou, N. D. (2020). Methods for flexible management of blockchain-based cryptocurrencies in electricity markets and smart grids. *IEEE Transactions on Smart Grid, 11*, 4227–4235.
11. Liu, Z., Wang, D., Wang, J., Wang, X., & Li, H. (2020). A blockchain-enabled secure power trading mechanism for smart grid employing wireless networks, special section on new advances in blockchain-based wireless networks. *IEEE Access, 8*, 177745–177756.
12. Andoni, M., Robu, V., Flynn, D., Abram, S., Geach, D., Jenkins, D., McCallum, P., & Peacock, A. (2019). Blockchain technology in the energy sector: A systematic review of challenges and opportunities. *Renewable and Sustainable Energy Reviews, 100*, 143–174.

13. Wang, Q., & Su, M. (2020). Integrating Blockchain technology into the energy sector from theory of blockchain to research and application of energy blockchain. *Computer Science Review, 37*, 100275.
14. Sedlmeir, J., Buhl, H. U., Fridgen, G., & Keller, R. (2020). The energy consumption of blockchain technology: Beyond Myth. *Business & Information Systems Engineering, 956*, 1–10.
15. Teufel, B., Sentic, A., & Barmet, M. (2019). Blockchain energy: Blockchain in future energy systems. *Journal of Electronic Science and Technology, 17*, 100011.
16. Afzal, M., Huang, Q., Amin, W., Umer, U., Raza, A., & Naeem, M. (2020). Blockchain enabled distributed demand side management in community energy system with smart homes. *IEEE Access, 8*, 37428–37439.
17. D'Oriano, L., Mastandrea, G., Rana, G., Raveduto, G., Croce, V., Verber, M., & Bertoncini, M. (2018). Decentralized blockchain flexibility system for Smart Grids: Requirements engineering and use cases. *IEEE (CANDO-EPE)*, 20–21.
18. Gur, A. O., Oksuzer, S., & Karaarslan, E. (2019). Blockchain based metering and billing system proposal with privacy protection for the electric network. *IEEE (ICSG)*, 25–26.
19. Lone, A. H., & Naaz, R. (2021). Applicability of Blockchain smart contracts in securing Internet and IoT: A systematic literature review. *Computer Science Review, 39*, 100360.
20. Liu, Y., & Zhang, S. (2020). Information security and storage of Internet of Things based on block chains. *Future Generation Computer Systems, 106*, 296–303.
21. Banerjee, M., Lee, J., & Choo, K. R. (2018). A blockchain future for internet of things security: a position paper. *Digital Communications and Networks, 4*, 149–160.
22. Reyna, A., Martín, C., Chen, J., Soler, E., & Diaz, M. (2018). On blockchain and its integration with IoT. *Challenges and Opportunities, Future Generation Computer Systems, 88*, 173–190.
23. Issaoui, Y., Khiat, A., Bahnasse, A., & Ouajji, H. (2019). Smart logistics: Study of the application of Blockchain technology, EUSPN. *Procedia Computer Science, 160*, 266–271.
24. Musleh, A. S., Yao, G., & Muyeen, S. M. (2019). Blockchain applications in smart grid–Review and frameworks. *IEEE Access, 7*, 86746–86757.
25. Hasankhani, A., Hakimi, S. M., Shafie-khah, M., & Asadolahi, H. (2021). Blockchain technology in the future smart grids: A comprehensive review and frameworks. *International Journal of Electrical Power & Energy Systems, 129*, 106811.
26. Suciu, G., Sachian, M. A., Dobrea, M., Istrate, C. I., Petrache, A. N., Vulpe, A., & Vochin, M. (2019). Securing the smart grid: A blockchain-based secure smart energy system. *IEEE International Universities Power Engineering Conference (UPEC)*.
27. Sun, M., & Zhang, J. (2020). Research on the application of block chain big data platform in the construction of new smart city for low carbon emission and green environment. *Computer Communications, 149*, 332–342.
28. Dutta, S. D., & Prasad, R. (2020). Digitalization of global cities and the smart grid. *Wireless Personal Communications., 113*, 1385–1395.
29. Liu, C., Chai, K. K., Zhang, X., & Chen, Y. (2019). Peer-to-peer electricity trading system: Smart contracts based proof-of-benefit consensus protocol. *Wireless Networks*, 1–12.
30. Agung, A. A. G., & Handayani, R. (2020). Blockchain for smart grid. *Journal of King Saud University – Computer and Information Sciences*.
31. Zhang, H., Wang, J., & Ding, Y. (2019). Blockchain-based decentralized and secure keyless signature scheme for smart grid. *Energy, 180*, 955–967.
32. Olivares-Rojas, J. C., Reyes-Archundia, E., Gutierrez-Gnecchi, J. A., Cerda-Jacobo, J., & Gonzalez-Murueta, J. W. (2020). A novel multitier blockchain architecture to protect data in smart metering systems. *IEEE Transactions on Engineering Management, 67*, 1271–1284.
33. Issaoui, Y., Khiat, A., Bahnasse, A., & Ouajji, H. (2019). Smart logistics: Study of the application of Blockchain technology, EUSPN. *Procedia Computer Science, 160*, 266–271.
34. Casado-Vara, R., Prieto, J., Prieta, F. D., & Corchado, J. M. (2018). How blockchain improves the supply chain: Case study alimentary supply chain. *IoTS, Procedia Computer Science, 134*, 393–398.

35. Awotunde, J. B., Ogundokun, R. O., Jimoh, R. G., Misra, S., & Aro, T. O. (2021). Machine learning algorithm for cryptocurrencies price prediction, artificial intelligence for cyber security: Methods, issues and possible horizons or opportunities. *Studies in Computational Intelligence book series, 972*, 421–447.
36. Misra, S. (2020). A step by step guide for choosing project topics and writing research papers in ICT related disciplines. In *Information and Communication Technology and Applications: Third International Conference, ICTA 2020*, Minna, Nigeria, November 24–27, Revised Selected Papers 3 (pp. 727–744). Springer.

Multifactor IoT Authentication System for Smart Homes Using Visual Cryptography, Digital Memory, and Blockchain Technologies

Aderonke Thompson, Adeola Abayomi, and Arome Junior Gabriel

1 Introduction

Internet of Things (IoT) is an emerging area of research, described as the network or organization of physical items, entities, or "things" embedded with electronics, sensors, software, and even network connection, which empowers these items/entities to access and trade information [1]. The "Things" in the IoT refer to any entity or object that can be allocated a unique identifier (IP address).

Recent studies reveal that breakthroughs in IoT emerged on the premise of wireless technology in conjunction with artificial intelligence (AI) and generated data. A large amount of sharable data between interconnected things necessitates the need for an adequate framework in data transfer management, privacy, and security, which is further initiated by appropriate authentication.

A robust IoT setup/architecture comprises of end-to-end solutions that offer cloud-enabled, efficient and innovative mechanisms for information exchange among physical objects, as well as provides for standardized connectivity with power to sustain interconnected devices and processes. The requirement of inherent power in IoT applications is an earnest need for capabilities to sustain years of operation.

The usage field of IoT is compartmented into three:

(a) Consumer IoT: These are connected devices, such as PCs, smartphones, cars, watches, entertainment gadgets, and other connected appliances.
(b) Commercial IoT: They are connected to medical equipment, inventory controls, and even device trackers.

A. Thompson · A. Abayomi · A. J. Gabriel (✉)
Federal University of Technology, Akure, Nigeria
e-mail: afthompson@futa.edu.ng; ajgabriel@futa.edu.ng

S. Misra, A. Kumar Tyagi (eds.), *Blockchain Applications in the Smart Era*,
EAI/Springer Innovations in Communication and Computing,
https://doi.org/10.1007/978-3-030-89546-4_14

(c) Industrial IoT: They include robots for manufacturing, monitoring devices for pipelines, flow gauges, wastewater monitors, electric meters, and other connected industrial gadgets.

1.1 Challenges Associated with IoT

Despite the aforementioned wide applications, IoT is confronted with some critical challenges, such as security, privacy, standards, regulation, and development. Out of these previous concerns, security and privacy have received notable research contributions. Security is a vast and crucial issue on the Internet and has often been regarded as the most significant concern in IoT networks [2, 3]. Other concerns highlighted by Banafa in [2] include connectivity, compatibility, and longevity. Although the current IoT ecosystems could connect a whole lot (thousands) of entities/items, decentralizing the currently centralized IoT setup will yield far better results or benefits. The compatibility issues in IoT setups arise from the absence of unified cloud services, the absence of normalized machine-to-machine (M2M) protocols, and varieties in firmware as well as operating systems in IoT devices. Several works of literature have even suggested integrating a private data center with the public cloud to create an extensible infrastructure as a possible solution to the security challenge. A hybrid cloud helps in data backup and disaster recovery by creating redundancy. It also has the ability to deploy IoT applications anytime and anywhere as the need arises.

Sequel to the importance of authentication in IoT's security and privacy preservation, Madsen in [4] stated that "providing for authentication in the IoT setup, has to do with recognizing the possibility for enabling new means of authenticating users through the devices and the neighboring entities or things." An authentication protocol is a communication protocol or cryptographic protocol intended for transmitting authentication data between two entities.

The article in [5] identified the users as the major issue in most IoT authentication mechanisms. This problem is further established in their requirement to memorize a plethora of diverse PINs, passwords, and even secrets. Usually, some accounts require more than one PIN or password for access. In the light of the above challenge, this current research work is proposing a digital memory-based authentication scheme in IoT using a lightweight encryption scheme with a view to enhancing the work of Shone *et al.*

Digital memories are an emerging area of interest for authentication. The emerging digital memory concept involves preserving the user's memories into audio, image, or video formats. Often, a cloud server is used to store embedded metadata, digital memories, which are used for user authentication. A user's multifactor digital memory is unique to an individual since it keeps or traces the online records of that person. These memories are potential means of flawless authentication [5]. One notable impetus for using digital memories in user verification is that users are expected to memorize detailed events or attributes of their lives. Digital memories come in an abundance of factors of authentication, viz, knowledge, possession, inheritance, location, and time factors. The digital memories can be encrypted using cryptography schemes to improve the confidentiality of information, such that the information is accessible strictly only by authorized parties.

Cryptography is a field of computing and mathematics that centers on methods for secure or private entity-entity communications even in the presence of an illegal, unauthorized third party [6, 7]. The main functions of cryptography are to offer confidentiality of information and ensure data integrity, authentication, and non-repudiation. Encryption schemes that require ciphertext to be sent through the communication channel are not immune from possible attacks. Furthermore, cryptography could be classified into symmetric encryption and asymmetric encryption. Establishing/setting up an encryption/communication key between a sender and a receiver in a secure/confidential manner remains a challenge/drawback of symmetric encryption schemes [8, 9], while large key size, high processing time and computational complexity, are drawbacks inherent in asymmetric encryption [10].

Visual cryptography scheme (VCS), a unique technique used to send data securely over the network, could be a possible solution to the shortcomings of conventional cryptography schemes. VCS provides for the encryption and decryption of secret information by the human visual system (HVS) in a manner devoid of complex mathematical computations. VCS is a more efficient and cost-effective encryption scheme to secure digital data over enterprise or open communication channels. Naor and Shamir in [11] were the first to come up with a visual secret information sharing strategy. Using VCS, a digital memory (image) is broken into n number of shares, and only an entity or individual in possession of all (or specified) n shares can reform the image (or decrypt the encoded digital memory). Image shares are encrypted via print on a separate sheet, such that superimposing the printed shares decrypts and reveals the image. VCS uses a simplified algorithm that makes it possible for digital information (or memories) to be encrypted so that reconstruction or decryption can be performed by the HVS in a simple, less complicated manner. In this research article, a (k, n) VCS model is adapted to a digital memory-based authentication to ensure efficient and cost-effective encryption of data over a channel. The values n and k represent the total number of shares produced via the VCS (used to encrypt the image) and the least number of shares needed to decrypt or reveal the image, respectively.

The use of the blockchain in this setup allows for the decentralization of the hitherto centralized conventional smart home authentication solutions. The identity management and data control at the smart home gateways allow for proper identification and necessary documentation of relevant information on blocks in the blockchain. This information is compared from time to time toward ensuring that the integrity of such information is preserved. Any integrity breach would be immediately discovered and flagged. The periodic updating of information on the blockchain blocks allows for the identification of the correct devices. Data leakage during transmission is also curtailed via encryption. Indeed, the cryptographic parameters are used by the blockchain.

Some related studies where machine techniques were adopted, including those related to digital memory-based authentication frameworks, have been reviewed, and in furtherance to the limitations identified from [2, 5, 12–14], an efficient authentication scheme is highly needed, and that is the major objective of this research article.

The remainder of this article is organized, such that Sect. 2 contains the appraisal of related research works in literature. The concept of VCS is briefly presented in Sect. 3. Section 4 presents a vivid description of the proposed system. Results from the experiments conducted are presented and discussed in Sect. 5. Finally, the conclusion, as well as suggestions for improvements in future research, is given in Sect. 6.

2 Related Works

With a view to identifying the existing research gaps and further providing enough justification for this current study, this article carried out a review of the most related literature by systematically exploring recognized reputable scientific databases, as recommended in [15]. Some of these related works are highlighted in the rest of Sect. 2.

The article in [16] presented "Modelling the choice users make in the PassPoint graphical password system." Therein, a digital image is partitioned into segments using a color-centered mean-shift segmentation procedure for identifying the most likely region at which users would make a click as their graphical password in the PassPoint system. Measures of saliency, such as the contrast in luminosity and hue between a fragment and its environment, are joined together to form a focus of attention (FoA) map for the image. The researchers sought to investigate the security of the PassPoint graphical password scheme and the sustainability of the essential images by offering a model for predicting the user's click points together with their saliency value. The authors opined that future research work could consider expanding the FoA map in order to add more to the image.

Elsewhere, the research in [17] reported on access control and authentication in IoT. This research was motivated by the need to improve Internet security and privacy issues in IoT. In a bid to improving the existing IoT access control and authentication strategies, this research in [17] proposed a more effective and viable method that adopts role-based access control (RBAC-based) and elliptic curve cryptography (ECC) techniques for access control and encryption, respectively. The methods proposed were analyzed and proven to be immune to the common cyber assaults. Still, there is a need to utilize digital memory in the authentication process due to the fact that the Public Key Cryptography based schemes incur computation overheads.

A more generic authorization framework for addressing both access control and authorization issues in IoT was presented in Seitz *et al.* [18]. The domain of focus in this work was the resource-controlled devices that require minimal or no human intervention or control. This research proposed a fine-grained and flexible access control that requires very little processing power and memory using an authorization engine to issue claims to unauthorized users and assertion profile denoted in a JSON notation. The scope of this framework is limited to being locally accessible in isolated places. However, one identified limitation is the issue of gateway requirement to read protocol messages. Hence, end-to-end security of protocol communications is not guaranteed.

The research work in [5] presented a mobile user authentication system for IoT using digital memory technology. The motivation behind this idea recognizes that the existing authentication strategies were highly susceptible to social engineering and phishing attacks. A two-factor authentication scheme using hybrid knowledge factor and possession factor was proposed. This scheme comprises three major entities in the mechanism: the digital memory authentication service (DMAS), the service provider (SP), and the user's smartphone (US). The user's smartphone has to be registered with the SP and the DMAS. The SP verifies if the identity of the smartphone has been registered when users make a request to access the SP's service. The SP states the number of questions and its difficulty for the preferred authentication challenge. The user provides responses to the questions, while the DMAS confirms their answer as correct or otherwise. If the user's answers correspond to the one saved in the DMAS during the registration, access is granted to the users by the SP. Protocols of this scheme were evaluated, and their results proved that the protocols are protected for authentication purposes. This work can be extended to incorporate multiple factor authentications at each link between the actors. The limitation posed by this research is that digital memory objects stored in the DMAS are not encrypted, and this introduces a possible privacy disclosure threat.

Another system based on the Datagram Transport Layer Security (DTLS)-based scheme and two-way authentication model for IoT is presented in [19]. The work was motivated by the need to achieve a tolerable level of security in IoT via solving the problem of end-to-end security solutions. The RSA public-key cryptography algorithm was proposed to work on the existing Internet standards. The DTLS protocol proposed for UDP/IPv6 networking was implemented and evaluated. Their evaluation results show that the proposed architecture in this research offers authentication, confidentiality, as well as message integrity, with manageable computation overhead. Suggestions for future work included adopting techniques that lessen packet headers for similar protocols to DTLS and reduce network overhead by applying an authenticated encryption with associated data (AEAD) modus operandi.

The article in [14] presented an improved user authentication scheme with privacy preservation, using digital memories. The users' authentication difficulty and confusion drove this research. Their work enhanced the scheme presented in [5] to preserve privacy, anti-tracking, and a multilevel security authentication system. The system proposed in this article achieved privacy-preserving authentication by adopting homomorphic encryption for the user's digital memory and leakproof user privacy; it improves data transmission security by adopting public-key encryption.

The system was able to provide improved privacy and security compared to the work in [5] and other schemes. However, the encryption approach used in this work encompasses several computational costs.

The work in [20] proposed a region of noninterest (RoNI) halftoned VCS toward achieving secure telemedicine. Specifically, their objective was to develop a scheme that will ensure secure remote healthcare delivery. Achieving this objective requires securing the large volume of information exchanged during doctor-patient consultation, MRI, X-ray, CT scan, disease diagnosis, and subsequent doctor's prescription, which are kept safe from unauthorized persons, either while in storage or while being shared over enterprise network channels. Although the VCS-based system yielded some efficiency, a multifactor strategy will yield better results and efficiency.

Giovanni and Valero in [21] proposed a fuzzy-based method that handles the sleeping time of smart home appliances in wireless sensor networks (WSNs). Their major motivation was the need for high-performing off-the-shelf hardware. Their results reveal that their method outperformed other existing methods.

The research in [22] implemented visual cryptography for face images in a biometric-based home security application to ensure that the biometric traits stored necessary for authentication are not compromised or altered by malicious attackers. Specifically, the authors proposed a system for preserving the privacy of face images stored and used for authentication in smart homes. The results of their proposed method show a very high level of accuracy (93%).

The quest for a free and fair voting system (VS) in Nigeria is an agelong problem. Abayomi-Zannu et al. in [23] suggested the adoption of blockchain technology for securing votes and, in general, ameliorating this quest. The article reported the development of a two-factor authentication mobile VS for voter's authentication. The results obtained on evaluating their system with the ISO 9241-11 usability model reveals that this system is suitable for use in a voting scenario.

Motivated by the agelong question of how to accurately predict the direction of flow of financial assets (cryptocurrency or stock) in terms of the prices, the article in [24] developed a machine learning model for cryptocurrency price prediction. Their research adapted the long short-term memory (LSTM) for developing this price prediction model. The results obtained revealed the superiority of LSTM over other models in terms of price prediction.

In [25], an extensive study on IoT technologies in the healthcare domain was carried out. The authors highlighted the enormous prospect of this combination. They opined that, in today's modern world healthcare systems, IoT finds application in a number of medical areas, such as blood information management, patient information management, real-time monitoring, and even medical emergency management. Their work was, however, a review, and no experiments were reported.

The article in [26] attempted to address the enormous energy crises that would arise due to the increasing demand for electricity in IoT networks. The paper proposed the bacteria foraging optimization technique for achieving demand-side management (DSM), an approach the author opined will ensure peak load reduction, hence, stability of the smart grid network. The results obtained from the experiments reveal that the proposed method was able to minimize electricity bills as well as the peak-to-average ratio.

There is a justifiable motivation for a lightweight authentication scheme to optimize time and memory complexity. This is expected to eliminate the shortcomings highlighted in the reviewed related works.

3 The Visual Cryptography Scheme (VCS)

VCS is a technique invented by Naor and Shamir in 1994 [8, 27]. This VCS allows a piece of given secret information to be encrypted as well as decrypted as the case may be, using the HVS in a fashion that is free of the relatively complex and usually mathematical computations inherent in conventional cryptography techniques [28, 29]. In the VCS scheme, a piece of the given secret information is broken into n number of shares, such that decryption is done only by an entity or individual in possession of all n shares. The implication is that someone with lesser (n-1) shares cannot reveal any information about the original image (except otherwise built into the particular VCS algorithm being used). Shares are printed on separate sheets so that carrying out decryption would imply overlaying the separate sheets or shares correctly. In most cases, once all the n shares are completely and properly overlaid, the original unencrypted information is recovered [30, 31].

Visual cryptography (VC) activities consist of two roles: a participant and a dealer. A participant holds the VC shares, while the dealer distributes the shares.

The encryption process of VC is accomplished on the basis of pixel expansion that varies according to the requirements of VC. The decryption process is of two kinds, namely, XOR and OR. Decryption using XOR operation is better in the act than that of OR [32, 33].

4 The Proposed System

The proposed system has three major actors or entities, viz, the digital memory authentication service (DMAS), the service provider (SP), as well as the user's smartphone (US).

(a) The US serves a dual function of an autonomous platform that allows for the exchange of information with other components and authentication factors as with its connection to the service provider using the user ID.
(b) The SP offers a specific service to the user. This service involves accepting and confirming the user ID for accessing the DMAS (where data pertain to the user's digital memory).
(c) The DMAS hosts or stores the user's digital memories in a cloud environment. It is based on these data that the DMAS offers part of the authentication.

4.1 Phases of the Proposed System

The proposed system has a number of phases. These phases are briefly highlighted in this subsection.

The Registration Phase

During registration, the user registers mainly with the SP as well as in the DMAS, with their email address being documented with the SP and DMAS. The user provides his or her security answers in response to security questions. Then, the user's digital memories (pictures) collected are fed as input into the (2, 2) VCS, where the picture is divided into two shares: one saved on the phone (user share) and the other saved on the database (database share).

The Login Phase

At the login phase, processing follows the following steps:

(i) A request for access to SP's service (where the user's email has been registered previously) is made by the user (US).
(ii) The SP's record is checked, in order to verify or authenticate the user's identity. Thereafter, the user's time stamp is registered (TS_U).
(iii) The SP sets up with the DMAS, an authentication session where the time stamp of the SP (TS_S) gets registered, the details of which are encrypted with public-key (kd).
(iv) The DMAS performs decryption on the data with kd^{-1} and verifies whether the email exists in DMAS records.
(v) The DMAS then forwards the question to the user which contains the encrypted digital memory object (EDMO). The details are all encrypted with DMAS private key (kd^{-1}).
(vi) After answering the questions, the user returns the response, which contains the email, time stamp, and their answer (ANS), to the challenge to the DMAS, and it is encrypted with the public-key (kd) of the DMAS.
(vii) The DMAS validates their answer, superimposes the user share on the database share to reveal the image, and delivers the result (RESULT) of the authentication question back to the SP after encrypting it with its (DMAS) private key (kd^{-1}).
(viii) If the authentication is successful (i.e., the result/response is correct), the SP grants the user access, i.e., the user will be granted access by the SP.
(ix) If unsuccessful (i.e., the result/response supplied is incorrect), then the user requests for new answer to the security question.
(x) The DMAS sends the answer (T_Pass) to the SP (which also sends it to the user's email address).

The user then starts the login process from step (i) to step (viii) and requests for the digital memory in the DMAS.

The algorithms for handling the stages in the proposed system are presented in threefold as follows:

First is the user registration. For this purpose, we have:

Begin

1. U → SP: {Email}
2. SP → DMAS: {Application}
3. DMAS → SP: {Kd, Email}
4. SP → U: {h(Email)}
5. U → DMAS: {EDMO, Email} kd

End

Second, we have the algorithm that handles user login:

Begin

1. U → SP: {Email, TS_U)
2. SP → DMAS: {Email, TS_S} kd
3. DMAS → U: {EDMO, Email, TS_D} kd^{-1}
4. U → DMAS: {Email, ANS, TS_U} kd
5. DMAS → SP: {RESULT, Email, TS_D} kd^{-1}
6. SP → U: {OK}

End

Then, third, if the user forgets the password, the following set of steps allows for the retrieval of the password:

Begin

7. U → SP: {Email, TS_U} kd
8. SP→ DMAS: {Email, TS_S} kd^{-1}
9. DMAS → SP: {Email, OTP, TS_D} kd^{-1}
10. SP → U: {Email, OTP, TS_S}

End

4.2 The Proposed System's Flowchart

In this application, a first-time user is required to register a valid email address, digital memory (image), and security answer to a predefined question. The system saves the security answer and compresses the image to a bitmap format in a byte. The image is split into two shares, and then, the two shares are encrypted. A share is saved on the smartphone (user share) and the other share on the database (database share). If the registration is successful, the user can log in by submitting the registered email address and security answer to allow for system authentication.

Next, the system validates or authenticates the email address and the security answer in the database. If there is a match, the system confirms the availability of the previously encrypted share that was saved on the user's smartphone and superimposes the user share with the database share. Decryption is done with reconstruction to reveal the image. If the image is revealed, access is granted to the user to either unlock or lock the door; otherwise, the user selects a new image for authentication via clicking the "Reset security image" button. Thus, the image undergoes the digital memory registration process.

On the other hand, if there is no match between the security answer provided by the user and the one saved on the database during registration, the user can request a reset of the security answer by clicking the "Reset security answer" button. Upon requesting a new security answer, the new answer is sent to the user's email; after that, the user can log in again. Figure 1 explains the various flow of activities in the application.

5 Results and Discussions

This research work used the Scyther tool to give a formal investigation of the system. Scyther is a programmed security convention verification instrument used to distinguish potential assaults and vulnerabilities.

Scyther was used to evaluate the following properties of our system:

(i) Reflection assault protection: To guarantee protection from assaults where validating users can be scammed into giving the solution to their own particular test.
(ii) Secrecy: To guarantee that the secrecy of qualifications, keys, tokens, and information is looked after, i.e., no gatecrashers can take them.
(iii) Replay assault protection: To guarantee protection from assaults.
(iv) Man-in-the-middle assault protection: In order to guarantee resilience to assaults, where malignant substances can catch and alter messages between two legal members of a communication session, without arousing suspicion.

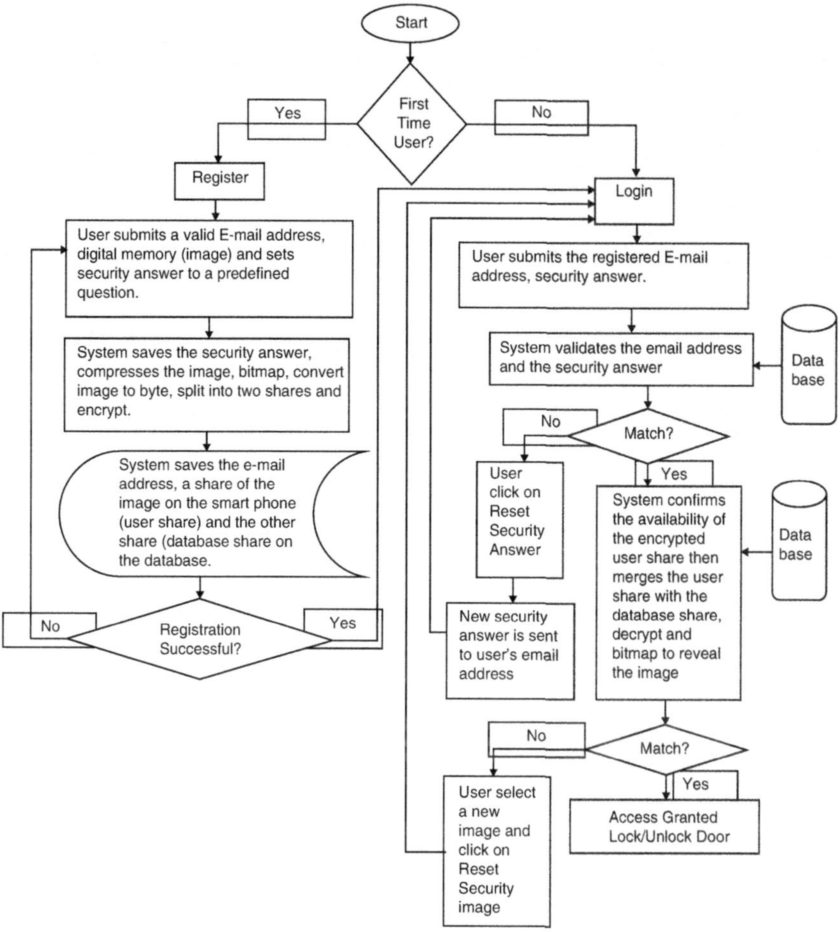

Fig. 1 The flowchart describing the functionality of the proposed system

First, Scyther was used to verify or evaluate the protocol that governs the registration of users by the DMAS. The results obtained show that there is no assault present in this scheme.

Figure 2 shows the results attained from Scyther's analysis. It was observed that there is no assault present in registration protocol of our scheme. The Scyther testing code is in Appendices A.

Secondly, Scyther was used to verify the login protocol that operates between the user and the DMAS. Figure 3 shows the results from Scyther's analysis. It was observed that there is no assault present in the login protocol in our scheme. Hence, our scheme obtained a good result.

By virtue of Figs. 2 and 3, we infer that Scyther could not establish any flaws or feasible assaults against this scheme. Scyther verification was reiterated both

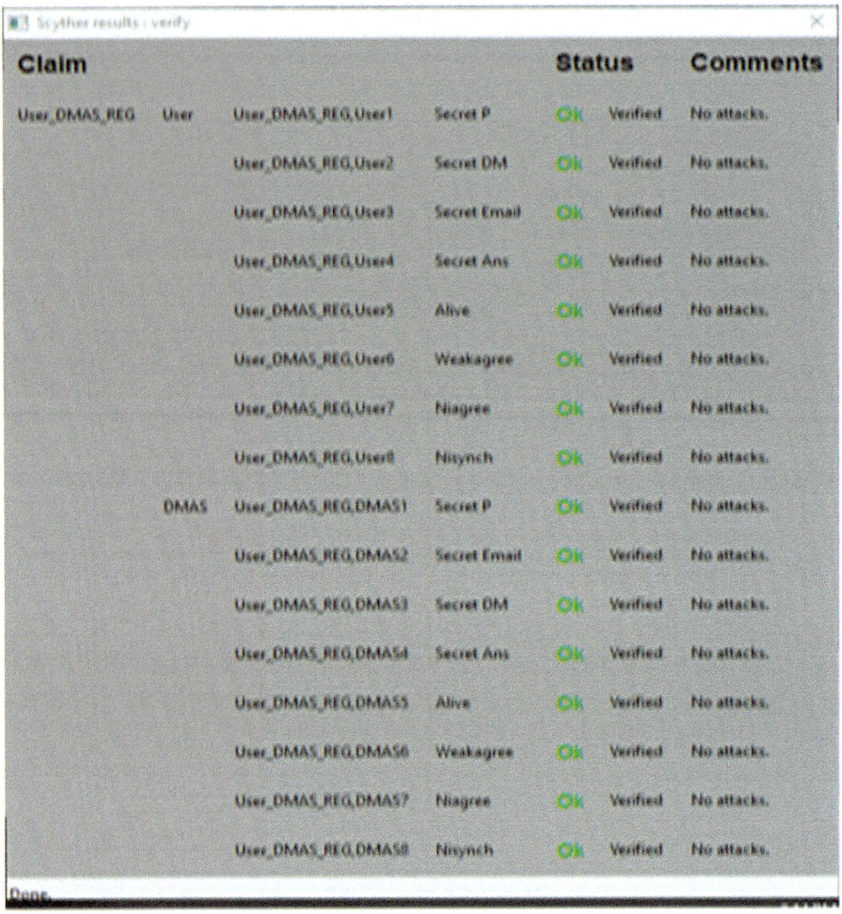

Fig. 2 Scyther results of formal analysis on the registration protocol

manually defined and Scyther's automatically generated claims, and the same results were generated.

5.1 Comparative Analysis of the Proposed System Versus Other Existing Works

A comparison is made between the proposed approach and the three existing most related research works, and this is summarized in Table 1. The proposed approach has the following advantages over the others:

Fig. 3 Scyther results of formal analysis on the login protocol and the DMAS

1. Lightweight authentication: The encryption method (VCS) used is not computationally intensive. Thus, it thereby optimizes time and memory complexity.
2. Different method: Asymmetric encryption (VCS) is used instead of the symmetric encryption to secure the communications among the actors, thereby reducing the risk of susceptibility to confidentiality and other common security attacks.
3. Multifactor authentication: Multifactor authentication is used in order to lessen the likelihood of a data breach via a compromised password.

Typically, a smart home is an aggregation of smart devices or objects that continuously or periodically interact or exchange information with themselves or some external gateways. Such a setup could leave gaps that pave way for privacy breaches, sensitive data leakage, and exposure or even deliberate tampering by unauthorized entities within or outside the network. These security issues are quite

Table 1 Comparative analysis of proposed system versus existing ones

Scheme	Digital memory	Privacy leakproof	Update digital memory	Multifactor authentication	Cost-effective	Lightweight authentication scheme	Decentralized	Encryption
PassPoint	✗	✗	✗	✗	✗	✗	✗	**N/A**
[5]	✓	✗	**N/A**	✓	✗	✗	✗	**Symmetric encryption**
[21]	✓	✓	✓	✗	✗	✗	✗	**Homomorphic encryption**
Proposed scheme	✓	✓	✓	✓	✓	✓	✓	**Visual cryptography encryption**

expedient and a huge requirement to the growth and general acceptability of the IoT paradigm.

This research proposes a multifactor strategy that involves the use of the digital memory, VCS, as well as blockchain technologies for achieving security. The blockchain with VCS encryption is used to ensure confidentiality within the smart home by permitting only authorized persons the privilege of access to data. Similarly, using our approach guarantees authentication in the smart home setup, as, via the network, the blockchain helps in verifying if an entity or object is a valid member or not. This governs the privileges the entity enjoys. Furthermore, via hashing, the integrity of records is preserved.

The blockchain technology brings decentralization into the existing centralized structure, hence, curbing the challenges, such as single point of failure, inherent in centralized systems. The unique identifiers (IDs) and other sensitive data about smart objects on the IoT network are first encrypted using the visual cryptography scheme (VCS) and, thereafter, securely recorded on the blockchain blocks in a manner that prevents data leakage and exposure. This information is compared from time to time to ensure that integrity of records is preserved.

6 Conclusion and Future Work

IoT is a growing area of interest in Internet technology, which accepts and sends information, receives and acts on the information, and can do both together. It is a means of taking resources and connecting them and can be controlled from any distance via the Internet. Moreover, the barriers and hurdles toward the rapid growth and adoption of IoT include data security, confidentiality, integrity, and privacy issues. Consequently, a scheme to protect information against unauthorized access using IoT authentication is desired. However, personal identification number (PIN)- or password-based authentication scheme has proven vulnerable to guessing, dictionary attack, and social engineering and shoulder surfing. These make authentication more difficult and confusing because users need to correctly recall complex and sometimes long piece of information or use extra hardware (such as a USB key). These issues were the motivation for digital memory in user authentication, the security of which mostly relies on trusted cryptography. In this study, the design of a secured digital memory-based authentication model for smart homes was successfully implemented in a mobile application and evaluated. This research adopted a two-factor authentication (2FA) using visual cryptography scheme for the encryption of digital memory, which lessens the likelihood of a data breach via a compromised password. In this scheme, the digital memory of the user is encrypted and split into two shares, one saved on the user's smartphone and the other saved on the DMAS. To authenticate, the exact two shares are needed. However, if only one share is available, the user cannot be granted access to control the smart door. Also, performance analysis and evaluation results were presented using the Scyther tool to give a formal investigation on potential assaults and vulnerabilities.

Scyther verification was repeated twelve times both manually defined and Scyther's automatically generated claims. The results proved that there was no assault present in the scheme. The speed of the entire system is about 50% of what is obtainable when conventional public-key cryptography schemes are used. Future work will consider the use of alternatives to power supply and biometrics as a multifactor authentication strategy.

References

1. Babu, B. S., Janeyulu, T. R., Narayana, I. L., & Srikanth, K. (2017). Trends of IoT. *International Journal of Engineering Trends and Technology, 43*(4), 185–188. https://doi.org/10.14445/22315381/ijett-v43p231
2. Banafa, A. (2017). *Three major challenges facing IoT - IEEE Internet of Things*. Retrieved April 10, 2018, from https://iot.ieee.org/newsletter/march-2017/three-major-challenges-facing-iot
3. Adeboje, O. T., Gabriel, A. J., & Adetunmbi, A. O. (2020). Development of an audio steganography system using discrete cosine transform and spread spectrum techniques. In O. Gervasi et al. (Eds.), *Lecture notes in computer science* (Vol. 12254, pp. 412–427). Springer. https://doi.org/10.1007/978-3030-58817-5_31
4. Madsen, P. (2015). *Authentication in the IoT – challenges and opportunities*, Secure ID News AVISIAN Publishing, Online Article. Retrieved April 12, 2018, from https://www.secureidnews.com/news-item/authentication-in-the-iot-challenges-and-opportunities/
5. Shone, N., Dobbins, C., Hurst, W., & Shi, Q. (2015). Digital memories based mobile user authentication for IoT. Proceedings - 15th IEEE International Conference on Computer and Information Technology, CIT 2015, 14th IEEE International Conference on Ubiquitous Computing and Communications, IUCC 2015, 13th IEEE International Conference on Dependable, Autonomic and Secure Computing, DASC 2015 and 13th IEEE International Conference on Pervasive Intelligence and Computing, PICom 2015 (pp. 1796–1802). https://doi.org/10.1109/CIT/IUCC/DASC/PICOM.2015.270
6. Baraket, M., Eder, C., & Hanke, T. (2010). *An introduction to cryptography. Information security management handbook* (6th edn, pp. 1121–1140). https://doi.org/10.1201/9781439833032.ch87.
7. Gabriel, A. J., Alese, B. K., Adetunmbi, A. O., & Adewale, O. S. (2013). Post-quantum crystography: A combination of post-quantum cryptography and steganography. The 8th International Conference for Internet Technology and Secured Transactions (ICITST-2013), Technically Co-sponsored by IEEE UK/RI Computer Chapter, 9th–12th December 2013, London, UK, pp. 454–457.
8. Daodu, M., Gabriel, A. J., Alese, B. K., & Adetunmbi, A. O. (2016). A Data Encryption Standard (DES) based Web Services Security Architecture. Anale. Seria Informatică. Vol. XIV fasc. 2 2016. 14(2):63–68
9. Alabi, O., Thompson, A. F., Alese, B. K., & Gabriel, A. J. (2020). Cloud application security using hybrid encryption. *Journal of Communications*. Foundation of Computer Science (FCS), NY, USA.
10. Gabriel, A. J., Alese, B. K., Adetunmbi, A. O., Adewale, O. S., & Sarumi, O. A. (2019). PostQuantum crystography system for secure electronic voting. *Open Computer Science, DeGruyter, 9*, 292–298. https://doi.org/10.1515/comp-2019-0018
11. Naor, M., & Shamir, A. (1995). Visual cryptography. In A. De Santis (Ed.), *Advances in Cryptology- EUROCRYPT'94. EUROCRYPT 1994* (Lecture notes in computer science) (Vol. 950). Springer. https://doi.org/10.1007/BFb0053419

12. Zhao, G., Si, X., Wang, J., Long, X., & Hu, T. (2011). A novel mutual authentication scheme for Internet of Things. In *Proceedings of 2011 international conference on modelling, identification and control* (pp. 563–566). IEEE. https://doi.org/10.1109/ICMIC.2011.5973767
13. Roussey, B. (2016). *5 challenges facing the Internet of Things*. Retrieved April 11, 2018, from http://techgenix.com/Internet-of-things-challenges/
14. Liu, J., Lyu, Q., Wang, Q., & Yu, X. (2017). A digital memories based user authentication scheme with privacy preservation. *PLOS ONE, 12*(11), e0186925. https://doi.org/10.1371/journal.pone.0186925
15. Misra, S. (2021). *A step by step guide for choosing project topics and writing research papers in ICT related disciplines, communications in computer and information science* (Vol. 1350, pp. 727–744). Springer.
16. Dirik, A. E. (2007). *Modeling user choice in the PassPoints graphical password scheme*. Retrieved July 16, 2019 from https://www.citeseerx.ist.psu.edu
17. Liu, J., Xiao, Y., & Chen, C. L. P. (2012). Authentication and access control in the Internet of Things. In *2012 32nd international conference on distributed computing systems workshops* (pp. 588–592). IEEE. https://doi.org/10.1109/ICDCSW.2012.23
18. Seitz, L., Selander, G., & Gehrmann, C. (2013). Authorization framework for the Internet-of-Things. In *2013 IEEE 14th international symposium on "A World of Wireless, Mobile and Multimedia Networks" (WoWMoM)* (pp. 1–6). IEEE. https://doi.org/10.1109/WoWMoM.2013.6583465
19. Kothmayr, T., Schmitt, C., Hu, W., Brünig, M., & Carle, G. (2013). DTLS based security and two-way authentication for the Internet of Things. https://doi.org/10.1016/j.adhoc.2013.05.003.
20. Bakshi, A., & Patel, A. K. (2019). Secure telemedicine using RONI halftoned visual cryptography without pixel expansion. *Journal of Information Security and Applications, 46*(2019), 281–295. www.elsevier.com/locate/jisa. https://doi.org/10.1016/j/jisa.2019.03.004
21. Giovanni, P., & Valerio, M. S. (2019). Wireless sensor networks for smart homes: A fuzzy-based solution for an energy-effective duty cycle. *Electronics-MDPI, 2019*(8), 131. https://doi.org/10.3390/electronics8020131
22. Mohan, J., & Rajesh, R. (2020). Enhancing home security through visual cryptography. *Microprocessors and Microsystems*. Available at Elsevier.com. https://doi.org/10.1016/j.micpro.2020.103355.
23. Abayomi-Zannu, T. P., Odun-Ayo, I., Tatama, B. F., & Misra, S. (2020). Implementing a mobile voting system utilizing blockchain technology and two-factor authentication in Nigeria. In *Proceedings of first international conference on computing, communications, and cyber-security (IC4S 2019)* (pp. 857–872). Springer.
24. Awotunde, J. B., Ogundokun, R. O., Jimoh, R. G., Misra, S., & Aro, T. O. (2021). Machine learning algorithm for cryptocurrencies price prediction. In S. Misra & A. Kumar Tyagi (Eds.), *Artificial intelligence for cyber security: Methods, issues, and possible horizons or opportunities* (Studies in computational intelligence) (Vol. 972). Springer. https://doi.org/10.1007/978-3-030-72236-4_17
25. Mohammed, M. N., Desyansah, S. F., Al-Zubaidi, S., & Yusuf, E. (2020). An internet of things-based smart homes and healthcare monitoring and management system. *Review. Journal of Physics: Conference Series, 1450 012079.*
26. Gabriel, A. J. (2020). Appliance scheduling towards energy management in IoT networks using Bacteria Foraging Optimization (BFO) algorithm. In A. E. Hassanien et al. (Eds.), *Artificial intelligence for sustainable development: Theory, practice and future applications* (Studies in computational intelligence) (Vol. 912, pp. 290–310). Springer. https://doi.org/10.1007/978-3-030-51920-9_15
27. Kukreja, S. Kasana, G., & Kasana, S. S. (2021). Copyright protection scheme for color images using extended visual cryptography. *Computers & Electrical Engineering, 91*, 106931. ISSN 0045-7906. https://doi.org/10.1016/j.compeleceng.2020.106931., https://www.sciencedirect.com/science/article/pii/S0045790620307801

28. Sridhar, S., & Sudha, G. F. (2021). Two in One Image Secret Sharing Scheme (TiOISSS) for extended progressive visual cryptography using simple modular arithmetic operations. *Journal of Visual Communication and Image Representation, 74*, 102996. ISSN 1047-3203. https://doi.org/10.1016/j.jvcir.2020.102996., https://www.sciencedirect.com/science/article/pii/S1047320320302121
29. Wang, L., Yan, B., Yang, H.-M., & Pan, J.-S. (2021). Flip extended visual cryptography for gray-scale and color cover images. *Symmetry, 13*(1), 65. https://doi.org/10.3390/sym13010065
30. Wu, X., & Peng, Y. N. (2021). Extended XOR-based visual cryptography schemes by integer linear program, signal processing, 2021, 108122, ISSN 0165-1684. https://doi.org/10.1016/j.sigpro.2021.108122., https://www.sciencedirect.com/science/article/pii/S0165168421001602
31. Rakshit, P., Ganguly, S., Pa, S., Aly, A. A., & Le, D. (2021). Securing technique using pattern-Based LSB audio steganography and intensity-based visual cryptography. *Computers, Materials & Continua, 67*. https://doi.org/10.32604/cmc.2021.014293A
32. Wang, G. (2014). *Content based authentication of visual cryptography attestation of authorship*. Retrieved from https://openrepository.aut.ac.nz/bitstream/handle/10292/8619/WangGY.pdf?sequence=3&isAllowed=y
33. Bhat, K., Uday, K. R., Ranjan, K. H. S., & Mahto, D. (2020). A novel scheme for lossless authenticated multiple secret images sharing using polynomials and extended visual cryptography. *IET Information Security, 15*(1), 13–22. https://doi.org/10.1049/ise2.12001

Index

S. Misra, A. Kumar Tyagi (eds.), *Blockchain Applications in the Smart Era*,
EAI/Springer Innovations in Communication and Computing,
https://doi.org/10.1007/978-3-030-89546-4

C

The manufacturer's authorised representative in the EU is Springer Nature Customer Service Centre GmbH, Europaplatz 3, 69115 Heidelberg, Germany. If you have any concerns regarding our products, please contact ProductSafety@springernature.com

Printed and bound by CPI Group (UK) Ltd, Croydon, CR0 4YY
28/09/2026
02224092-0004